McGraw-Hill

Illustrative Mathematics™

Course 1

Mc
Graw
Hill

Cover credit: (l)Karan Limwattanapiboon/EyeEm/Getty Images, (tr, br)McGraw-Hill Education, (cr)Liyao Xie/Moment/ Getty Images

mheducation.com/prek-12

Send all inquiries to:
McGraw-Hill Education
STEM Learning Solutions Center
8787 Orion Place
Columbus, OH 43240

ISBN: 978-0-07-687510-8
MHID: 0-07-687510-5

Illustrative Mathematics, Course 1
Student Edition, Volume 1

Printed in the United States of America.

7 8 9 10 11 12 LMN 28 27 26 25 24 23 22 21

Contents in Brief

Image Source, Yana Gayvoronskaya/123RF, oatawa/Shutterstock, Pavel Nesvadba/Shutterstock, Vlad61/Shutterstock, Andrea Marzorati/123RF, f11photo/Shutterstock

McGraw-Hill

Illustrative Mathematics

(t)Pobytov/DigitalVision Vectors/Getty Images, (b)Liyao Xie/Moment/Getty Images

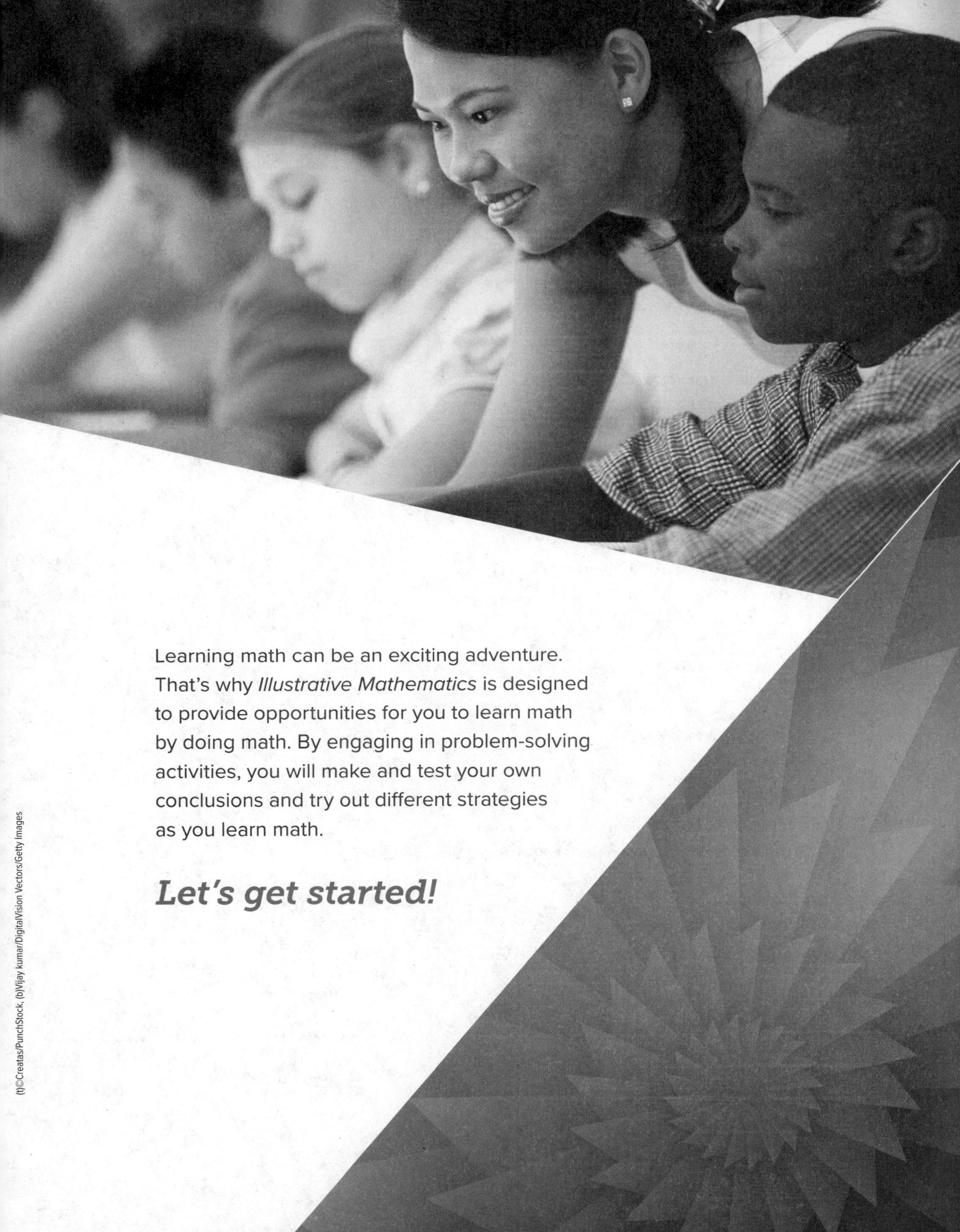

Learning math can be an exciting adventure. That's why *Illustrative Mathematics* is designed to provide opportunities for you to learn math by doing math. By engaging in problem-solving activities, you will make and test your own conclusions and try out different strategies as you learn math.

Let's get started!

Unit 1

Area and Surface Area

Michael DeYoung/Blend Images LLC

Unit 2

Introducing Ratios

What Are Ratios?

Equivalent Ratios

Representing Equivalent Ratios

Solving Ratio and Rate Problems

Part-Part-Whole Ratios

Let's Put It to Work

Unit 3

Unit Rates and Percentages

Easy Production/Image Source

Unit 4

Dividing Fractions

Yana Gayvoronskaya/123RF

Unit 5

Arithmetic in Base Ten

oatawa/Shutterstock

Unit 6

Expressions and Equations

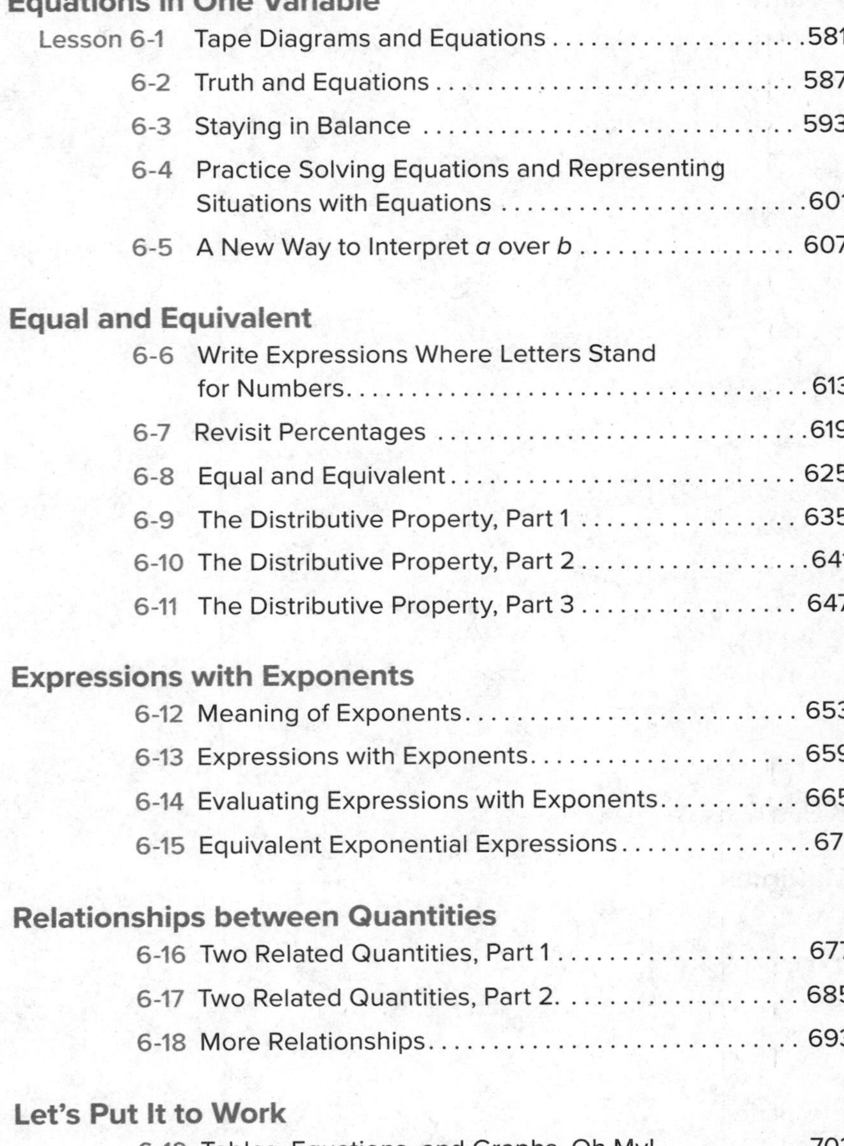

Pavel Nesvadba/Shutterstock

Unit 7

Rational Numbers

Vlad61/Shutterstock

Negative Numbers and Absolute Value

Inequalities

The Coordinate Plane

Common Factors and Common Multiples

Let's Put It to Work

Unit 8

Data Sets and Distributions

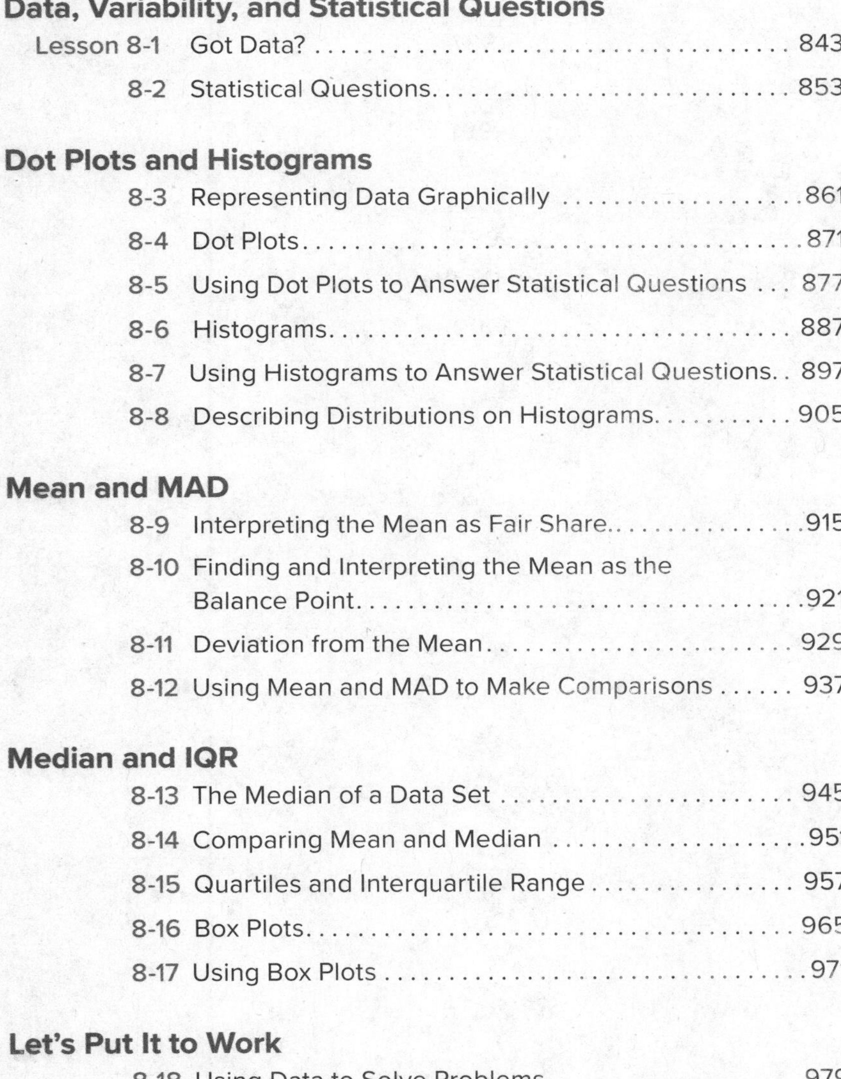

Andrea Marzorati/123RF

Unit 9

Putting It All Together

f1photo/Shutterstock

Making Connections

Voting

Unit 1

Area and Surface Area

Michael DeYoung/Blend Images LLC

At the end of this unit, you'll apply what you learned about area and surface area to design your own tent.

Topics

- Reasoning to Find Area
- Parallelograms
- Triangles
- Polygons
- Surface Area
- Squares and Cubes
- Let's Put It to Work

Unit 1

Area and Surface Area

Reasoning to Find Area

Parallelograms

Triangles

Polygons

Surface Area

Squares and Cubes

Let's Put It to Work

Lesson 1-1

Tiling the Plane

NAME ______________________ DATE ____________ PERIOD ____________

Learning Goal Let's look at tiling patterns and think about area.

Warm Up

1.1 Which One Doesn't Belong: Tilings

Which pattern doesn't belong?

Pattern A

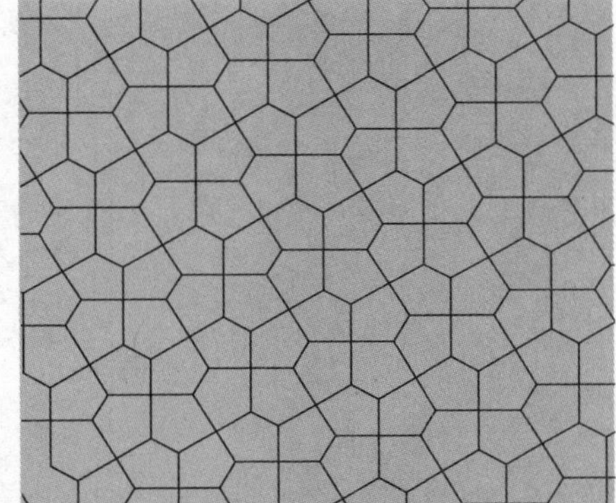

Pattern B

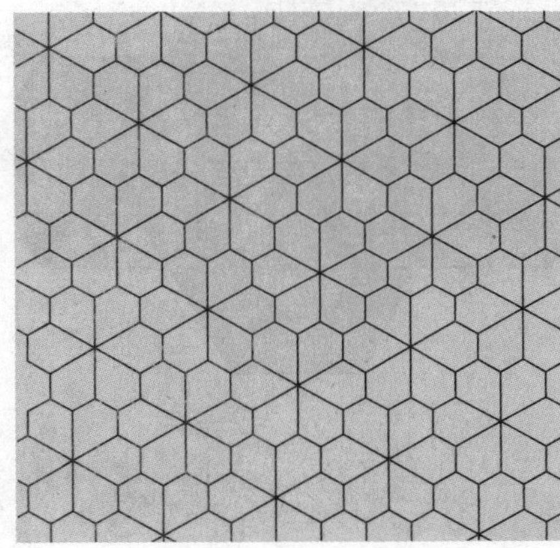

Pattern C

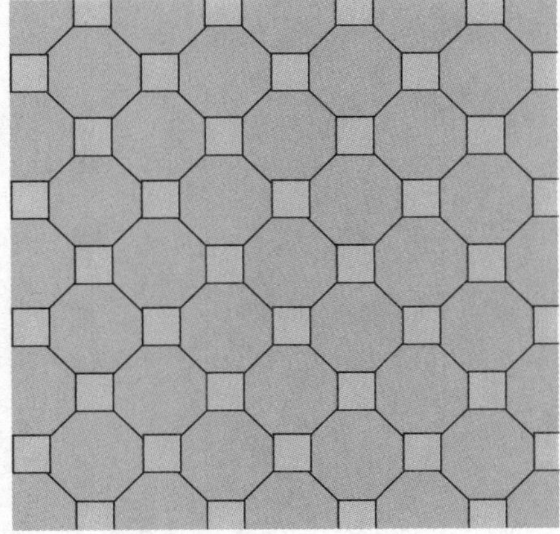

Pattern D

Activity
1.2 More Red, Green, or Blue?

Your teacher will assign you to look at Pattern A or Pattern B.

In your pattern, which shape covers more of the plane: blue rhombuses, red trapezoids, or green triangles? Explain how you know.

Pattern A

Pattern B

NAME _______________ DATE _______________ PERIOD _______________

Are you ready for more?

On graph paper, create a tiling pattern so that:

- The pattern has at least two different shapes.
- The same amount of the plane is covered by each type of shape.

Summary

Tiling the Plane

In this lesson, we learned about *tiling* the plane, which means covering a two-dimensional region with copies of the same shape or shapes such that there are no gaps or overlaps.

Then, we compared tiling patterns and the shapes in them. In thinking about which patterns and shapes cover more of the plane, we have started to reason about **area**.

We will continue this work, and to learn how to use mathematical tools strategically to help us do mathematics.

Glossary

area

region

Practice
Tiling the Plane

1. Which square—large, medium, or small—covers more of the plane? Explain your reasoning.

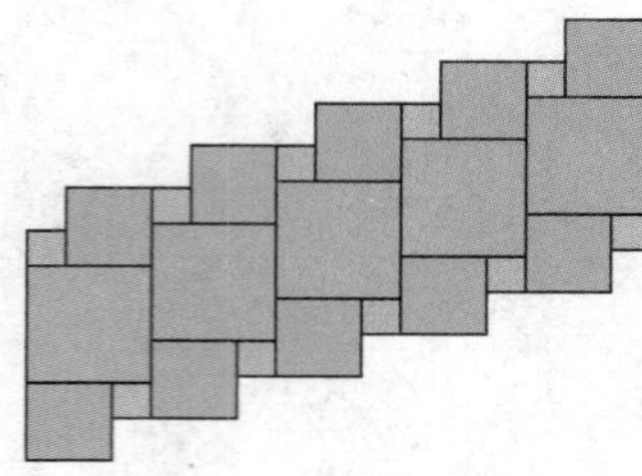

2. Draw three different quadrilaterals, each with an area of 12 square units.

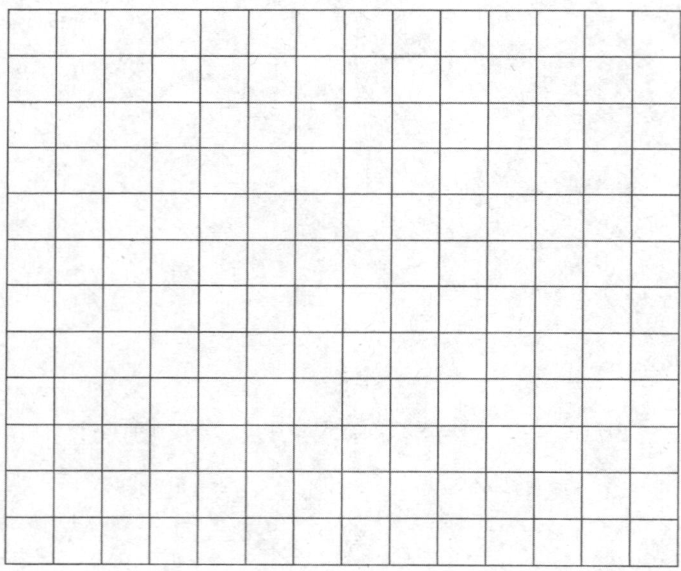

NAME ______________________ DATE ____________ PERIOD ________

3. Use copies of the rectangle to show how a rectangle could:

a. tile the plane.

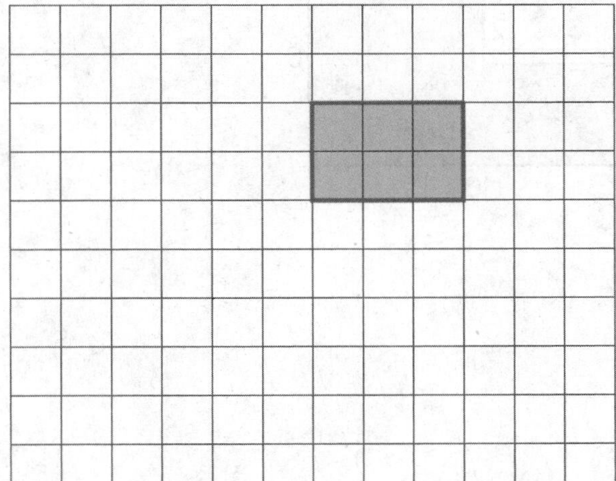

b. *not* tile the plane.

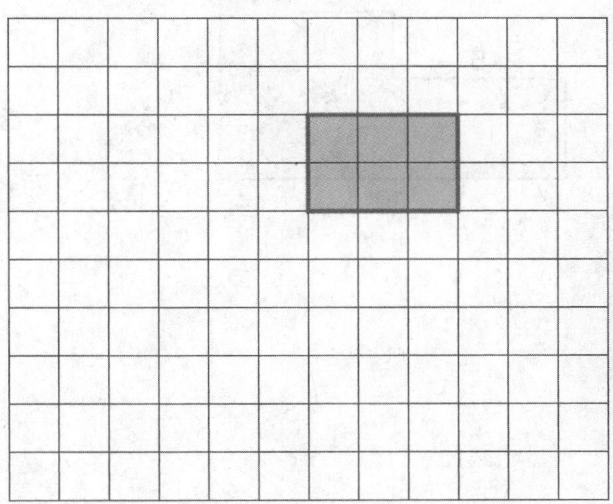

4. The area of this shape is 24 square units. Which of these statements is true about the area? Select **all** that apply.

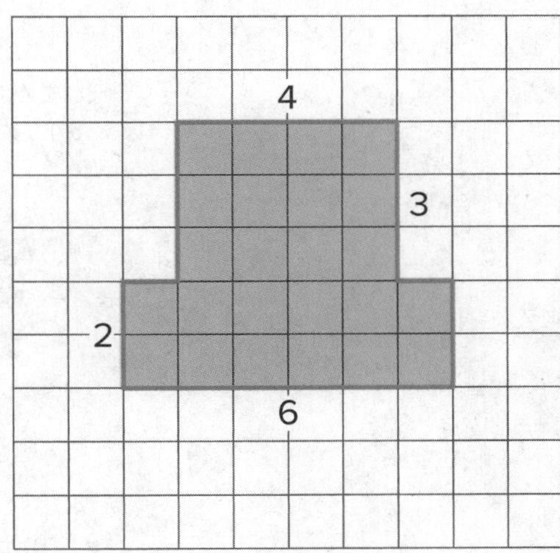

(A) The area can be found by counting the number of squares that touch the edge of the shape.

(B) It takes 24 grid squares to cover the shape without gaps and overlaps.

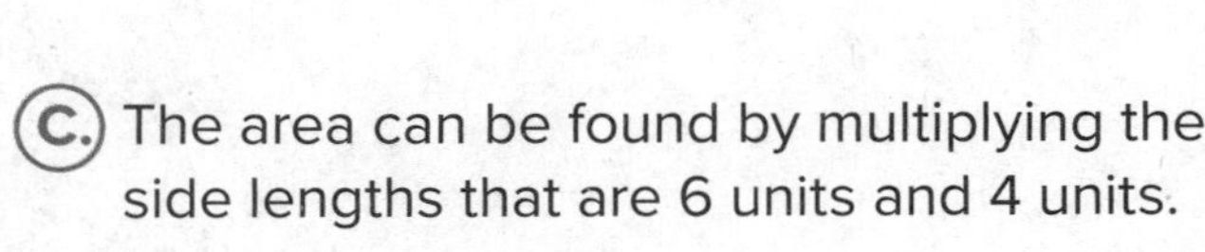

(C) The area can be found by multiplying the side lengths that are 6 units and 4 units.

(D) The area can be found by counting the grid squares inside the shape.

(E) The area can be found by adding 4×3 and 6×2.

5. Here are two copies of the same figure. Show two different ways for finding the area of the shaded region. All angles are right angles.

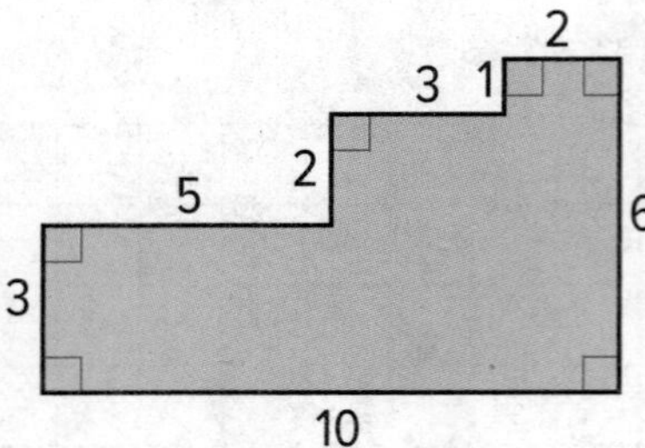

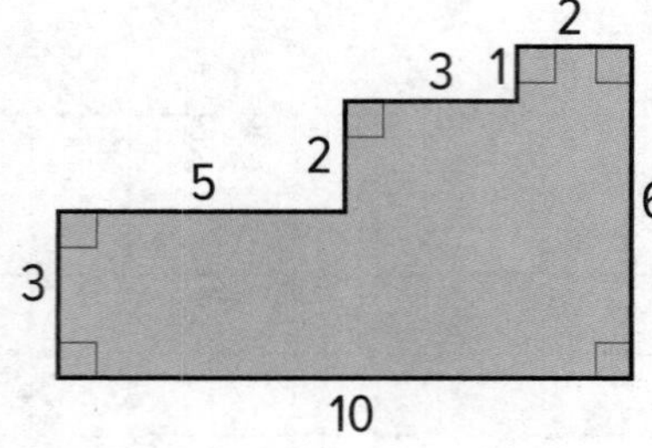

6. Which shape has a larger area: a rectangle that is 7 inches by $\frac{3}{4}$ inch, or a square with side length of $2\frac{1}{2}$ inches? Show your reasoning.

Lesson 1-2

Finding Area by Decomposing and Rearranging

NAME ______________________ DATE ____________ PERIOD ____________

Learning Goal Let's create shapes and find their areas.

Warm Up

2.1 What Is Area?

You may recall that the term **area** tells us something about the number of squares inside a two-dimensional shape.

1. Here are four drawings that each show squares inside a shape. Select **all** drawings whose squares could be used to find the area of the shape. Be prepared to explain your reasoning.

Drawing A

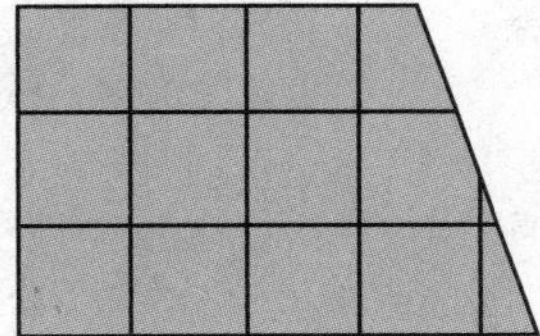

Drawing B

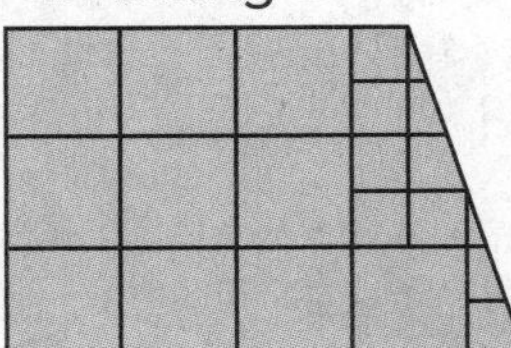

Drawing C

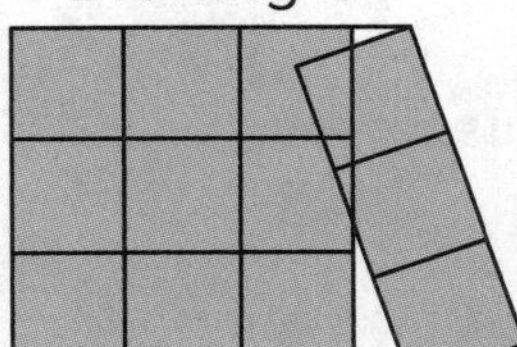

Drawing D

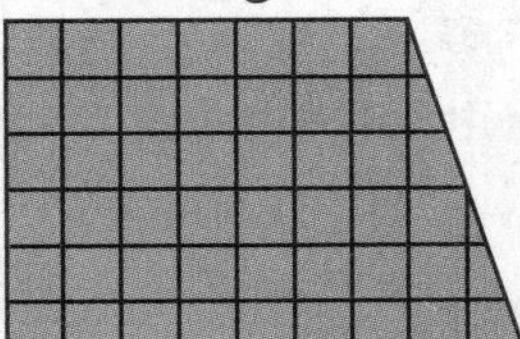

2. Write a definition of area that includes all the information that you think is important.

Activity

2.2 Composing Shapes

Your teacher will give you one square and some small, medium, and large right triangles. The area of the square is 1 square unit.

1. Notice that you can put together two small triangles to make a square. What is the area of the square composed of two small triangles? Be prepared to explain your reasoning.

2. Use your shapes to create a new shape with an area of 1 square unit that is not a square. Trace your shape.

3. Use your shapes to create a new shape with an area of 2 square units. Trace your shape.

NAME ________________ DATE ________ PERIOD ________

4. Use your shapes to create a *different* shape with an area of 2 square units. Trace your shape.

5. Use your shapes to create a new shape with an area of 4 square units. Trace your shape.

Are you ready for more?

Find a way to use all of your pieces to compose a single large square. What is the area of this large square?

Activity

2.3 Tangram Triangles

Recall that the area of the square you saw earlier is 1 square unit. Complete each statement and explain your reasoning.

1. The area of the small triangle is __________ square units. I know this because . . .

2. The area of the medium triangle is __________ square units. I know this because . . .

3. The area of the large triangle is __________ square units. I know this because . . .

NAME _______________ DATE _______________ PERIOD _______________

Summary
Finding Area by Decomposing and Rearranging

Here are two important principles for finding **area**.

1. If two figures can be placed one on top of the other so that they match up exactly, then they have the *same area*.
2. We can **decompose** a figure (break a figure into pieces) and **rearrange** the pieces (move the pieces around) to find its area.

Here are illustrations of the two principles.

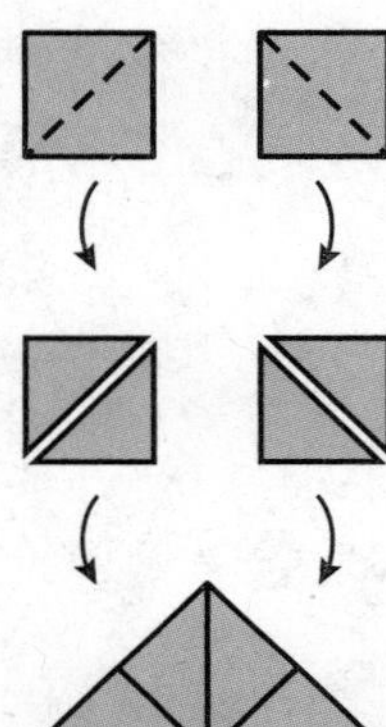

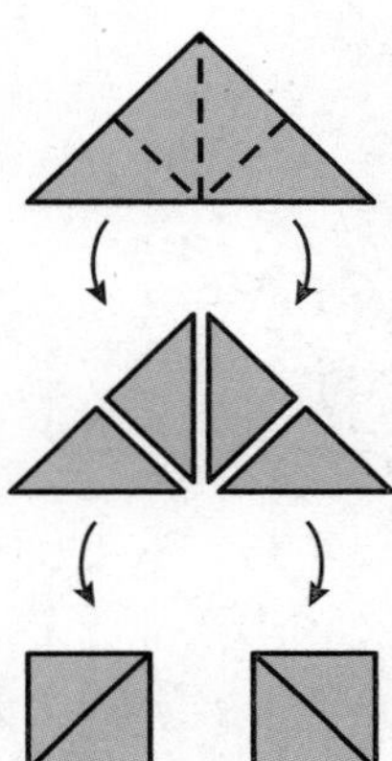

- Each square on the left can be decomposed into 2 triangles. These triangles can be rearranged into a large triangle. So the large triangle has the *same area* as the 2 squares.
- Similarly, the large triangle on the right can be decomposed into 4 equal triangles. The triangles can be rearranged to form 2 squares. If each square has an area of 1 square unit, then the area of the large triangle is 2 square units.

  We also can say that each small triangle has an area of $\frac{1}{2}$ square unit.

Glossary

compose

decompose

Practice

Finding Area by Decomposing and Rearranging

1. The diagonal of a rectangle is shown.

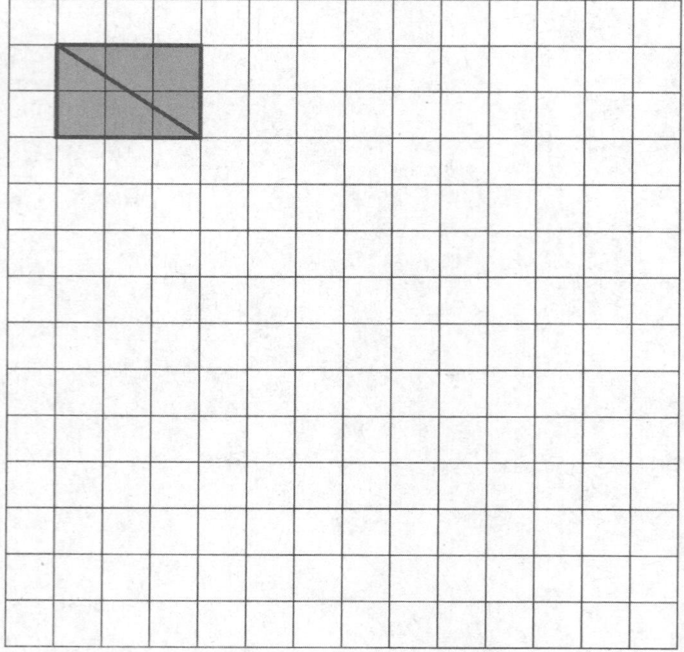

   a. Decompose the rectangle along the diagonal and recompose the two pieces to make a *different* shape.

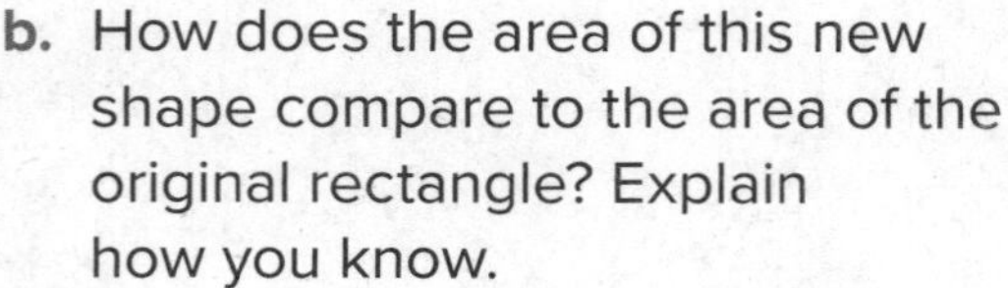

   b. How does the area of this new shape compare to the area of the original rectangle? Explain how you know.

2. Priya decomposed a square into 16 smaller, equal-size squares and then cut out 4 of the small squares and attached them around the outside of original square to make a new figure. How does the area of her new figure compare with that of the original square?

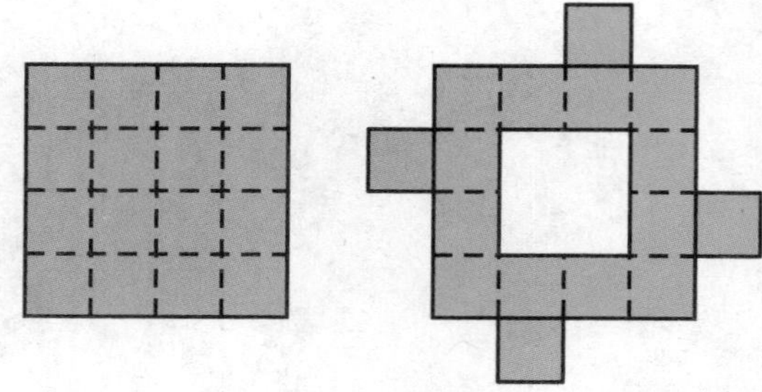

   (A.) The area of the new figure is greater.

   (B.) The two figures have the same area.

   (C.) The area of the original square is greater.

   (D.) We don't know because neither the side length nor the area of the original square is known.

NAME ______________________ DATE ______________ PERIOD ____________

3. The area of the square is 1 square unit. Two small triangles can be put together to make a square or to make a medium triangle.

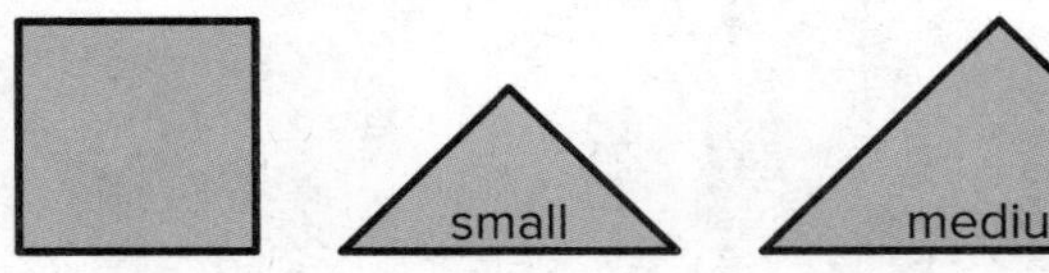

Which figure(s) has an area of $1\frac{1}{2}$ square units? Select **all** that apply.

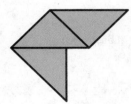

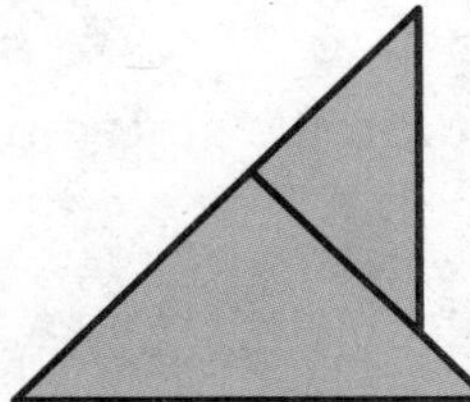

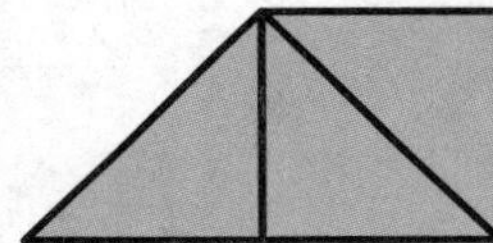

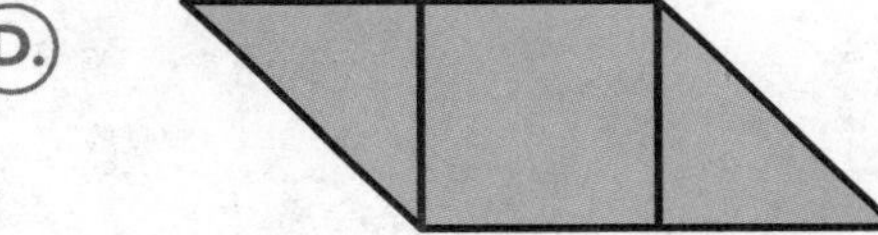

4. The area of a rectangular playground is 78 square meters. If the length of the playground is 13 meters, what is its width? **(Lesson 1-1)**

5. A student said, "We can't find the area of the shaded region because the shape has many different measurements, instead of just a length and a width that we could multiply."

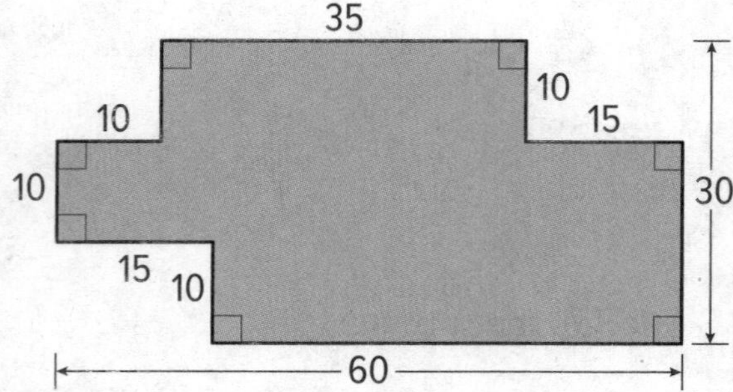

Explain why the student's statement about area is incorrect. **(Lesson 1-1)**

Lesson 1-3

Reasoning to Find Area

NAME ______________________ DATE ____________ PERIOD ____________

Learning Goal Let's decompose and rearrange shapes to find their areas.

Warm Up

3.1 Comparing Regions

Is the area of Figure A greater than, less than, or equal to the area of the shaded region in Figure B? Be prepared to explain your reasoning.

Figure A

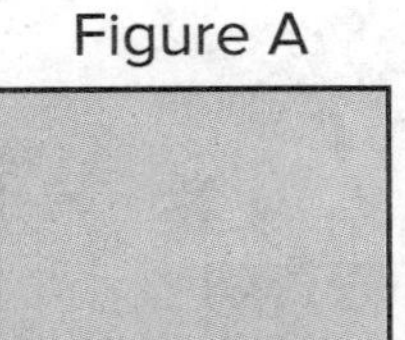

Figure B

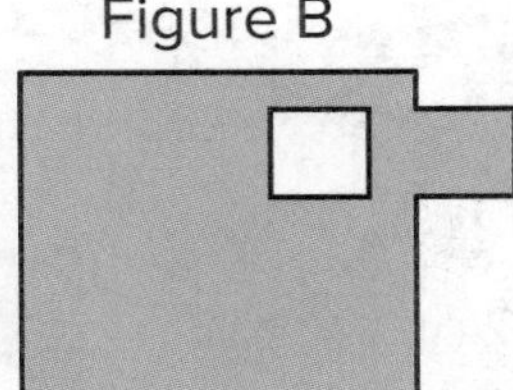

Activity

3.2 On the Grid

Each grid square is 1 square unit. Find the area, in square units, of each shaded region without counting every square. Be prepared to explain your reasoning.

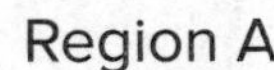

Region A

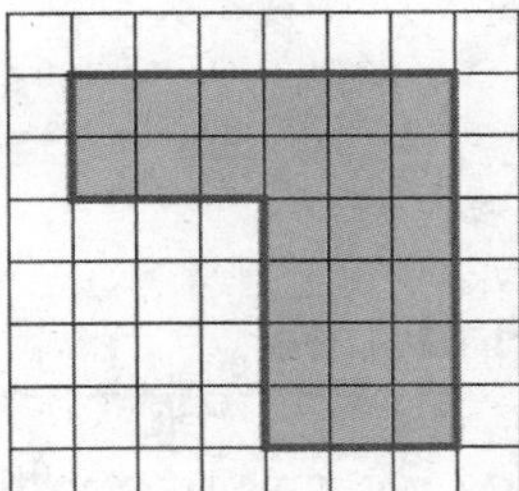

Region B

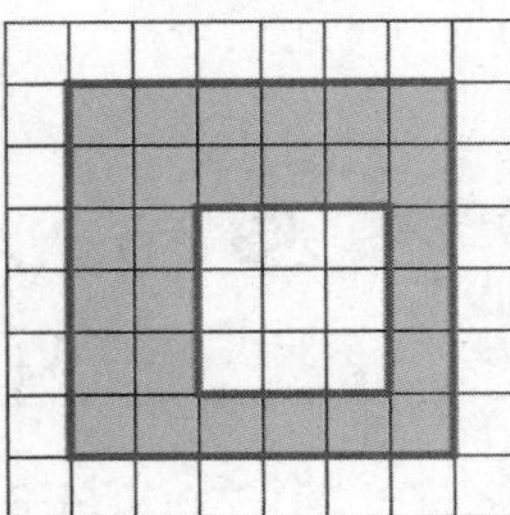

Region C

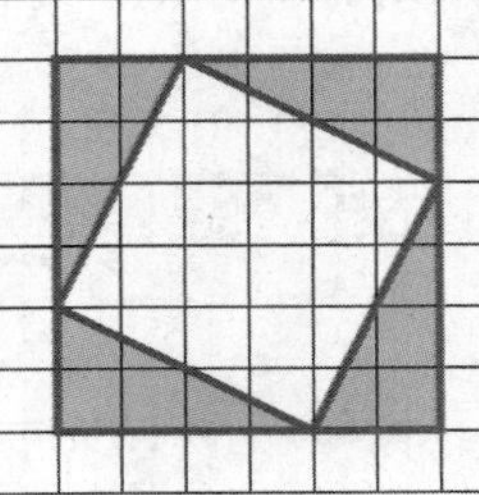

Region D

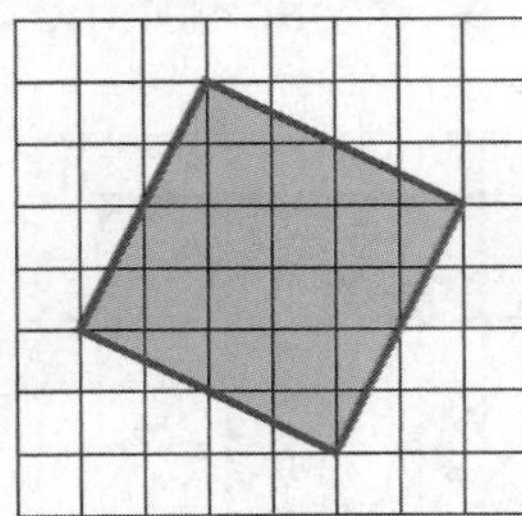

Are you ready for more?

Rearrange the triangles from Figure C so they fit inside Figure D. Draw and color a diagram of your work.

Activity
3.3 Off the Grid

Find the area of the shaded region(s) of each figure. Explain or show your reasoning.

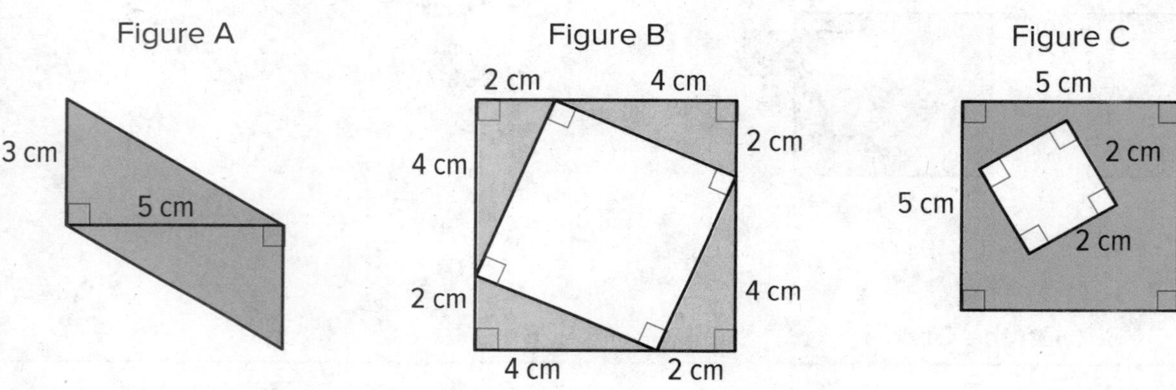

Summary
Reasoning to Find Area

There are different strategies we can use to find the area of a region. We can:

1. Decompose it into shapes whose areas you know how to calculate; find the area of each of those shapes, and then add the areas.

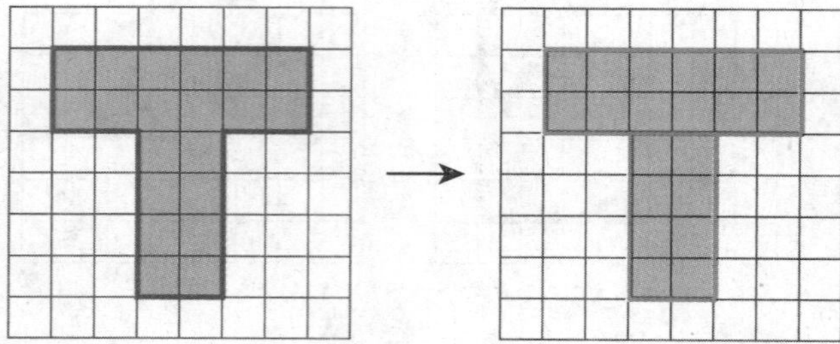

NAME ______________________________ DATE ______________ PERIOD ____________

2. Decompose it and rearrange the pieces into shapes whose areas you know how to calculate; find the area of each of those shapes, and then add the areas.

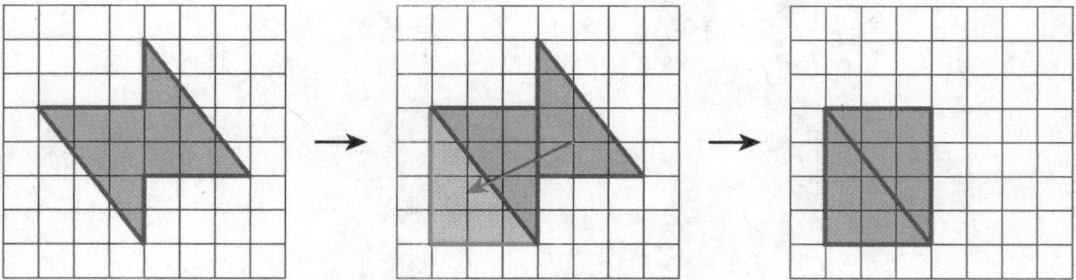

3. Consider it a shape with a missing piece; calculate the area of the shape and the missing piece, and then subtract the area of the piece from the area of the shape.

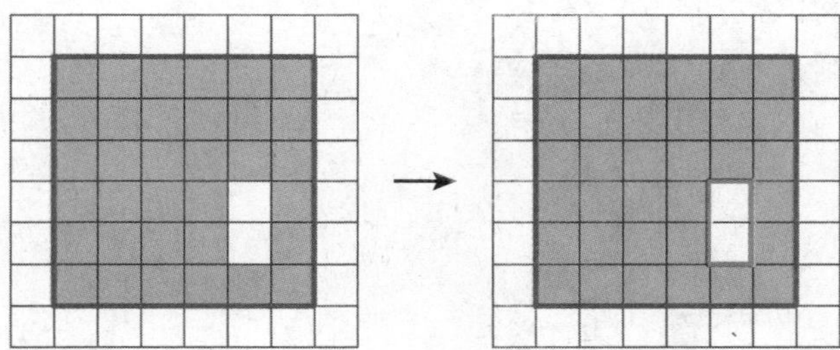

The area of a figure is always measured in square units. When both side lengths of a rectangle are given in centimeters, then the area is given in square centimeters.

For example, the area of this rectangle is 32 square centimeters.

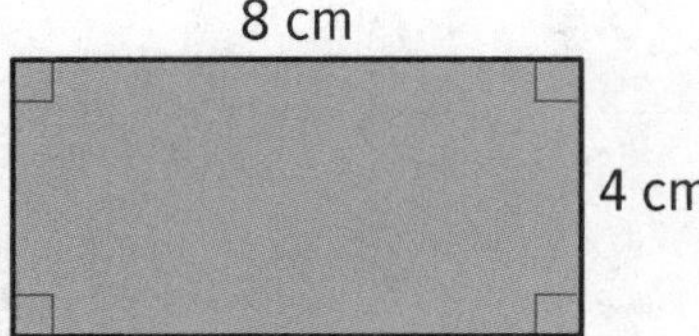

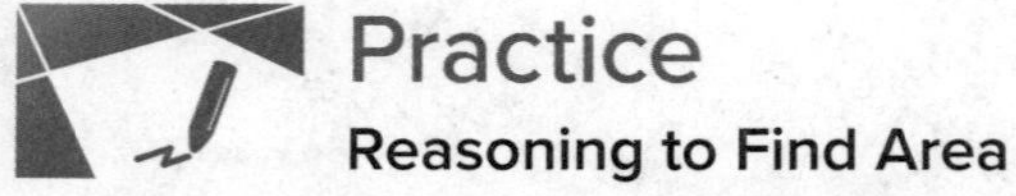

Practice
Reasoning to Find Area

1. Find the area of each shaded region. Show your reasoning.

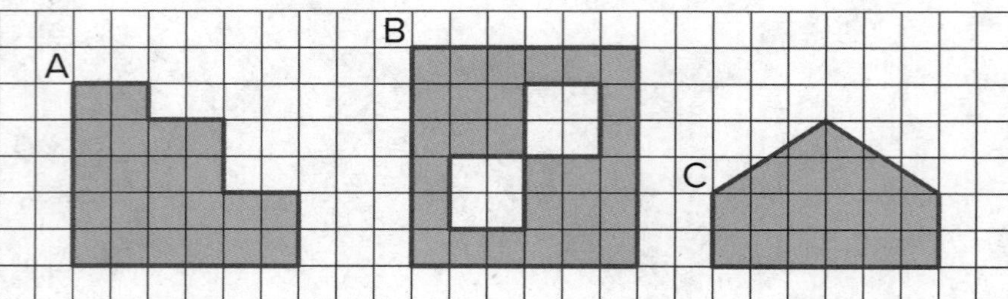

NAME ______________________ DATE ____________ PERIOD ____________

2. Find the area of each shaded region. Show or explain your reasoning.

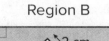

Region A

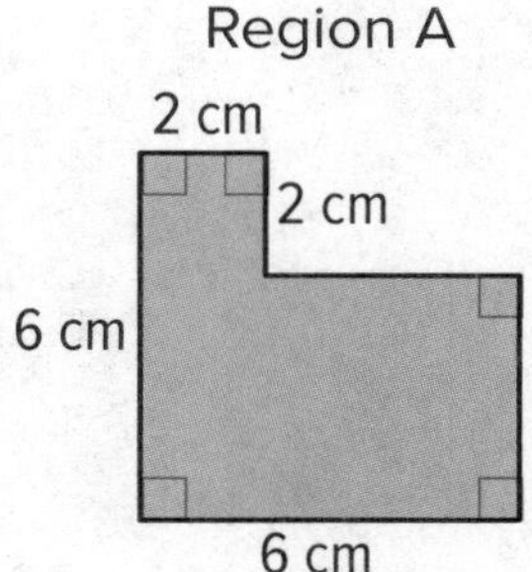

Region B

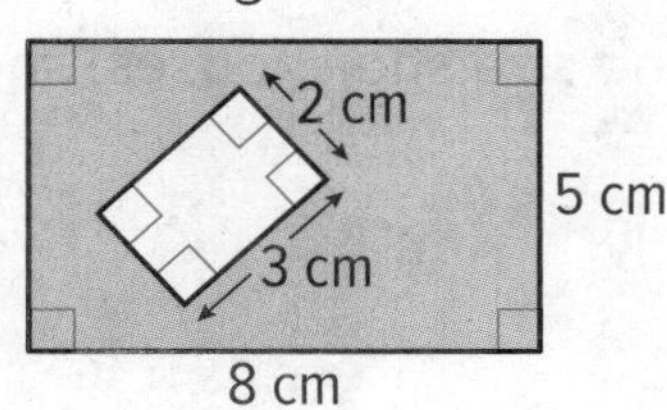

Region C

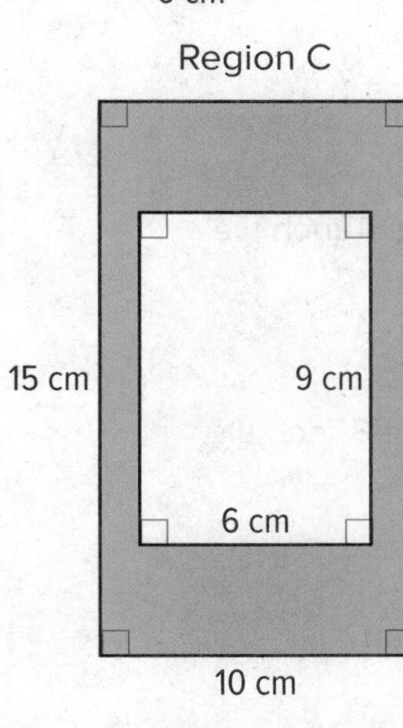

Region D

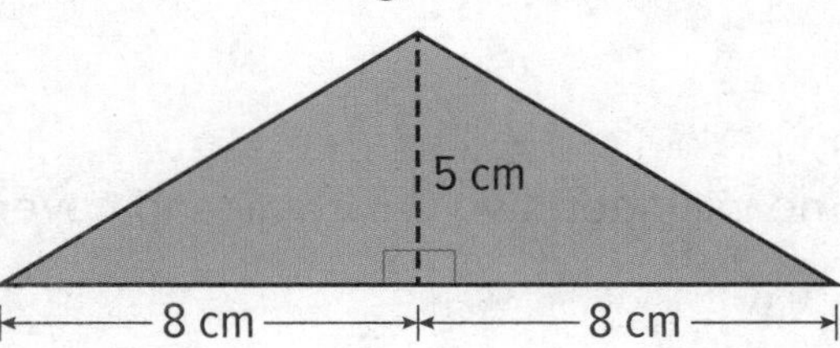

3. Two plots of land have very different shapes. Noah said that both plots of land have the same area. Do you agree with Noah? Explain your reasoning.

Plot A

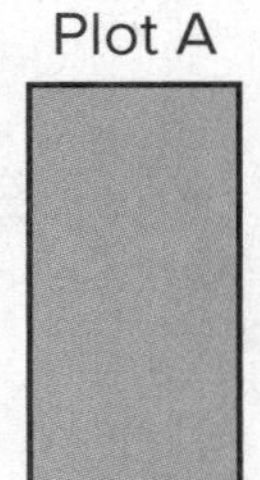

Plot B

4. A homeowner is deciding on the size of tiles to use to fully tile a rectangular wall in her bathroom that is 80 inches by 40 inches. The tiles are squares and come in three side lengths: 8 inches, 4 inches, and 2 inches. State if you agree with each statement about the tiles. Explain your reasoning. **(Lesson 1-2)**

   a. Regardless of the size she chooses, she will need the same number of tiles.

   b. Regardless of the size she chooses, the area of the wall that is being tiled is the same.

   c. She will need two 2-inch tiles to cover the same area as one 4-inch tile.

   d. She will need four 4-inch tiles to cover the same area as one 8-inch tile.

   e. If she chooses the 8-inch tiles, she will need a quarter as many tiles as she would with 2-inch tiles.

Lesson 1-4

Parallelograms

NAME ______________________ DATE ____________ PERIOD ________

Learning Goal Let's investigate the features and area of parallelograms.

Warm Up

4.1 Features of a Parallelogram

Figures A, B, and C are **parallelograms**. Figures D, E, and F are *not* parallelograms.

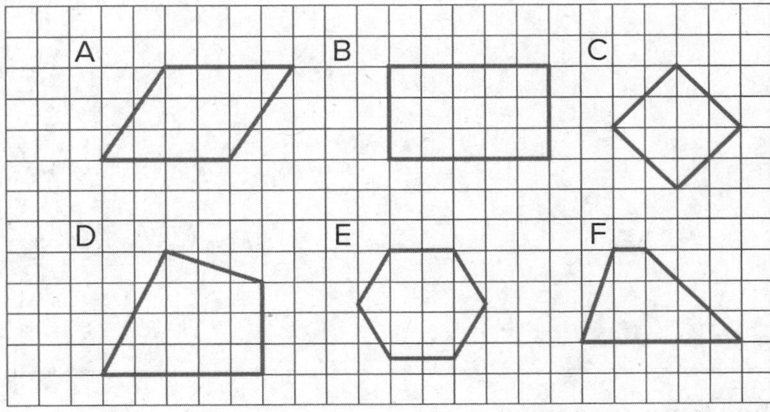

Study the examples and non-examples. What do you notice about:

1. the number of sides that a parallelogram has?

2. opposite sides of a parallelogram?

3. opposite angles of a parallelogram?

Activity
4.2 Area of a Parallelogram

Find the area of each parallelogram. Show your reasoning.

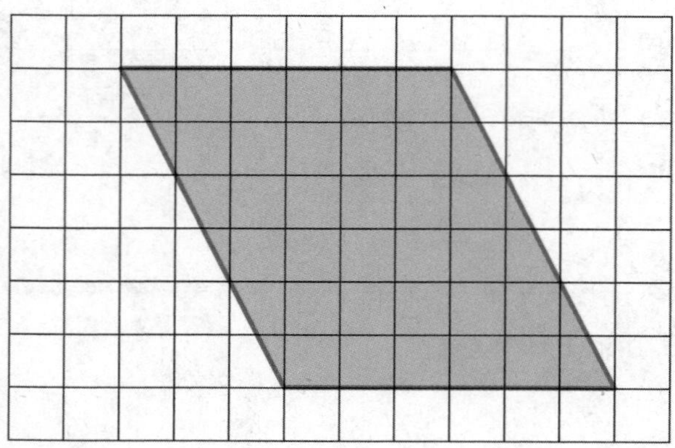

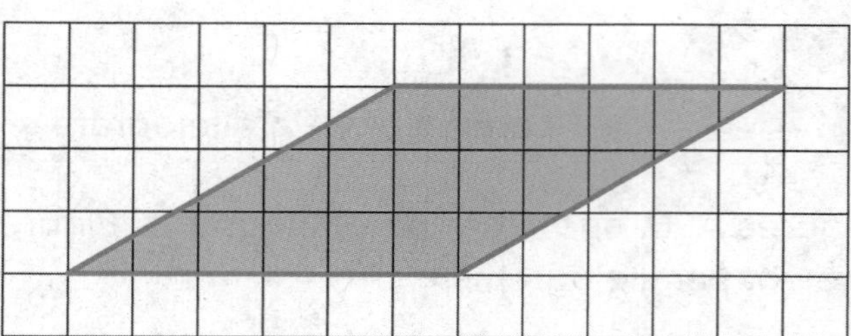

Activity
4.3 Lots of Parallelograms

Find the area of each parallelogram. Show your reasoning.

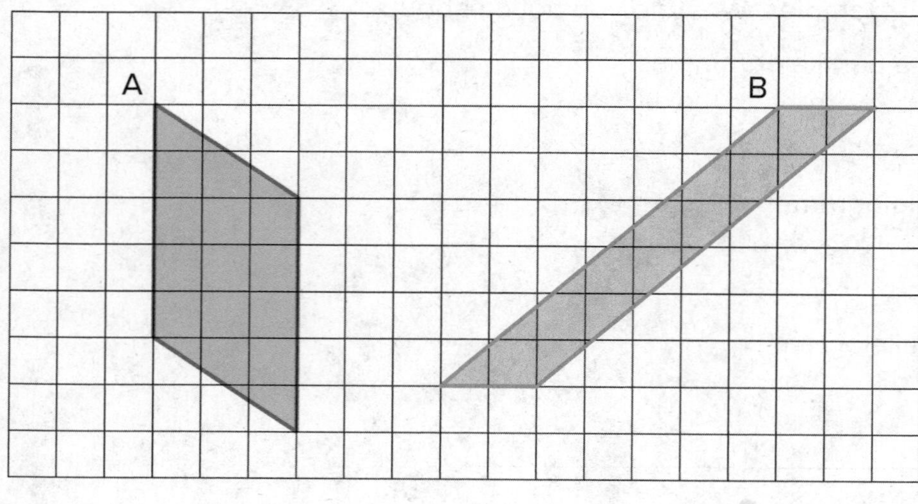

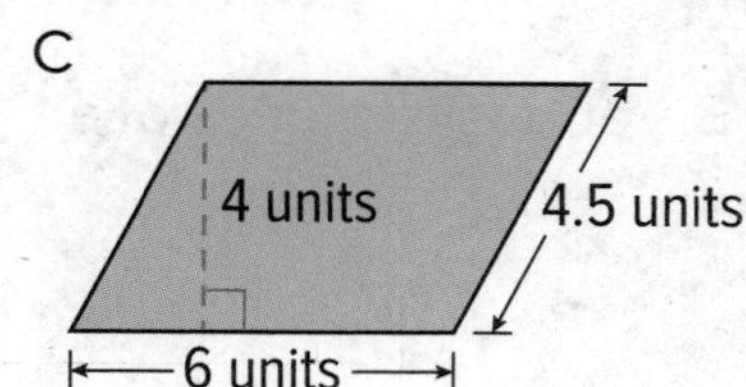

NAME _______________ DATE _______________ PERIOD _______________

Summary
Parallelograms

A **parallelogram** is a quadrilateral (it has four sides). The opposite sides of a parallelogram are parallel. It is also true that:

- The opposite sides of a parallelogram have equal length.
- The opposite angles of a parallelogram have equal measure.

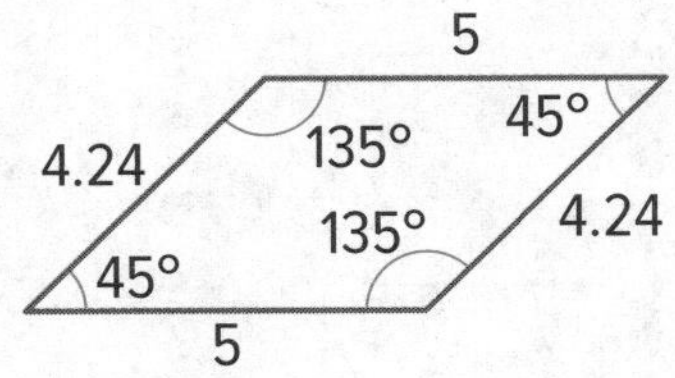

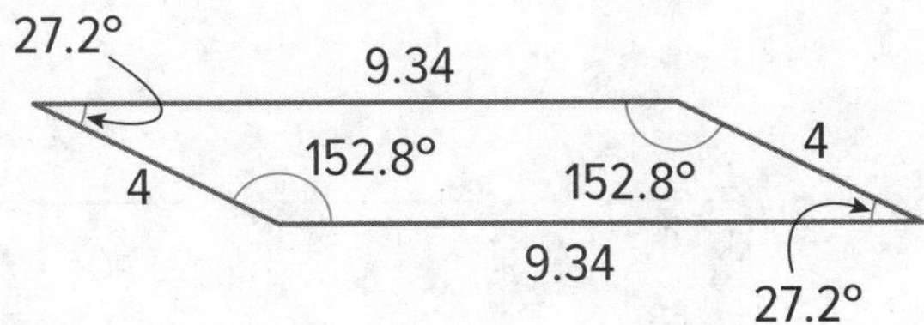

There are several strategies for finding the area of a **parallelogram**.

1. We can decompose and rearrange a parallelogram to form a rectangle. Here are three ways:

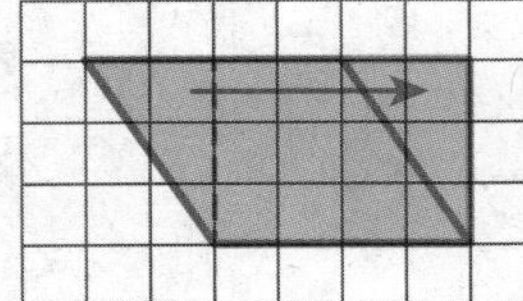 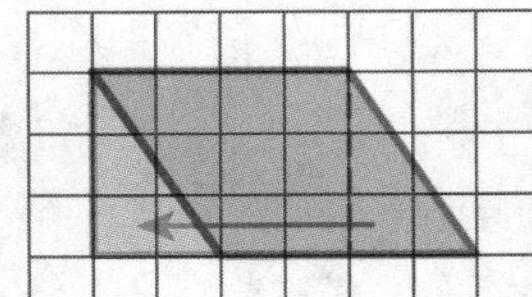 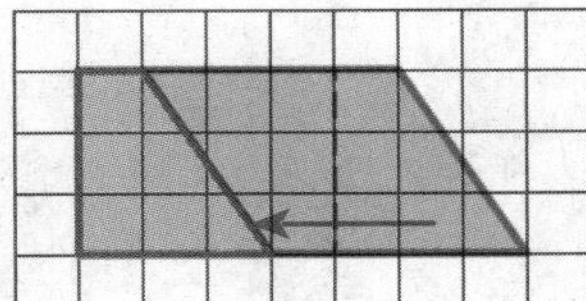

2. We can enclose the parallelogram and then subtract the area of the two triangles in the corner.

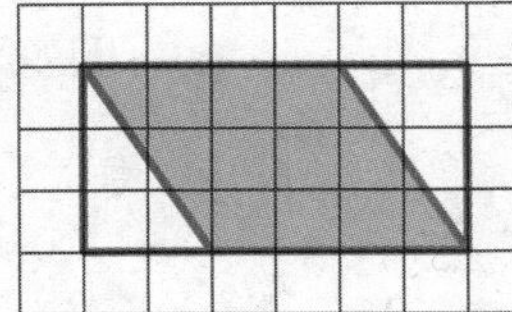 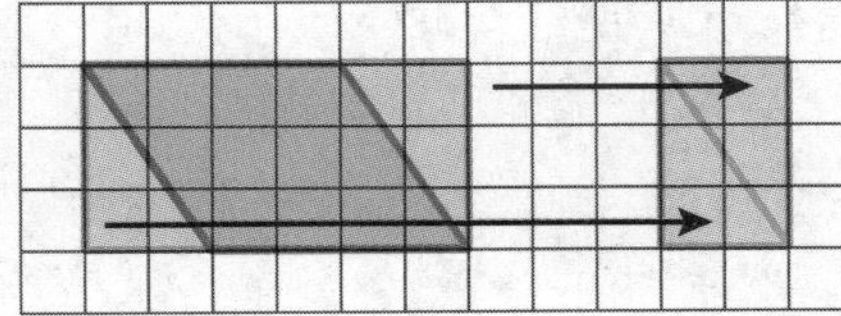

Both of these ways will work for any parallelogram.

However, for some parallelograms, the process of decomposing and rearranging requires a lot more steps than if we enclose the parallelogram with a rectangle and subtract the combined area of the two triangles in the corners.

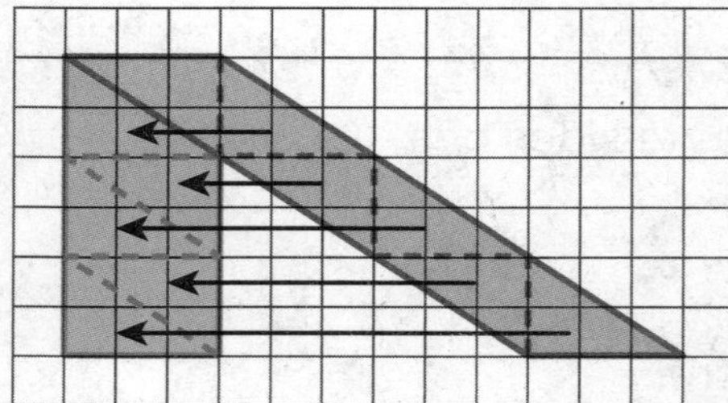

Glossary

parallelogram

quadrilateral

NAME ______________________ DATE ____________ PERIOD ________

Practice

Parallelograms

1. Select **all** of the parallelograms. For each figure that is *not* selected, explain how you know it is not a parallelogram.

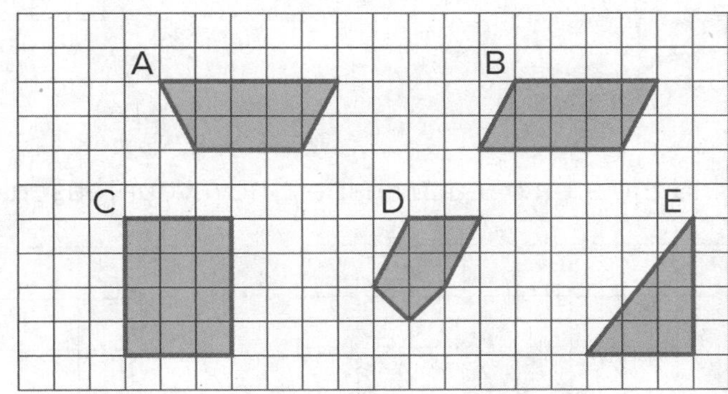

2. a. Decompose and rearrange this parallelogram to make a rectangle.

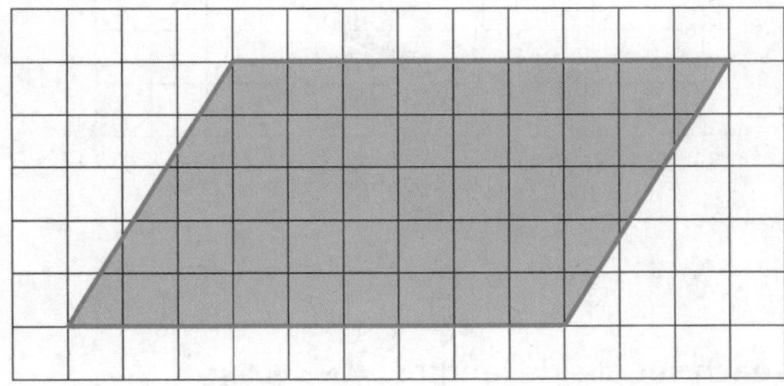

b. What is the area of the parallelogram? Explain your reasoning.

3. Find the area of the parallelogram.

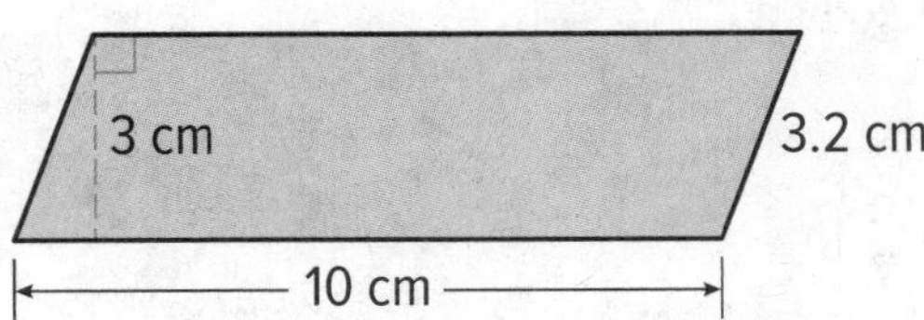

4. Explain why this quadrilateral is *not* a parallelogram.

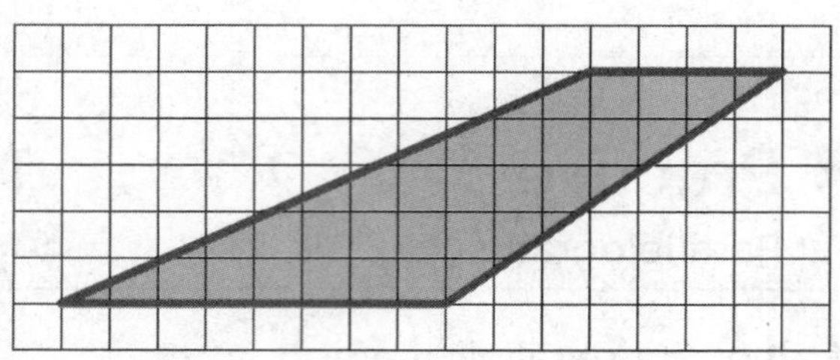

5. Find the area of each shape. Show your reasoning. **(Lesson 1-3)**

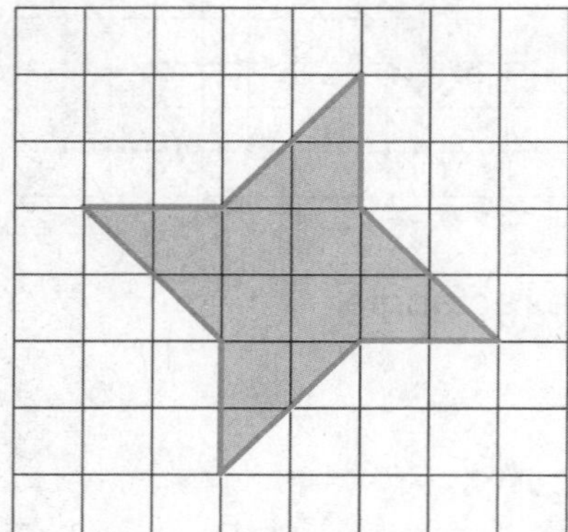

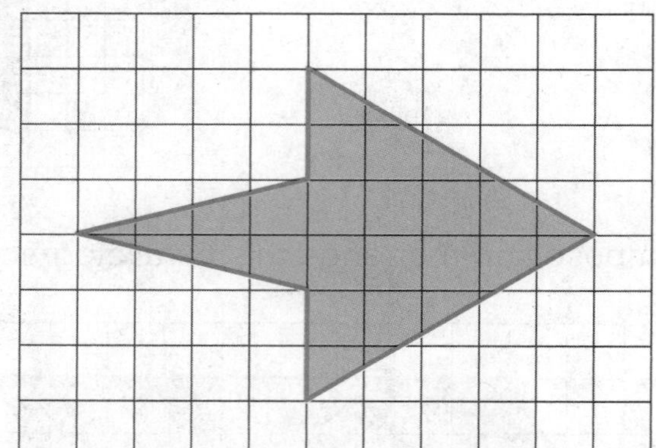

6. Find the area of the rectangle with each set of side lengths. **(Lesson 1-1)**

   a. 5 in and $\frac{1}{3}$ in

   b. 5 in and $\frac{4}{3}$ in

   c. $\frac{5}{2}$ in and $\frac{4}{3}$ in

   d. $\frac{7}{6}$ in and $\frac{6}{7}$ in

Lesson 1-5

Bases and Heights of Parallelograms

NAME ________________________ DATE ____________ PERIOD ____________

Learning Goal Let's investigate the area of parallelograms some more.

Warm Up

5.1 A Parallelogram and Its Rectangles

Elena and Tyler were finding the area of this parallelogram.

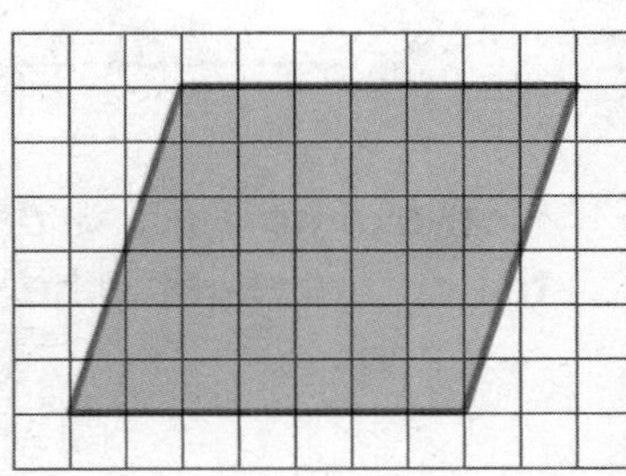

Here is how Elena did it.

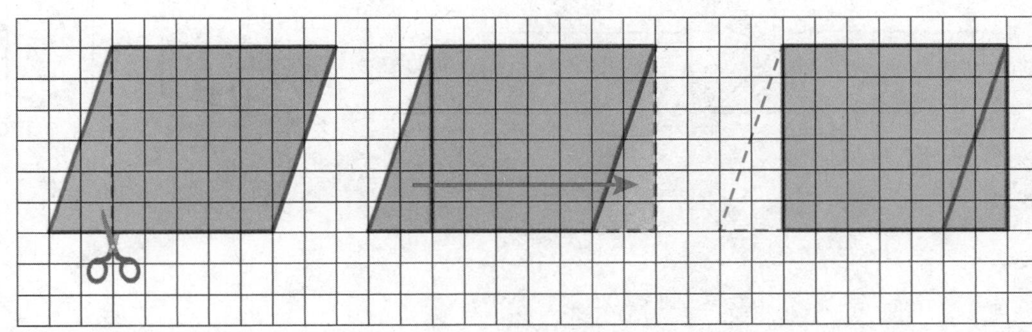

Here is how Tyler did it.

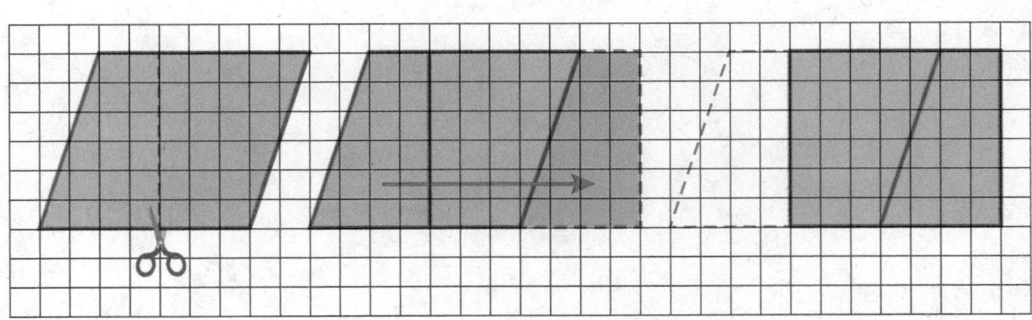

How are the two strategies for finding the area of a parallelogram the same? How they are different?

Activity

5.2 The Right Height?

Study the examples and non-examples of **bases** and **heights** of parallelograms.

- **Examples:** The dashed segments in these drawings represent the corresponding height for the given base.

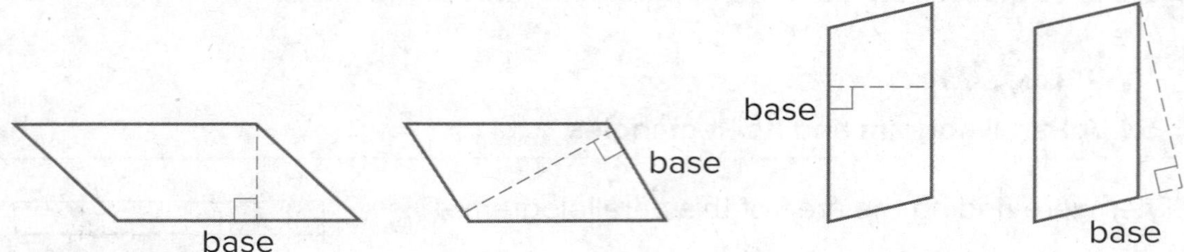

- **Non-examples:** The dashed segments in these drawings do *not* represent the corresponding height for the given base.

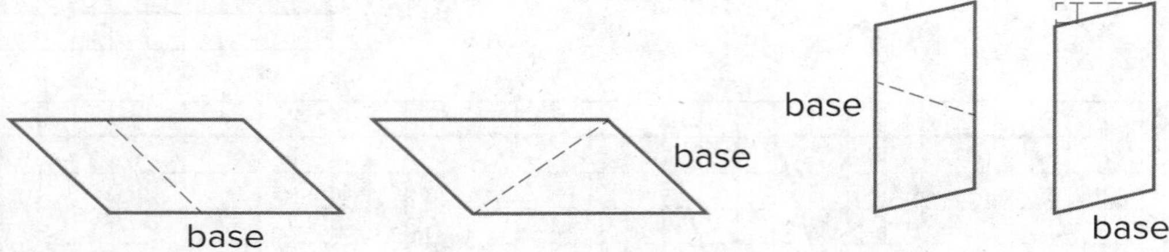

NAME ______________________ DATE ______________ PERIOD ______________

1. Select **all** the statements that are true about bases and heights in a parallelogram.

   A. Only a horizontal side of a parallelogram can be a base.

   B. Any side of a parallelogram can be a base.

   C. A height can be drawn at any angle to the side chosen as the base.

   D. A base and its corresponding height must be perpendicular to each other.

   E. A height can only be drawn inside a parallelogram.

   F. A height can be drawn outside of the parallelogram, as long as it is drawn at a 90-degree angle to the base.

   G. A base cannot be extended to meet a height.

2. Five students labeled a base b and a corresponding height h for each of these parallelograms. Are all drawings correctly labeled? Explain how you know.

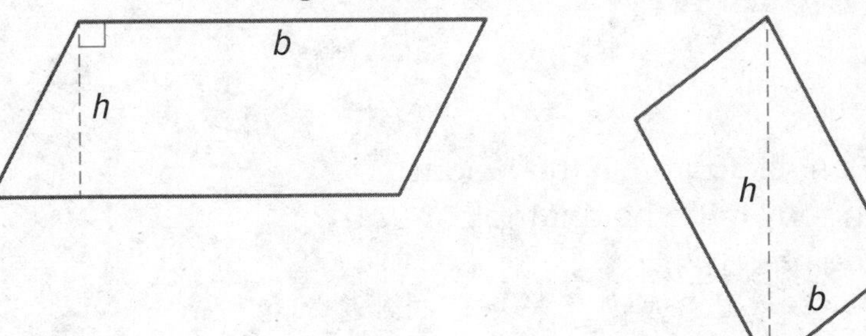

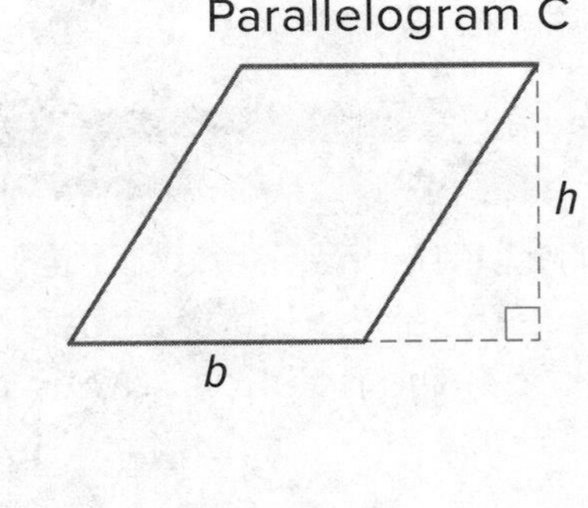

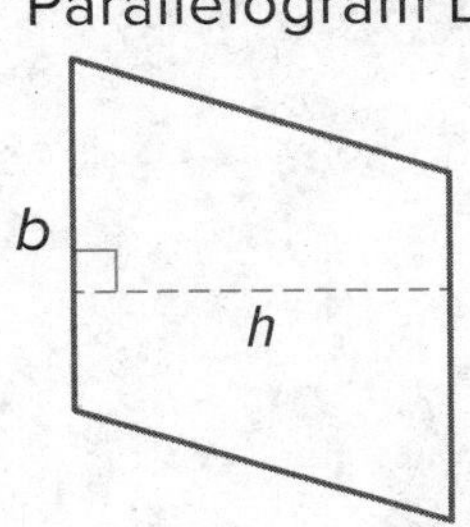

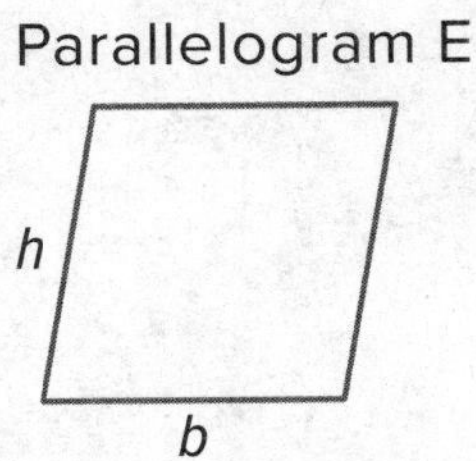

Activity
5.3 Finding the Formula for Area of Parallelograms

For each parallelogram:

- Identify a base and a corresponding height. Record their lengths in the table.
- Find the area of the parallelogram. Record it in the last column of the table.

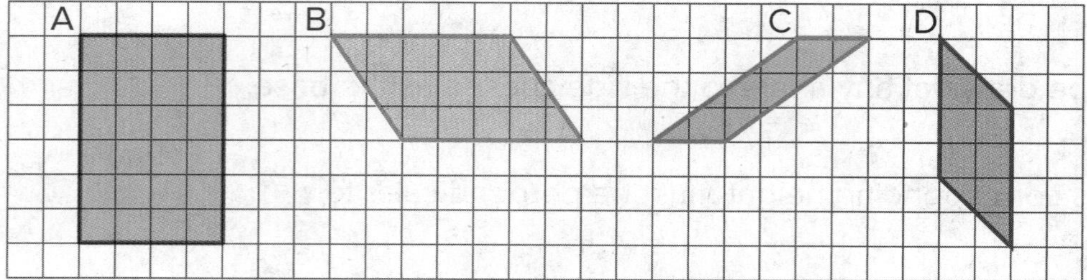

Parallelogram	Base (units)	Height (units)	Area (sq units)
A			
B			
C			
D			
Any Parallelogram	b	h	

In the last row, write an expression for the area of any parallelogram, using b and h.

Are you ready for more?

1. What happens to the area of a parallelogram if the height doubles but the base is unchanged? If the height triples? If the height is 100 times the original?

2. What happens to the area if both the base and the height double? Both triple? Both are 100 times their original lengths?

NAME _______________ DATE _______________ PERIOD _______________

Summary

Bases and Heights of Parallelograms

We can choose any of the four sides of a parallelogram as the **base**. Both the side (the segment) and its length (the measurement) are called the base.

If we draw any perpendicular segment from a point on the base to the opposite side of the parallelogram, that segment will always have the same length. We call that value the **height**. There are infinitely many segments that can represent the height!

Here are two copies of the same parallelogram.

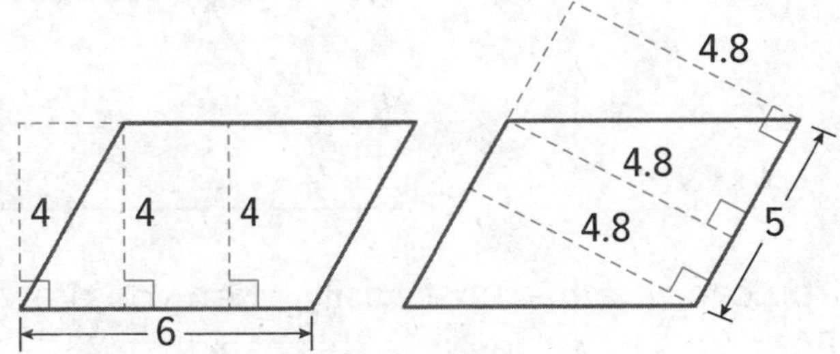

On the left, the side that is the base is 6 units long. Its corresponding height is 4 units.

On the right, the side that is the base is 5 units long. Its corresponding height is 4.8 units.

For both, three different segments are shown to represent the height. We could draw in many more!

No matter which side is chosen as the base, the area of the parallelogram is the product of that base and its corresponding height. We can check this:

$$4 \times 6 = 24 \quad \text{and} \quad 4.8 \times 5 = 24$$

We can see why this is true by decomposing and rearranging the parallelograms into rectangles.

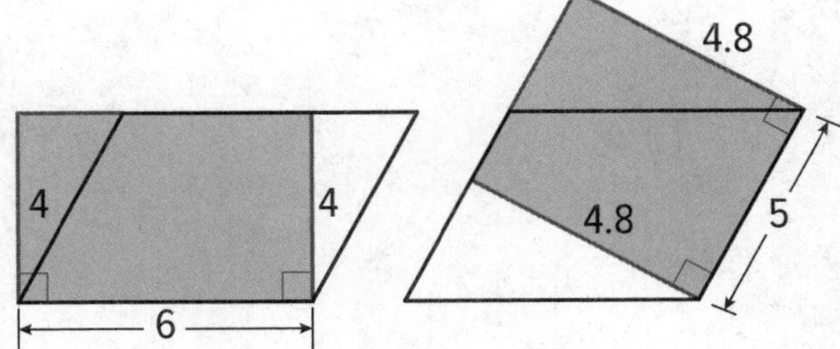

Notice that the side lengths of each rectangle are the base and height of the parallelogram.

Even though the two rectangles have different side lengths, the products of the side lengths are equal, so they have the same area!

And both rectangles have the same area as the parallelogram.

We often use letters to stand for numbers. If b is base of a parallelogram (in units), and h is the corresponding height (in units), then the area of the parallelogram (in square units) is the product of these two numbers, $b \cdot h$.

Notice that we write the multiplication symbol with a small dot instead of a $\times$ symbol. This is so that we don't get confused about whether $\times$ means multiply, or whether the letter x is standing in for a number.

In high school, you will be able to prove that a perpendicular segment from a point on one side of a parallelogram to the opposite side will always have the same length.

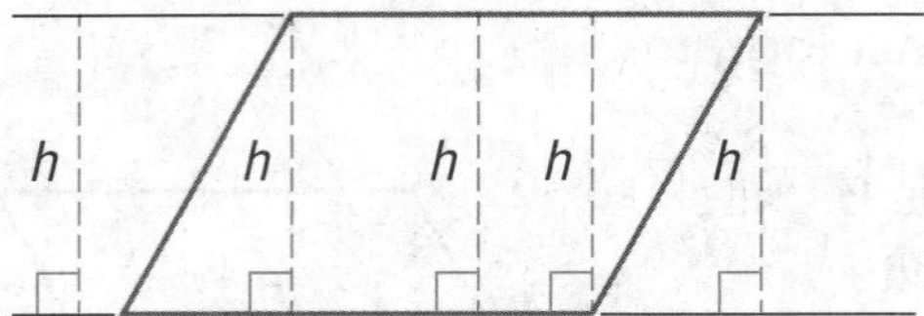

You can see this most easily when you draw a parallelogram on graph paper. For now, we will just use this as a fact.

Glossary

base (of a parallelogram or triangle)

height (of a parallelogram or triangle)

NAME _______________________ DATE ____________ PERIOD ________

Practice
Bases and Heights of Parallelograms

1. Select **all** parallelograms that have a correct height labeled for the given base.

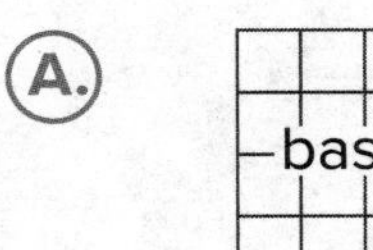

Ⓐ

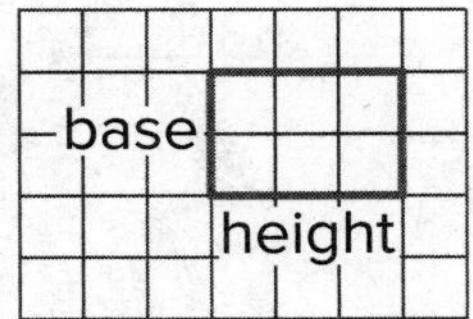

Ⓑ

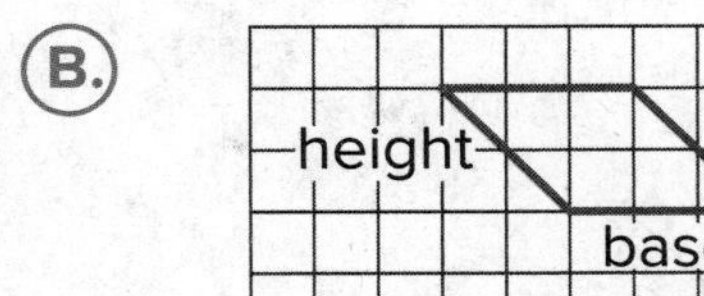

Ⓒ

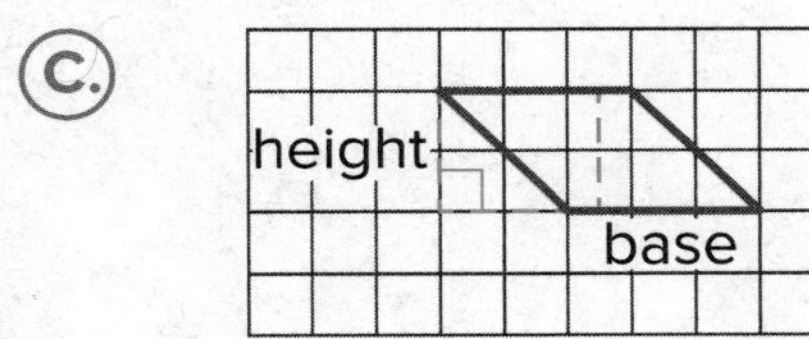

Ⓓ

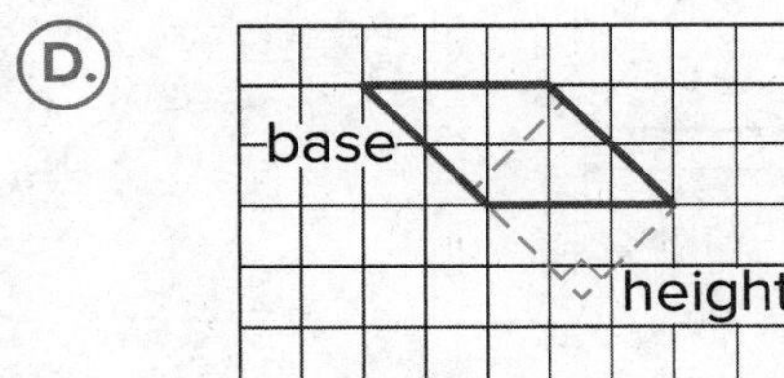

2. The side labeled *b* has been chosen as the base for this parallelogram. Draw a segment showing the height corresponding to that base.

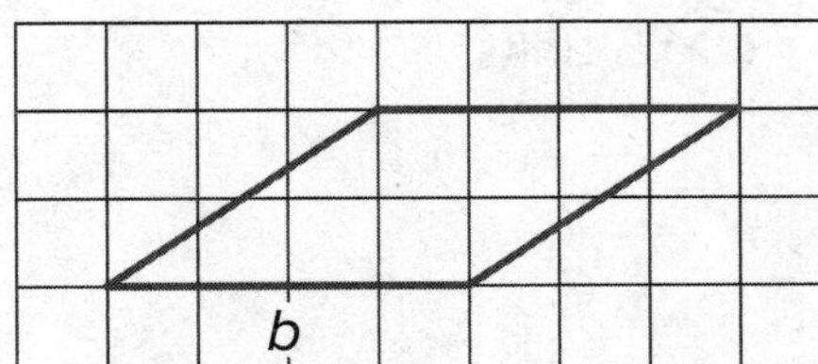

3. Find the area of each parallelogram.

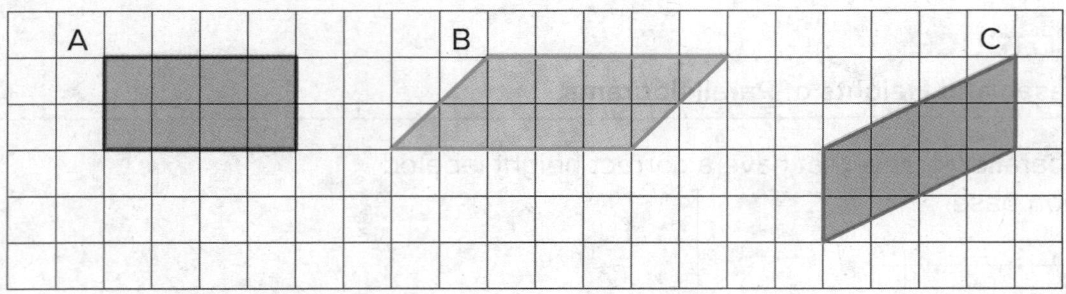

4. If the side that is 6 units long is the base of this parallelogram, what is its corresponding height?

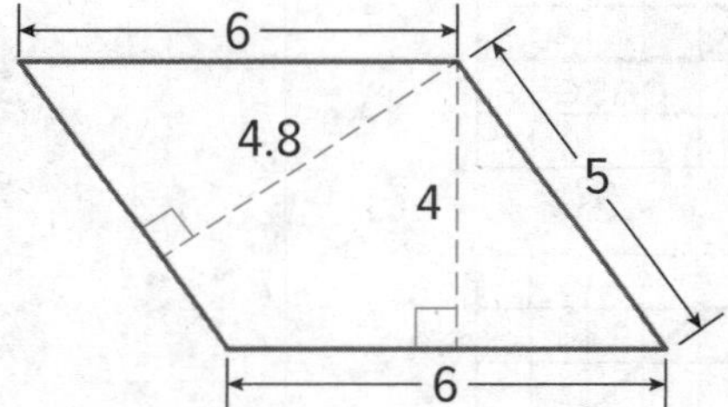

A. 6 units

B. 4.8 units

C. 4 units

D. 5 units

NAME ______________________ DATE ____________ PERIOD ________

5. Find the area of each parallelogram.

Parallelogram A

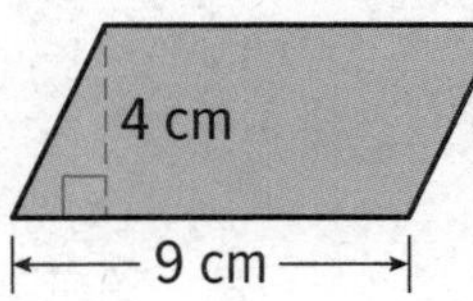

Parallelogram B

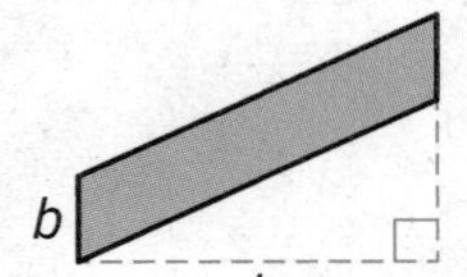

Parallelogram C

b

h

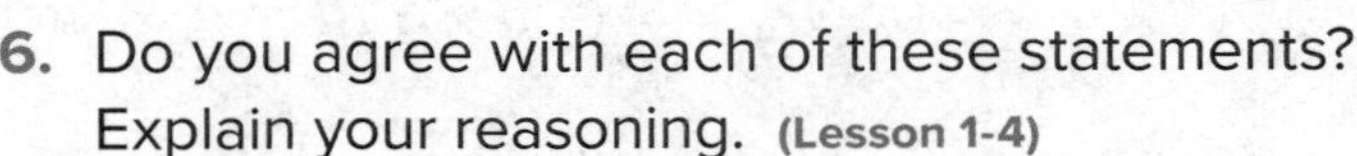

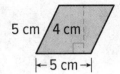

6. Do you agree with each of these statements? Explain your reasoning. **(Lesson 1-4)**

a. A parallelogram has six sides.

b. Opposite sides of a parallelogram are parallel.

c. A parallelogram can have one pair or two pairs of parallel sides.

d. All sides of a parallelogram have the same length.

e. All angles of a parallelogram have the same measure.

7. A square with an area of 1 square meter is decomposed into 9 identical small squares. Each small square is decomposed into two identical triangles. **(Lesson 1-2)**

   a. What is the area, in square meters, of 6 triangles? If you get stuck, consider drawing a diagram.

   b. How many triangles are needed to compose a region that is $1\frac{1}{2}$ square meters?

Lesson 1-6

Area of Parallelograms

NAME ________________ DATE ________ PERIOD ________

Learning Goal Let's practice finding the area of parallelograms.

Warm Up

6.1 Missing Dots

How many dots are in the image?

How do you see them?

Activity

6.2 More Areas of Parallelograms

1. Find the area of each parallelogram. Show your reasoning.

Parallelogram A

Parallelogram B

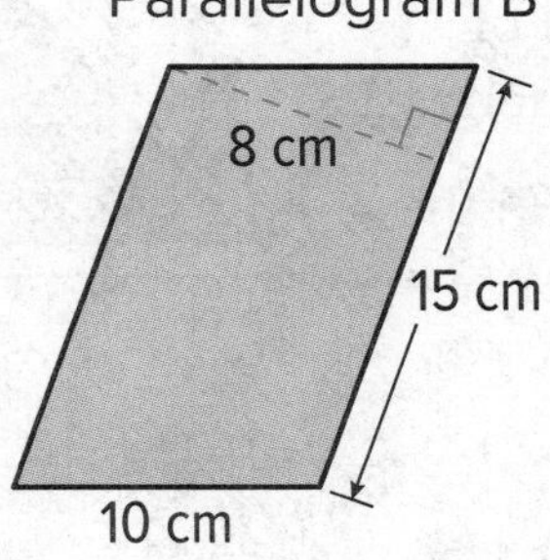

Parallelogram C

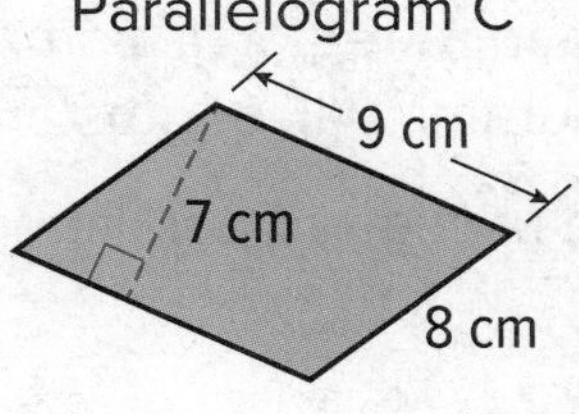

Parallelogram D

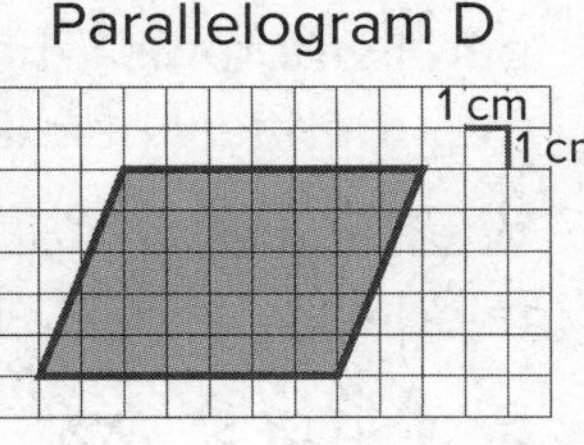

2. In Parallelogram B, what is the corresponding height for the base that is 10 cm long? Explain or show your reasoning.

3. Two different parallelograms P and Q both have an area of 20 square units. Neither of the parallelograms are rectangles.

On the grid, draw two parallelograms that could be P and Q.

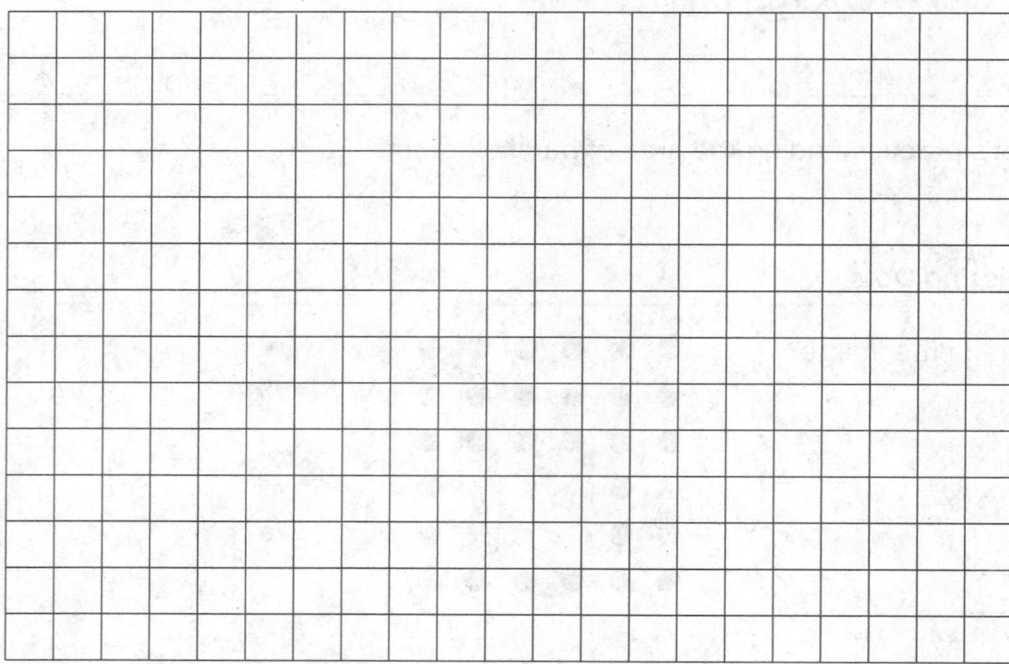

Are you ready for more?

Here is a parallelogram composed of smaller parallelograms. The shaded region is composed of four identical parallelograms. All lengths are in inches.

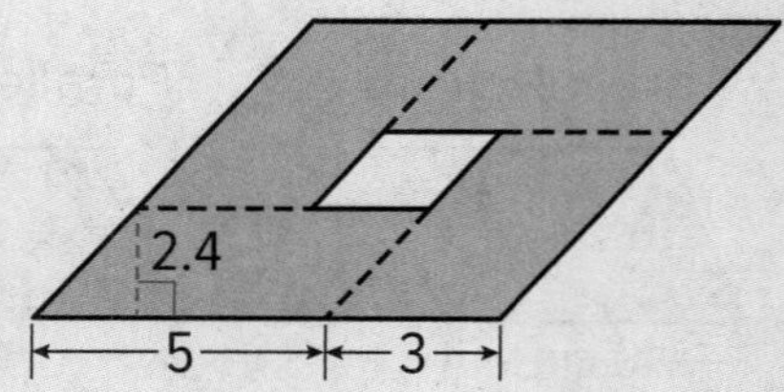

What is the area of the unshaded parallelogram in the middle?

Explain or show your reasoning.

NAME _______________ DATE _______________ PERIOD _______________

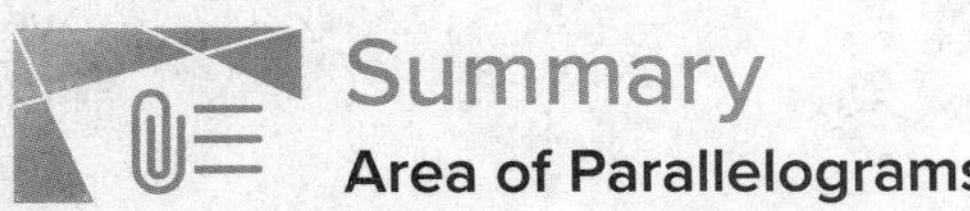

Summary
Area of Parallelograms

Any pair of base and corresponding height can help us find the area of a parallelogram, but some base-height pairs are more easily identified than others.

- When a parallelogram is drawn on a grid and has *horizontal* sides, we can use a horizontal side as the base. When it has *vertical* sides, we can use a vertical side as the base. The grid can help us find (or estimate) the lengths of the base and of the corresponding height.

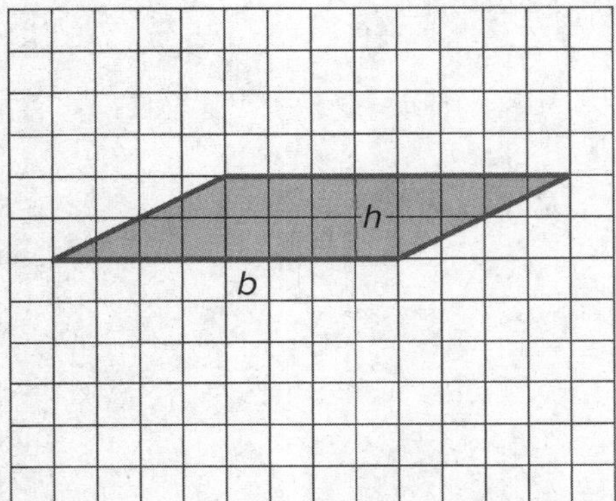

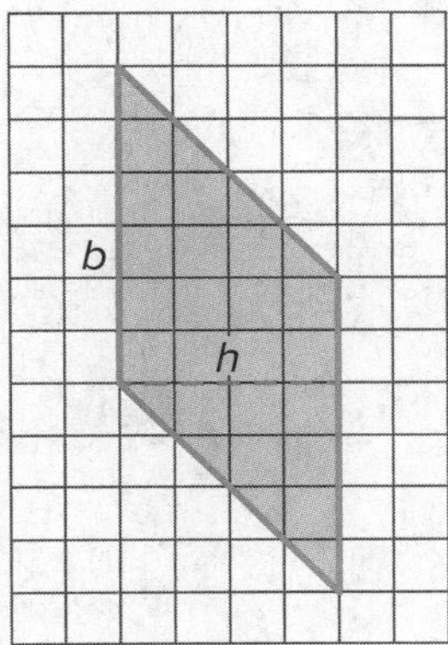

- When a parallelogram is *not* drawn on a grid, we can still find its area if a base and a corresponding height are known.

In this parallelogram, the corresponding height for the side that is 10 units long is not given, but the height for the side that is 8 units long is given. This base-height pair can help us find the area.

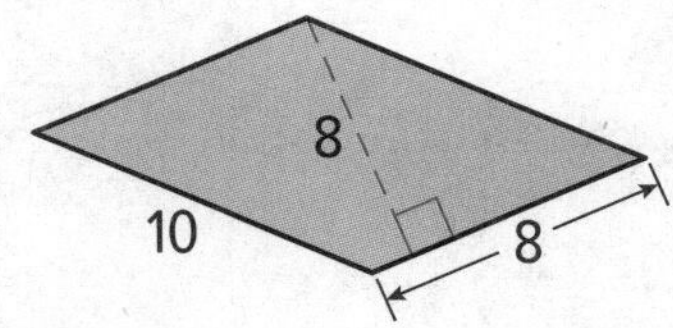

Regardless of their shape, parallelograms that have the same base and the same height will have the same area; the product of the base and height will be equal. Here are some parallelograms with the same pair of base-height measurements.

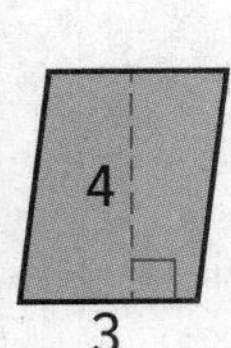

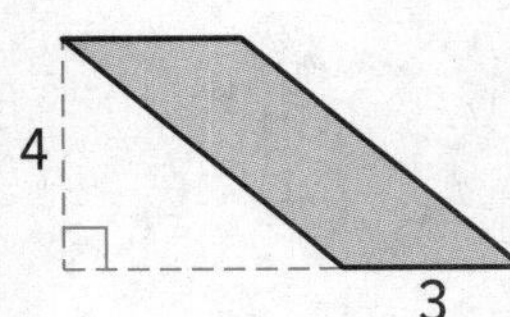

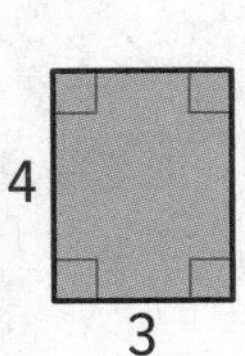

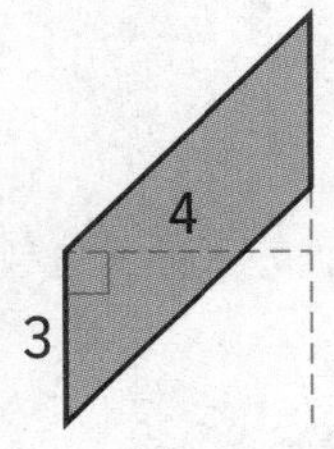

Practice

Area of Parallelograms

1. Which **three** of these parallelograms have the same area as each other? Each grid is the same size.

   A. 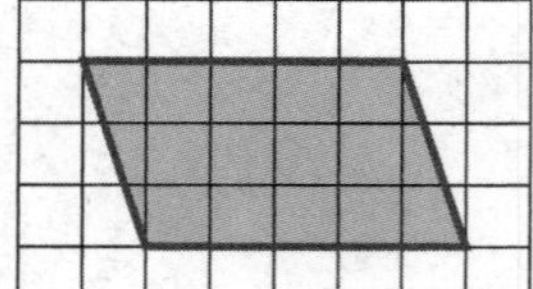

   B. 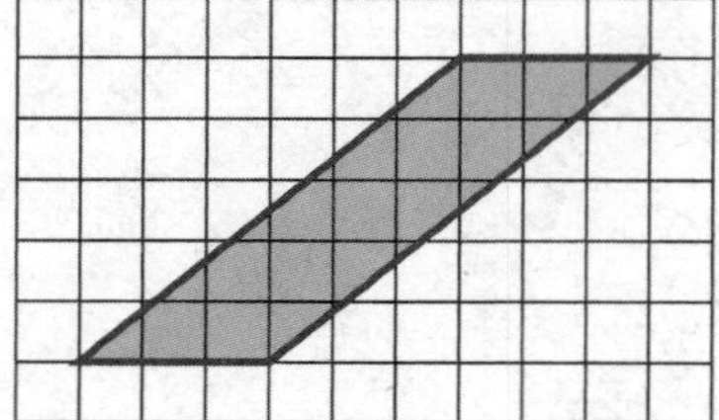

   C. 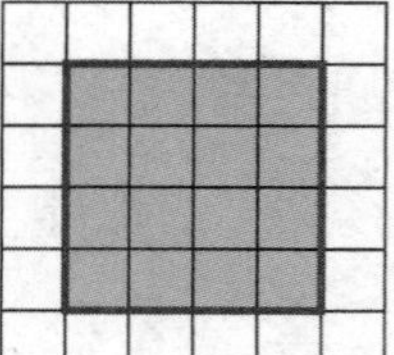

   D. 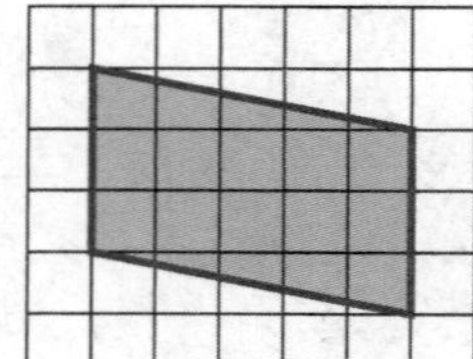

2. Which pair of base and height produces the greatest area? All measurements are in centimeters.

   A. $b = 4$, $h = 3.5$

   B. $b = 0.8$, $h = 20$

   C. $b = 6$, $h = 2.25$

   D. $b = 10$, $h = 1.4$

NAME ______________________ DATE ____________ PERIOD ____________

3. Here are the areas of three parallelograms. Use them to find the missing length (labeled with a "?") on each parallelogram.

Parallelogram A

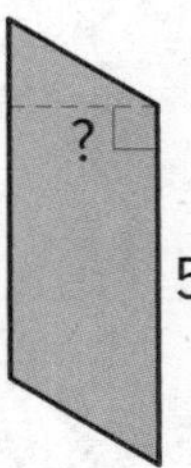

Area = 10 square units

Parallelogram B

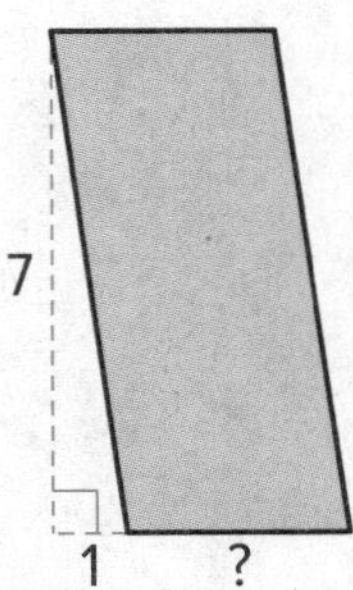

Area = 21 square units

Parallelogram C

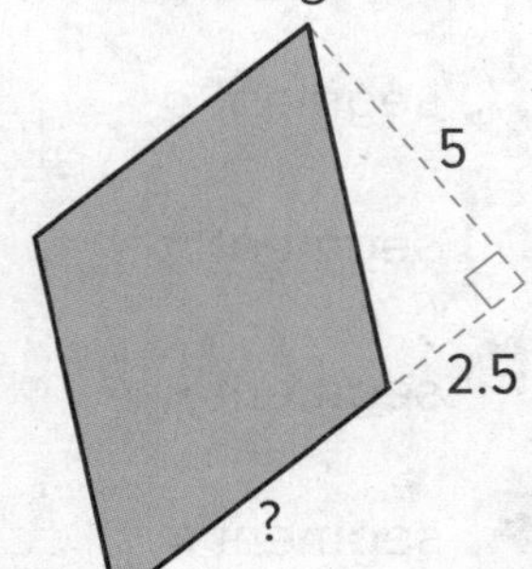

Area = 25 square units

Gerhard1302/Alamy Stock Photo

4. The Dockland Building in Hamburg, Germany is shaped like a parallelogram. If the length of the building is 86 meters and its height is 55 meters, what is the area of this face of the building?

5. Select **all** segments that could represent a corresponding height if the side m is the base. **(Lesson 1-5)**

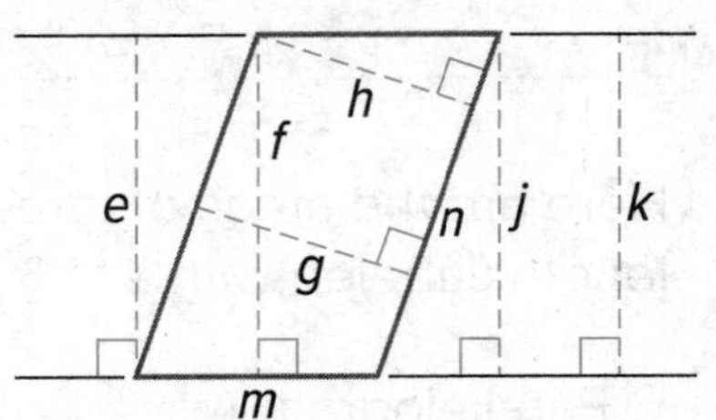

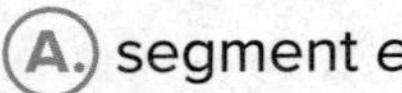

A. segment e

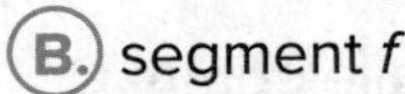

B. segment f

C. segment g

D. segment h

E. segment j

F. segment k

G. segment n

6. Find the area of the shaded region. All measurements are in centimeters. Show your reasoning. **(Lesson 1-3)**

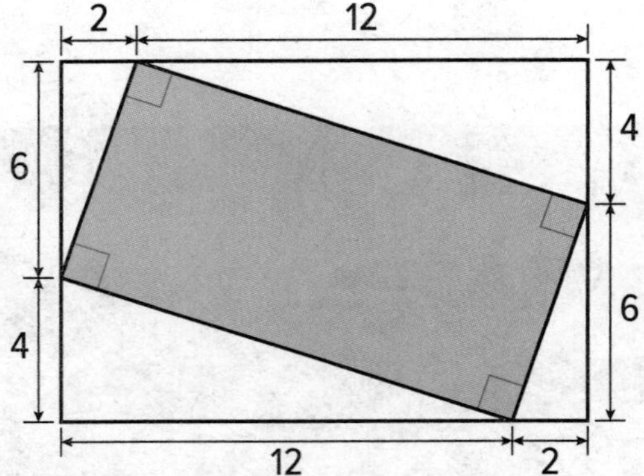

Lesson 1-7

From Parallelograms to Triangles

NAME ____________________ DATE __________ PERIOD __________

Learning Goal Let's compare parallelograms and triangles.

Warm Up

7.1 Same Parallelograms, Different Bases

Here are two copies of a parallelogram. Each copy has one side labeled as the base *b* and a segment drawn for its corresponding height and labeled *h*.

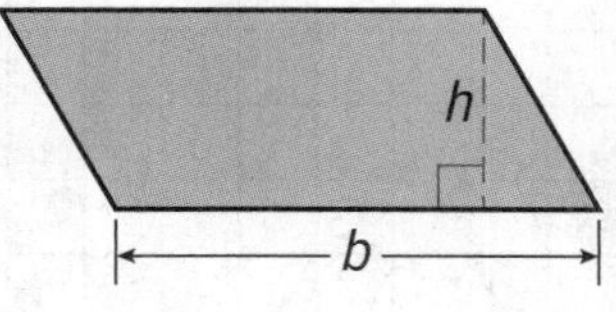

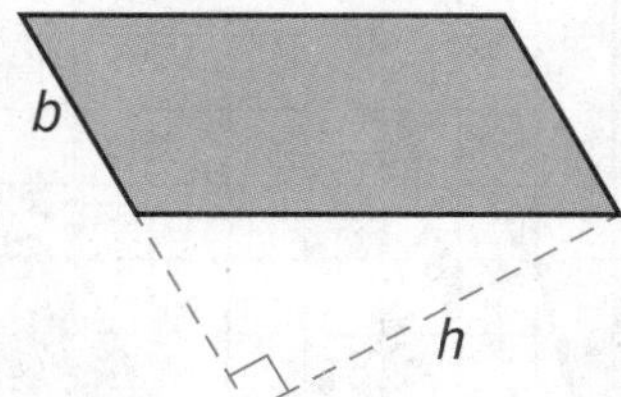

1. The base of the parallelogram on the left is 2.4 centimeters; its corresponding height is 1 centimeter. Find its area in square centimeters.

2. The height of the parallelogram on the right is 2 centimeters. How long is the base of that parallelogram? Explain your reasoning.

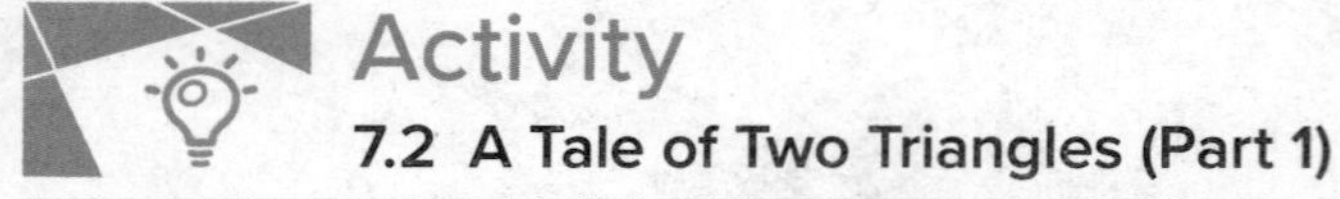

Activity
7.2 A Tale of Two Triangles (Part 1)

Two polygons are identical if they match up exactly when placed one on top of the other.

1. Draw *one* line to decompose each polygon into two identical triangles, if possible. Use a straightedge to draw your line.

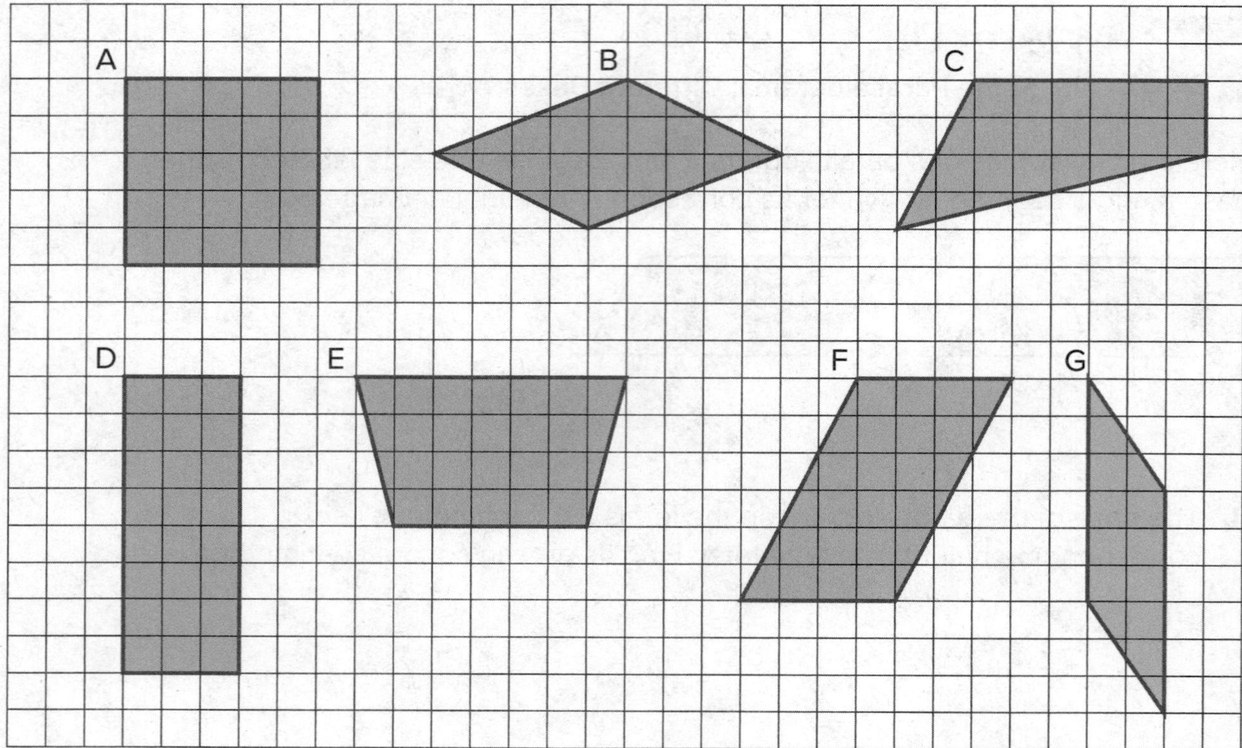

2. Which quadrilaterals can be decomposed into two identical triangles?

Pause here for a small-group discussion.

3. Study the quadrilaterals that can, in fact, be decomposed into two identical triangles. What do you notice about them? Write a couple of observations about what these quadrilaterals have in common.

NAME ______________________ DATE ______________ PERIOD ____________

Are you ready for more?

On the grid, draw some other types of quadrilaterals that are not already shown. Try to decompose them into two identical triangles. Can you do it?

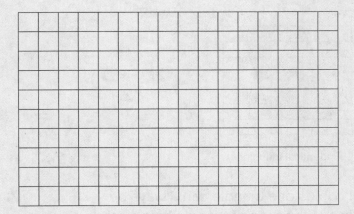

Come up with a rule about what must be true about a quadrilateral for it to be decomposed into two identical triangles.

Activity

7.3 A Tale of Two Triangles (Part 2)

Your teacher will give your group several pairs of triangles. Each group member should take 1 or 2 pairs.

1. **a.** Which pair(s) of triangles do you have?

   b. Can each pair be composed into a rectangle? A parallelogram?

2. Discuss with your group your responses to the first question. Then, complete each statement with *All*, *Some*, or *None*. Sketch 1–2 examples in the space at the right to illustrate each completed statement.

   a. ________________ of these pairs of identical triangles can be composed into a *rectangle*.

   b. ________________ of these pairs of identical triangles can be composed into a *parallelogram*.

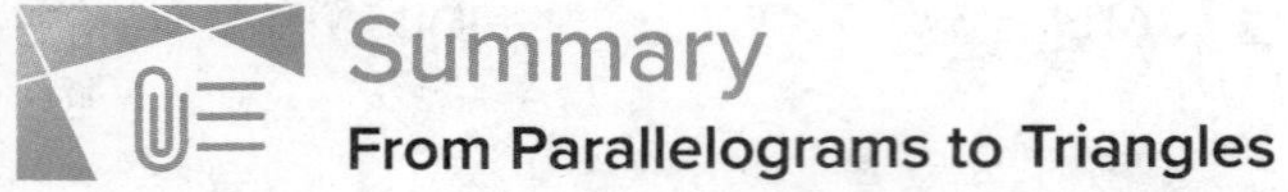

Summary

From Parallelograms to Triangles

A parallelogram can always be decomposed into two identical triangles by a segment that connects opposite vertices.

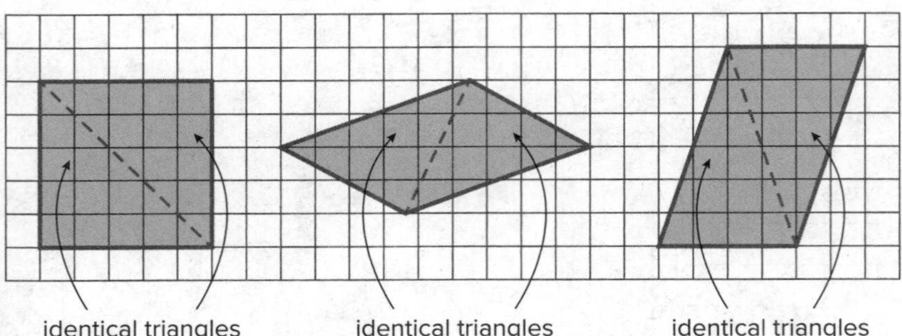

Going the other way around, two identical copies of a triangle can always be arranged to form a parallelogram, regardless of the type of triangle being used.

To produce a parallelogram, we can join a triangle and its copy along any of the three sides, so the same pair of triangles can make different parallelograms.

Here are examples of how two copies of both Triangle A and Triangle F can be composed into three different parallelograms.

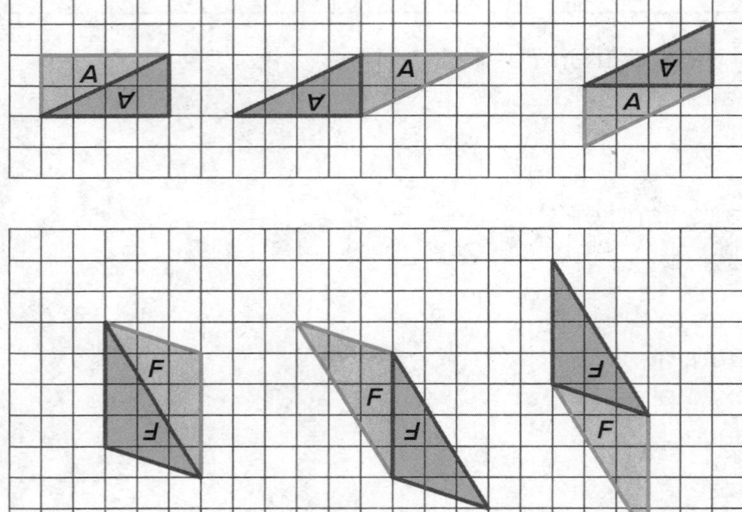

This special relationship between triangles and parallelograms can help us reason about the area of any triangle.

NAME ______________________ DATE ____________ PERIOD ________

Practice

From Parallelograms to Triangles

1. To decompose a quadrilateral into two identical shapes, Clare drew a dashed line as shown in the diagram.

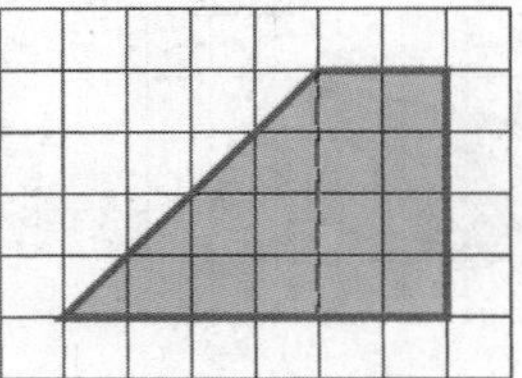

   a. She said that the two resulting shapes have the same area. Do you agree? Explain your reasoning.

   b. Did Clare partition the figure into two identical shapes? Explain your reasoning.

2. Triangle R is a right triangle. Can we use two copies of Triangle R to compose a parallelogram that is not a square? If so, explain how or sketch a solution. If not, explain why not.

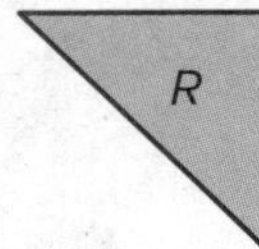

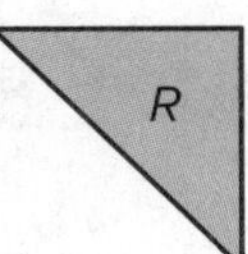

3. Two copies of this triangle are used to compose a parallelogram. Which parallelogram *cannot* be a result of the composition? If you get stuck, consider using tracing paper.

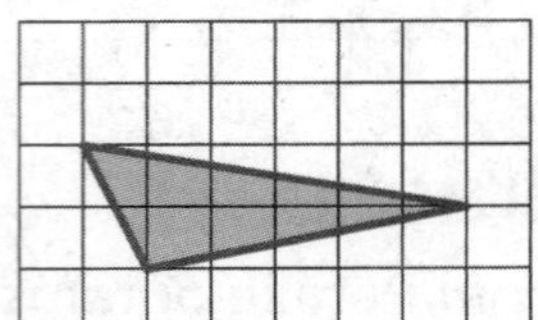

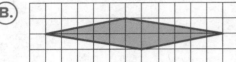

A.

C.

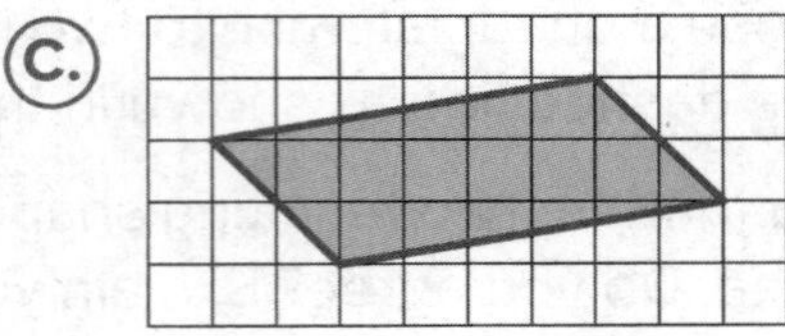

B.

D.

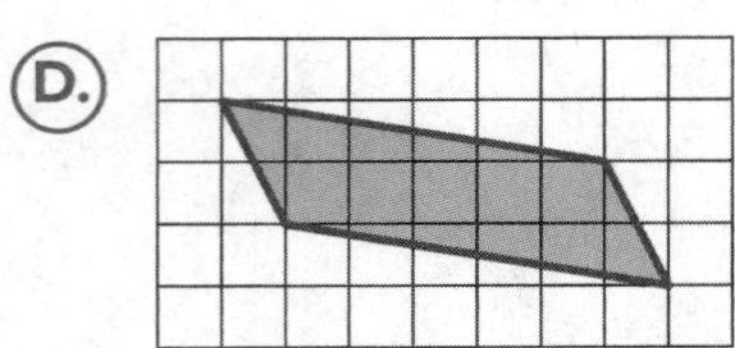

4. a. On the grid, draw at least three different quadrilaterals that can each be decomposed into two identical triangles with a single cut (show the cut line). One or more of the quadrilaterals should have non-right angles.

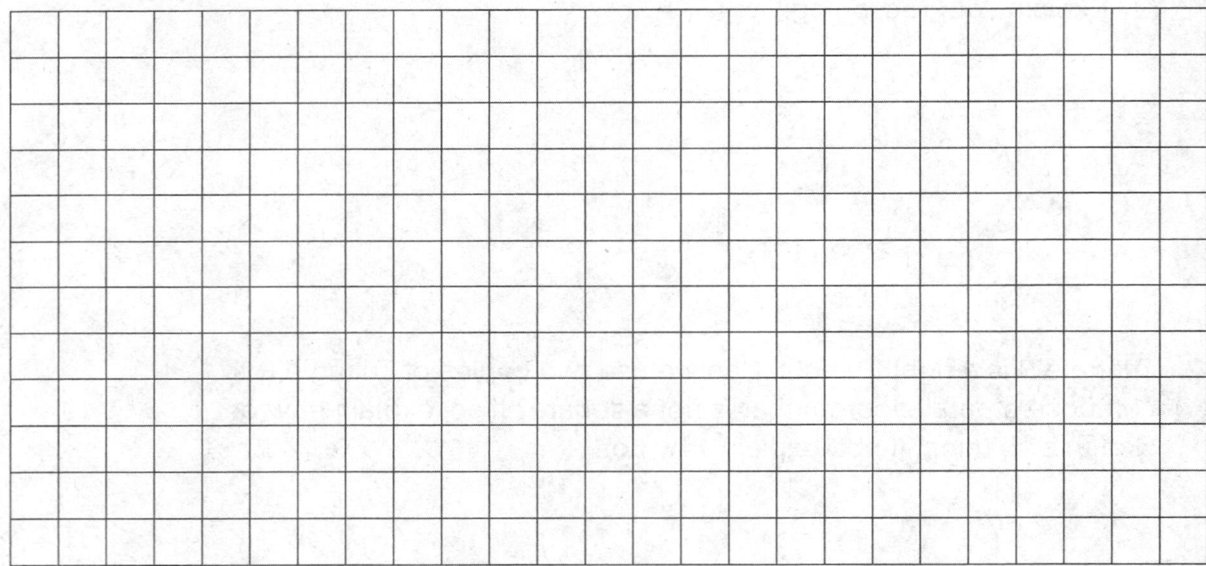

b. Identify the type of each quadrilateral.

NAME ______________________ DATE ____________ PERIOD ____________

5. Respond to each question. **(Lesson 1-6)**

a. A parallelogram has a base of 9 units and a corresponding height of $\frac{2}{3}$ units. What is its area?

b. A parallelogram has a base of 9 units and an area of 12 square units. What is the corresponding height for that base?

c. A parallelogram has an area of 7 square units. If the height that corresponds to a base is $\frac{1}{4}$ unit, what is the base?

6. Select **all** the segments that could represent the height if side n is the base. **(Lesson 1-5)**

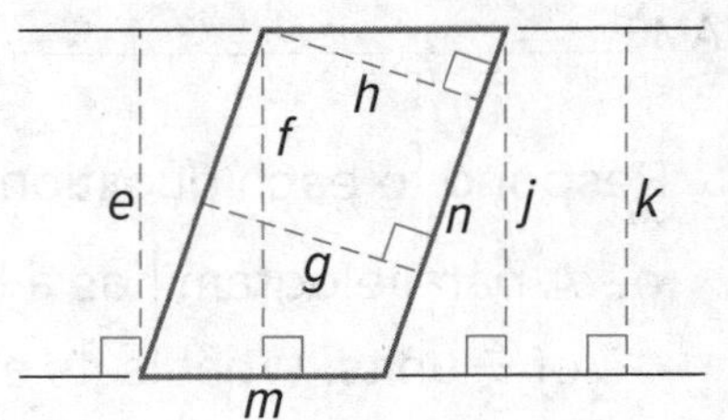

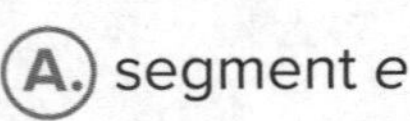
A. segment e

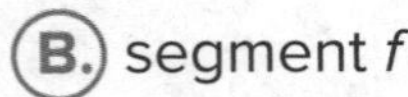
B. segment f

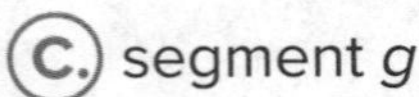
C. segment g

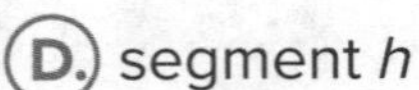
D. segment h

E. segment m

F. segment n

G. segment j

H. segment k

Lesson 1-8

Area of Triangles

NAME ______________________ DATE ____________ PERIOD ____________

Learning Goal Let's use what we know about parallelograms to find the area of triangles.

Warm Up

8.1 Composing Parallelograms

Here is Triangle *M*. Han made a copy of Triangle *M* and composed three different parallelograms using the original *M* and the copy, as shown here.

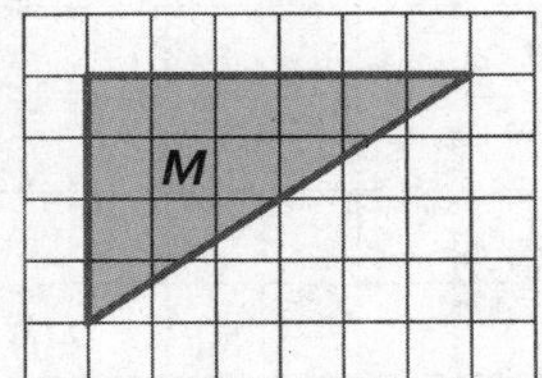

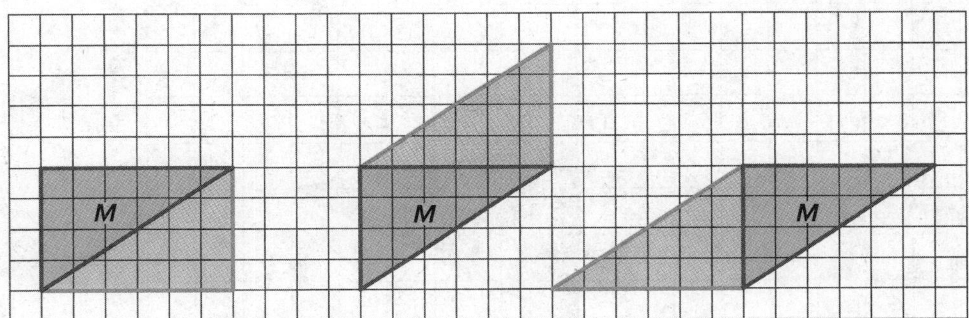

1. For each parallelogram Han composed, identify a base and a corresponding height, and write the measurements on the drawing.

2. Find the area of each parallelogram Han composed. Show your reasoning.

Activity

8.2 More Triangles

Find the areas of at least two of these triangles. Show your reasoning.

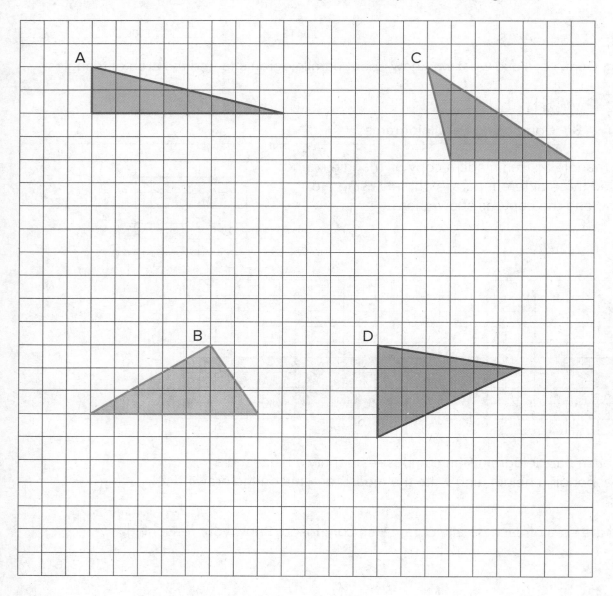

A

C

B

D

NAME ______________________ DATE ______________ PERIOD ____________

Activity

8.3 Decomposing a Parallelogram

1. Your teacher will give you two copies of a parallelogram. Glue or tape *one* copy of your parallelogram here and find its area. Show your reasoning.

2. Decompose the second copy of your parallelogram by cutting along the dotted lines. Take *only* the small triangle and the trapezoid, and rearrange these two pieces into a different parallelogram. Glue or tape the newly composed parallelogram on your paper.

3. Find the area of the new parallelogram you composed. Show your reasoning.

4. What do you notice about the relationship between the area of this new parallelogram and the original one?

5. How do you think the area of the large triangle compares to that of the new parallelogram: Is it larger, the same, or smaller? Why is that?

6. Glue or tape the remaining large triangle to your paper. Use any part of your work to help you find its area. Show your reasoning.

Are you ready for more?

Can you decompose this triangle and rearrange its parts to form a rectangle? Describe how it could be done.

NAME ______________________ DATE ______________ PERIOD __________

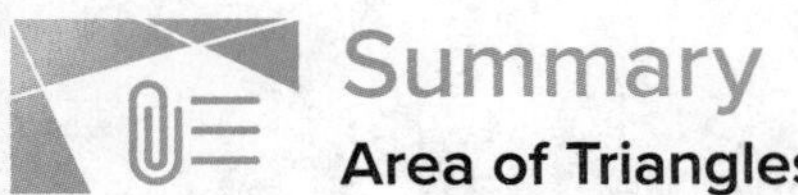

Summary
Area of Triangles

We can reason about the area of a triangle by using what we know about parallelograms. Here are three general ways to do this.

1. Make a copy of the triangle and join the original and the copy along an edge to create a parallelogram. Because the two triangles have the same area, one copy of the triangle has half the area of that parallelogram.

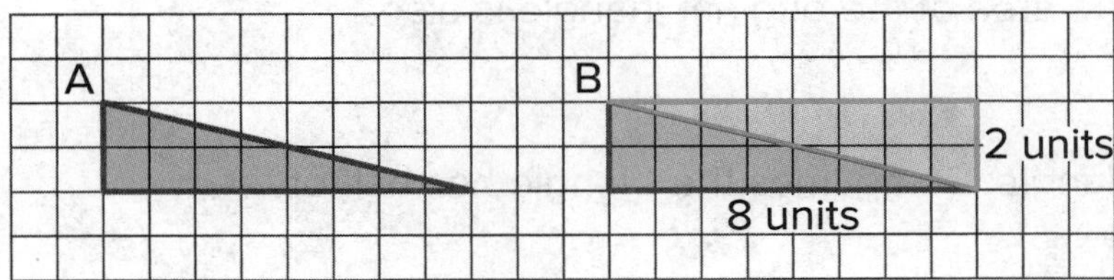

The area of Parallelogram B is 16 square units because the base is 8 units and the height 2 units. The area of Triangle A is half of that, which is 8 square units.

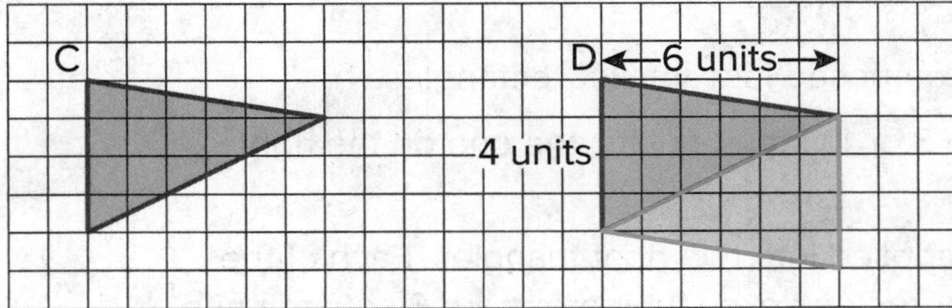

The area of Parallelogram D is 24 square units because the base is 4 units and the height 6 units. The area of Triangle C is half of that, which is 12 square units.

2. Decompose the triangle into smaller pieces and compose them into a parallelogram.

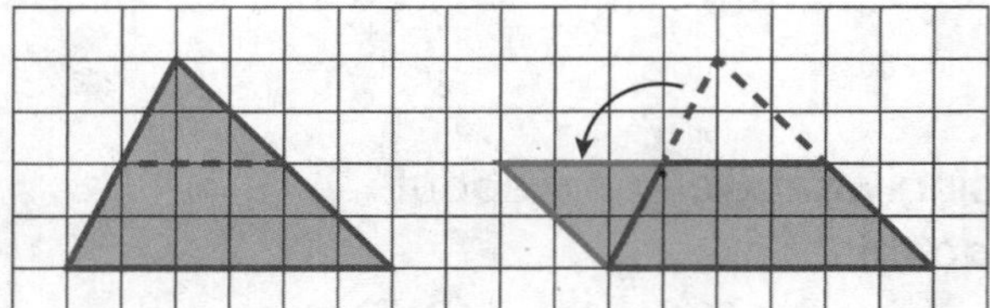

In the new parallelogram, $b = 6$, $h = 2$, and $6 \cdot 2 = 12$, so its area is 12 square units. Because the original triangle and the parallelogram are composed of the same parts, the area of the original triangle is also 12 square units.

3. Draw a rectangle around the triangle. Sometimes the triangle has half of the area of the rectangle.

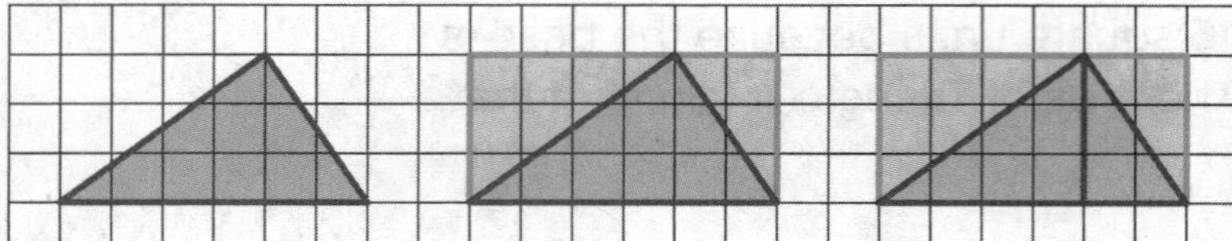

The large rectangle can be decomposed into smaller rectangles.

The one on the left has area $4 \cdot 3$ or 12 square units; the one on the right has area $2 \cdot 3$ or 6 square units.

The large triangle is also decomposed into two right triangles. Each of the right triangles is half of a smaller rectangle, so their areas are 6 square units and 3 square units. The large triangle has area 9 square units.

Sometimes, the triangle is half of what is left of the rectangle after removing two copies of the smaller right triangles.

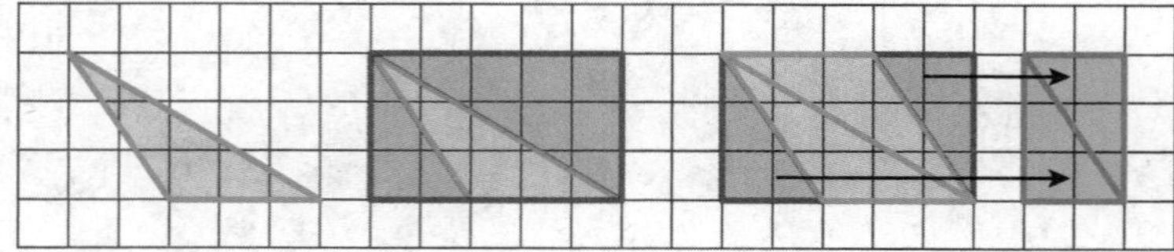

The right triangles being removed can be composed into a small rectangle with area $(2 \cdot 3)$ square units.

What is left is a parallelogram with area $5 \cdot 3 - 2 \cdot 3$, which equals $15 - 6$ or 9 square units. Notice that we can compose the same parallelogram with two copies of the original triangle! The original triangle is half of the parallelogram, so its area is $\frac{1}{2} \cdot 9$ or 4.5 square units.

NAME ______________________ DATE ______________ PERIOD ____________

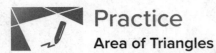

Practice

Area of Triangles

1. To find the area of this right triangle, Diego and Jada used different strategies.

Diego drew a line through the midpoints of the two longer sides, which decomposes the triangle into a trapezoid and a smaller triangle. He then rearranged the two shapes into a parallelogram.

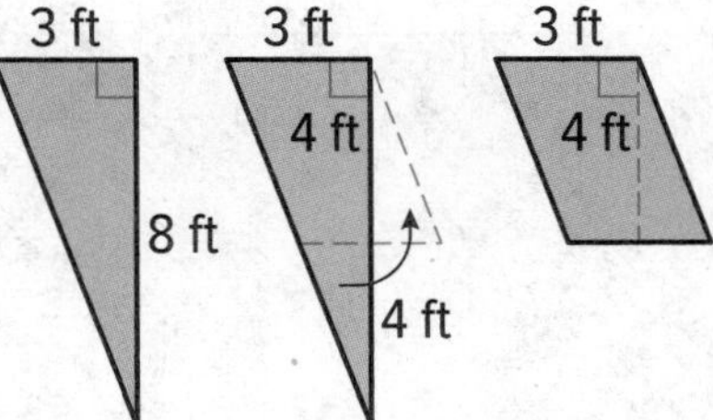

Jada made a copy of the triangle, rotated it, and lined it up against one side of the original triangle so that the two triangles make a parallelogram.

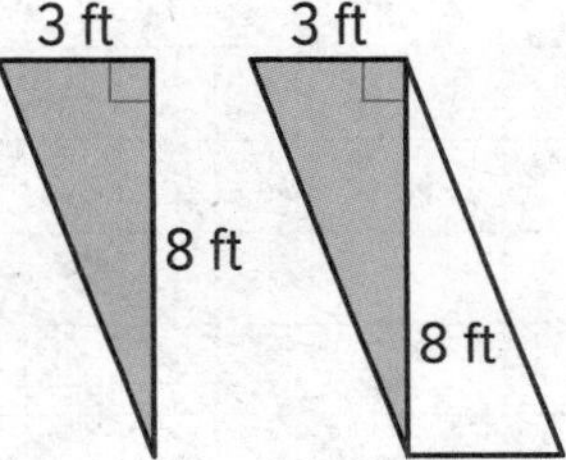

a. Explain how Diego might use his parallelogram to find the area of the triangle.

b. Explain how Jada might use her parallelogram to find the area of the triangle.

2. Find the area of each triangle. Explain or show your reasoning.

a.

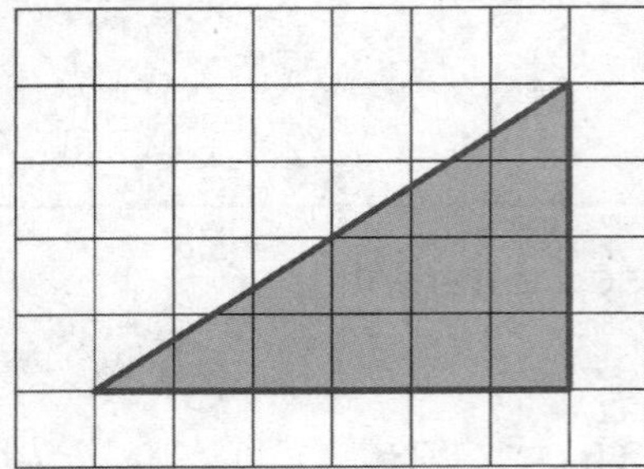

b.

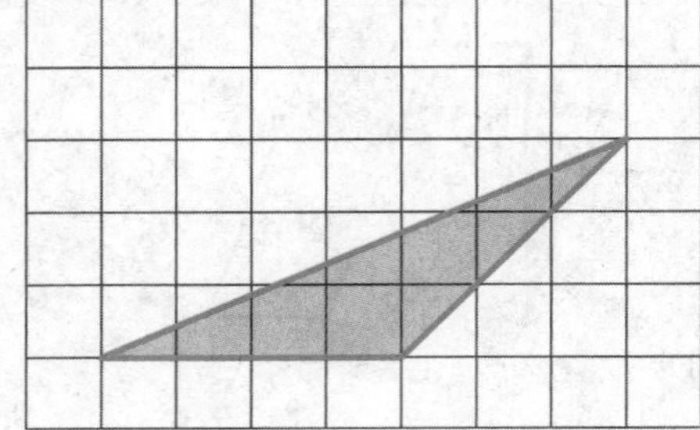

NAME ______________________ DATE ____________ PERIOD ________

3. Which of the three triangles has the greatest area? Show your reasoning. If you get stuck, try using what you know about the area of parallelograms.

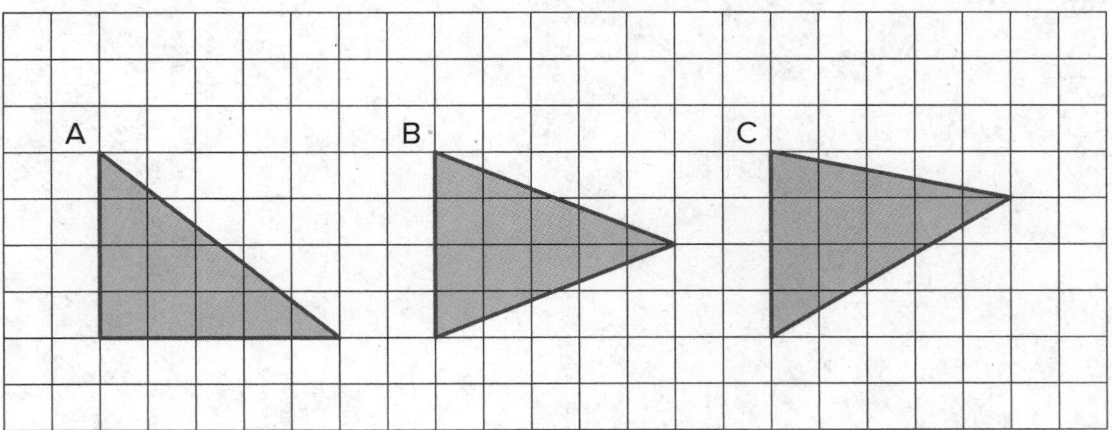

4. Draw an identical copy of each triangle such that the two copies together form a parallelogram. If you get stuck, consider using tracing paper. **(Lesson 1-7)**

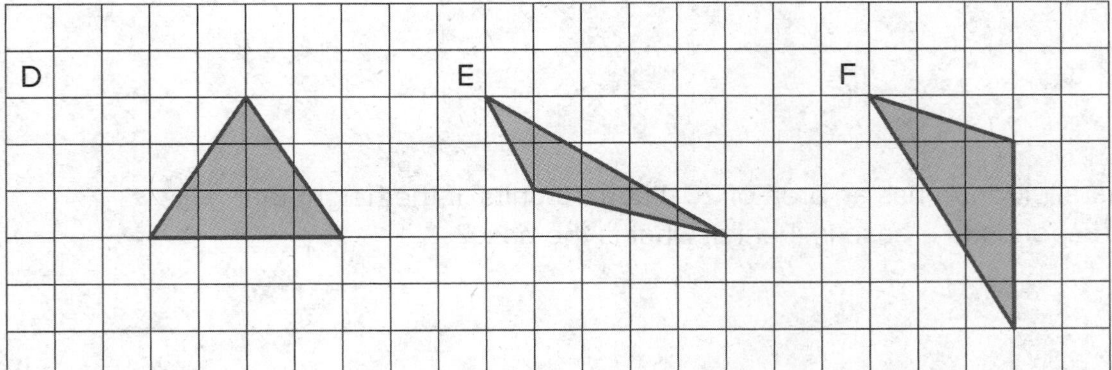

5. Respond to each question. **(Lesson 1-6)**

   a. A parallelogram has a base of 3.5 units and a corresponding height of 2 units. What is its area?

   b. A parallelogram has a base of 3 units and an area of 1.8 square units. What is the corresponding height for that base?

   c. A parallelogram has an area of 20.4 square units. If the height that corresponds to a base is 4 units, what is the base?

Lesson 1-9

Formula for the Area of a Triangle

NAME ______________________ DATE ______________ PERIOD ______________

Learning Goal Let's write and use a formula to find the area of a triangle.

Warm Up

9.1 Bases and Heights of a Triangle

Study the examples and non-examples of **bases** and **heights** in a triangle.

- **Examples:** These dashed segments represent heights of the triangle.

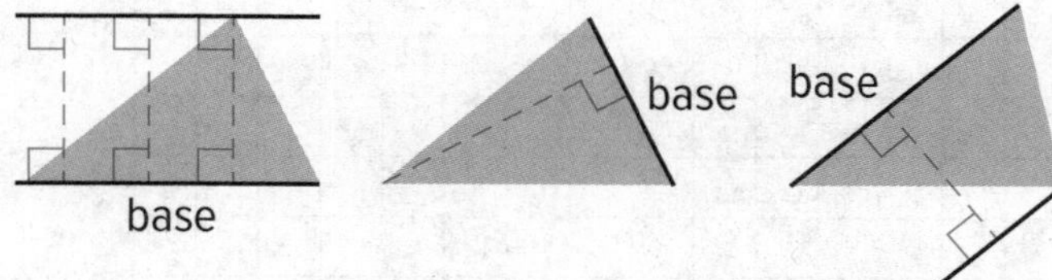

- **Non-examples:** These dashed segments do *not* represent heights of the triangle.

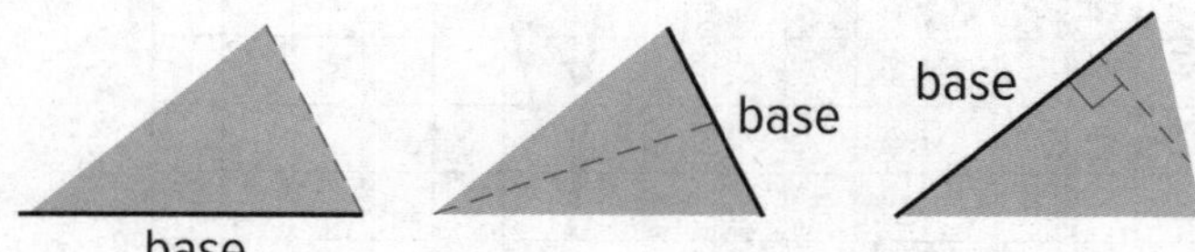

Select **all** the statements that are true about bases and heights in a triangle.

A. Any side of a triangle can be a base.

B. There is only one possible height.

C. A height is always one of the sides of a triangle.

D. A height that corresponds to a base must be drawn at an acute angle to the base.

E. A height that corresponds to a base must be drawn at a right angle to the base.

F. Once we choose a base, there is only one segment that represents the corresponding height.

G. A segment representing a height must go through a vertex.

Activity

9.2 Finding a Formula for Area of a Triangle

For each triangle:

- Identify a base and a corresponding height and record their lengths in the table.
- Find the area of the triangle and record it in the last column of the table.

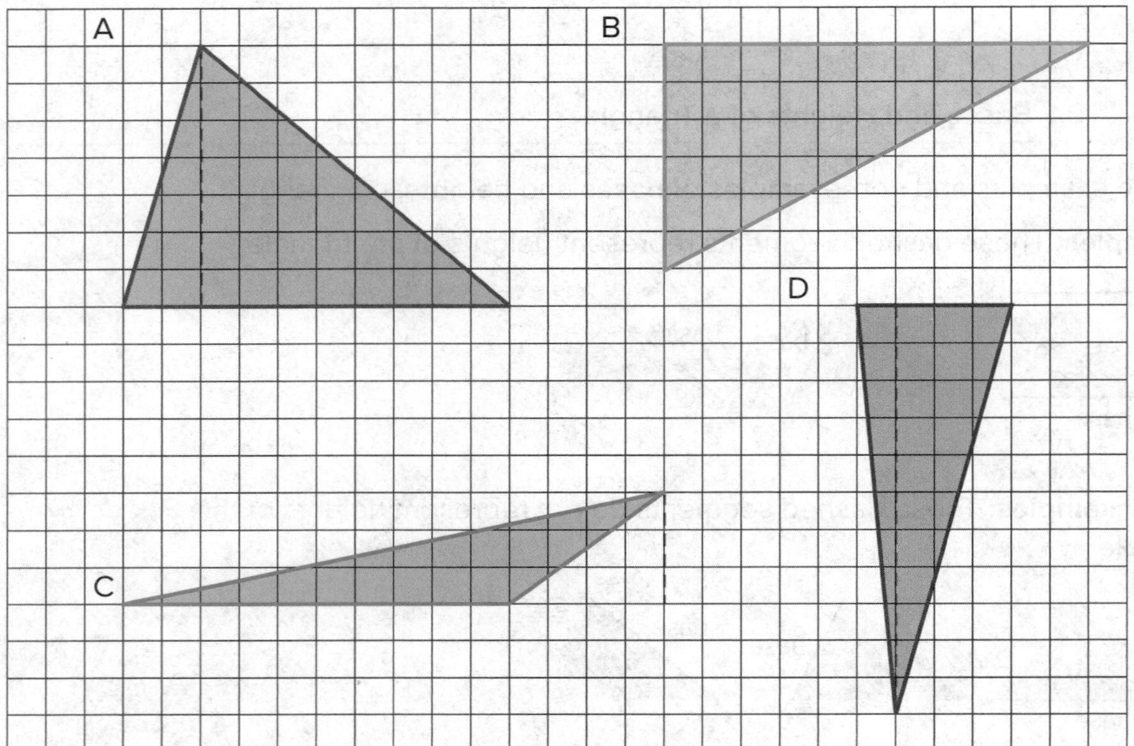

Triangle	Base (units)	Height (units)	Area (square units)
A			
B			
C			
D			
Any Triangle	b	h	

In the last row, write an expression for the area of any triangle, using b and h.

NAME ______________________ DATE ____________ PERIOD ________

Activity

9.3 Applying the Formula for Area of Triangles

For each triangle, circle a base measurement that you can use to find the area of the triangle. Then, find the area of any *three* triangles. Show your reasoning.

Triangle A

6 cm

5 cm

Triangle B

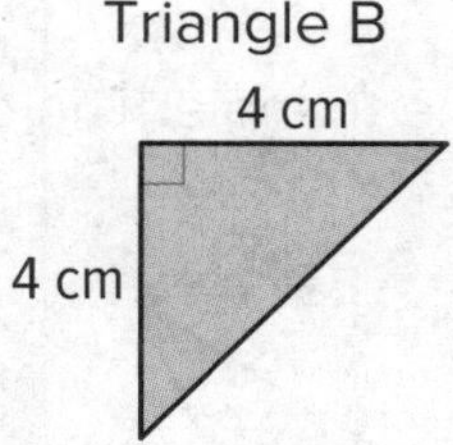

Triangle C

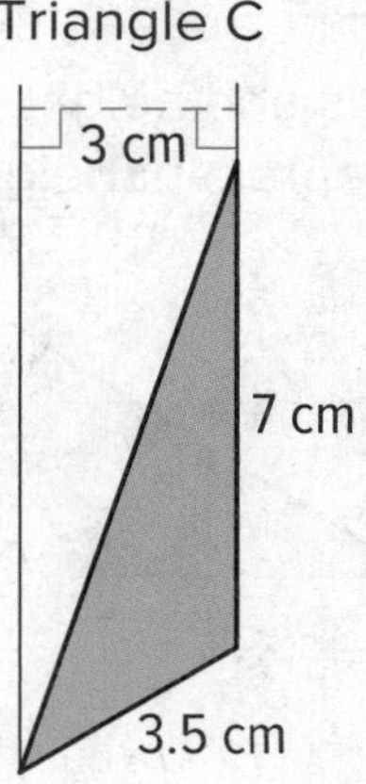

Triangle D

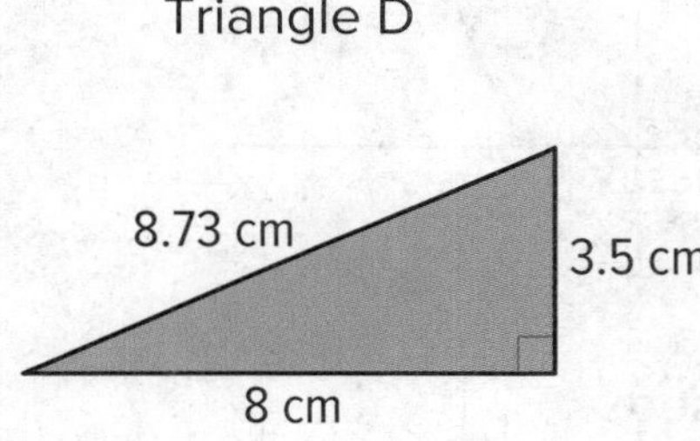

Triangle E

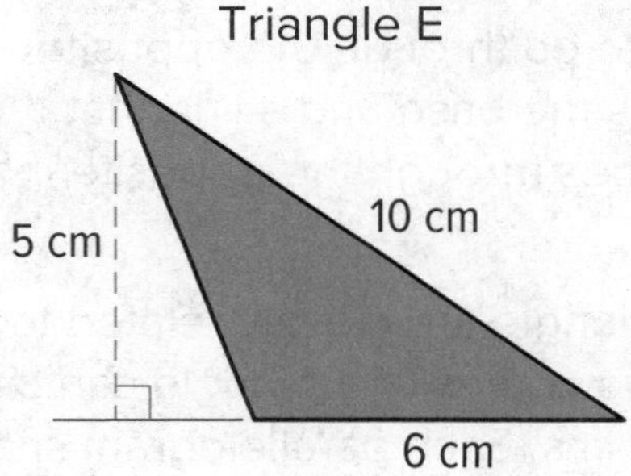

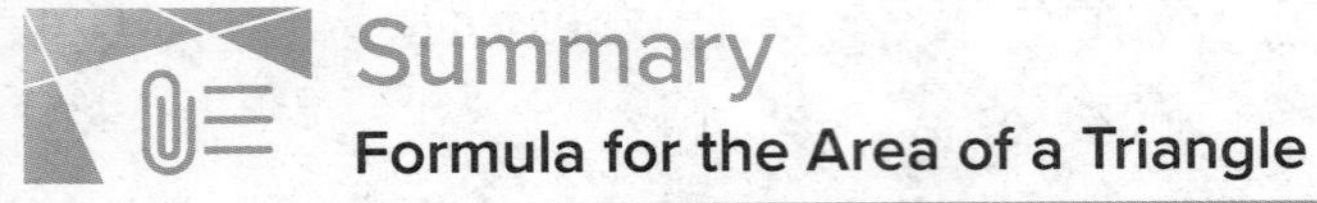

Summary

Formula for the Area of a Triangle

We can choose any of the three sides of a triangle to call the **base**. The term "base" refers to both the side and its length (the measurement).

The corresponding **height** is the length of a perpendicular segment from the base to the vertex opposite of it. The **opposite vertex** is the vertex that is *not* an endpoint of the base.

Here are three pairs of bases and heights for the same triangle. The dashed segments in the diagrams represent heights.

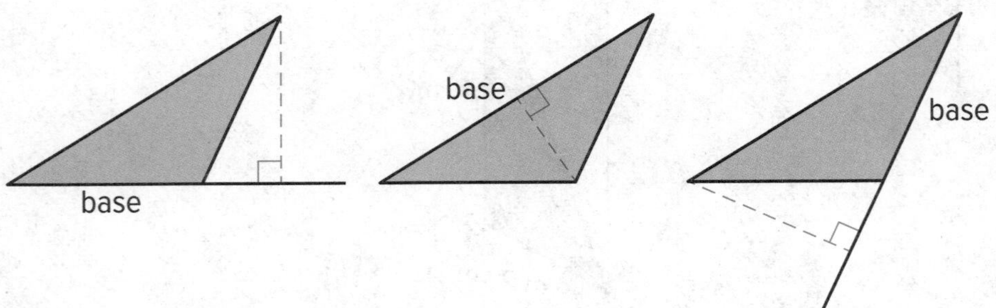

A segment showing a height must be drawn at a right angle to the base, but it can be drawn in more than one place. It does not have to go through the opposite vertex, as long as it connects the base and a line that is parallel to the base and goes through the opposite vertex, as shown here.

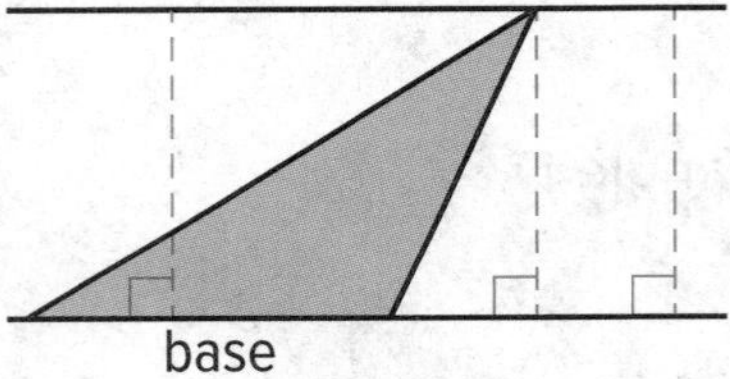

The base-height pairs in a triangle are closely related to those in a parallelogram. Recall that two copies of a triangle can be composed into one or more parallelograms. Each parallelogram shares at least one base with the triangle.

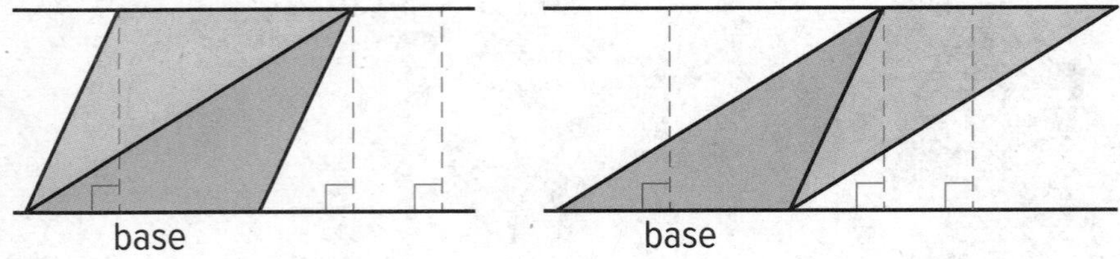

For any base that they share, the corresponding height is also shared, as shown by the dashed segments.

NAME ______________________ DATE ______________ PERIOD ____________

We can use the base-height measurements and our knowledge of parallelograms to find the area of any triangle.

- The formula for the area of a parallelogram with base b and height h is $A = b \cdot h$.
- A triangle takes up half of the area of a parallelogram with the same base and height. We can therefore express the area A of a triangle as: $A = \frac{1}{2} \cdot b \cdot h$

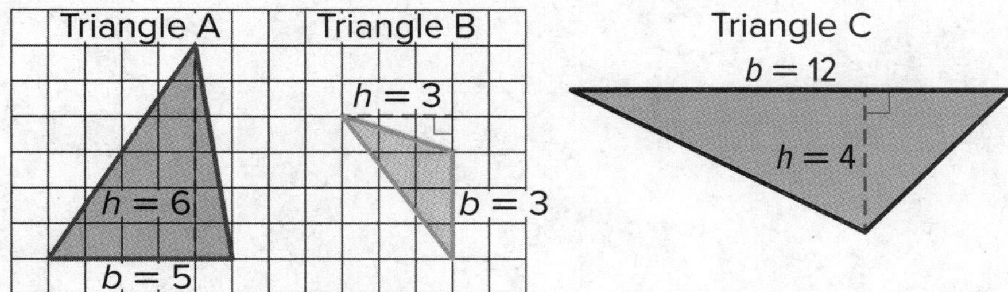

- The area of Triangle A is 15 square units because $\frac{1}{2} \cdot 5 \cdot 6 = 15$.
- The area of Triangle B is 4.5 square units because $\frac{1}{2} \cdot 3 \cdot 3 = 4.5$.
- The area of Triangle C is 24 square units because $\frac{1}{2} \cdot 12 \cdot 4 = 24$.

In each case, one side of the triangle is the base but neither of the other sides is the height. This is because the angle between them is not a right angle.

In right triangles, however, the two sides that are perpendicular can be a base and a height.

The area of this triangle is 18 square units whether we use 4 units or 9 units for the base.

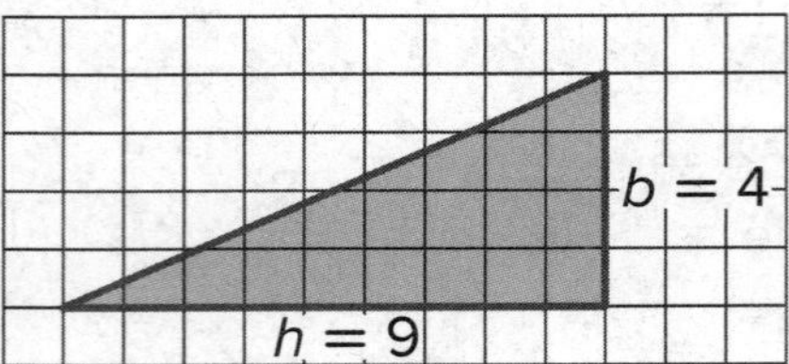

Glossary

opposite vertex

NAME ______________________ DATE ____________ PERIOD ________

Practice

Formula for the Area of a Triangle

1. Select **all** drawings in which a corresponding height *h* for a given base *b* is correctly identified.

Ⓐ 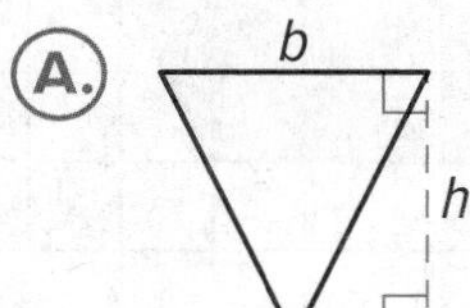

Ⓑ 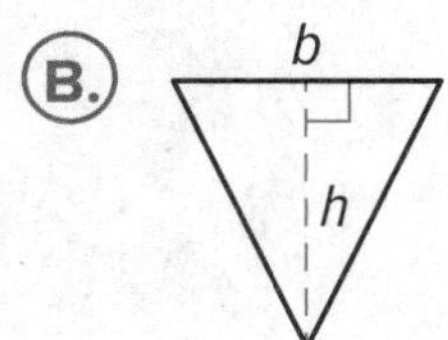

Ⓒ 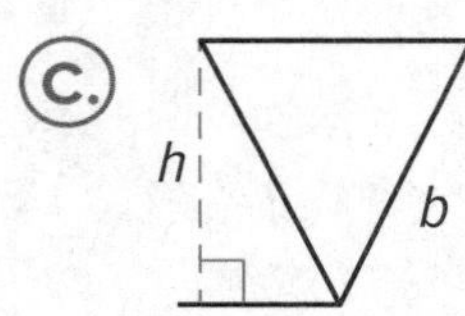

Ⓓ 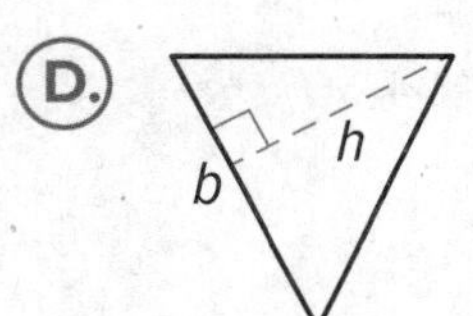

Ⓔ 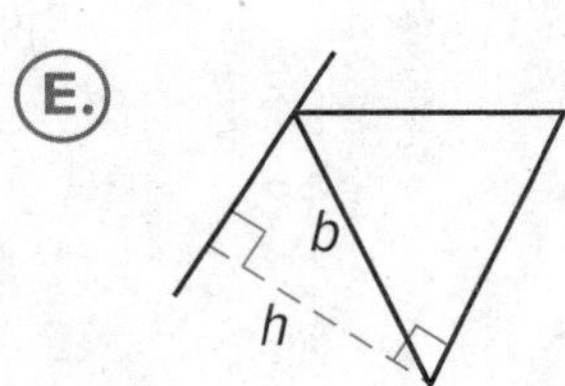

Ⓕ 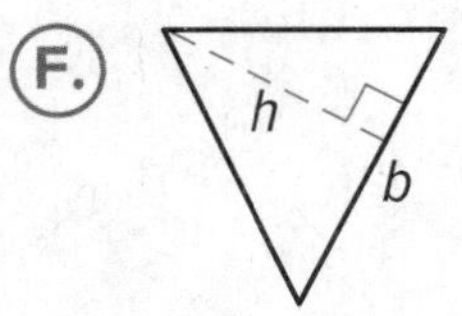

2. For each triangle, a base and its corresponding height are labeled.

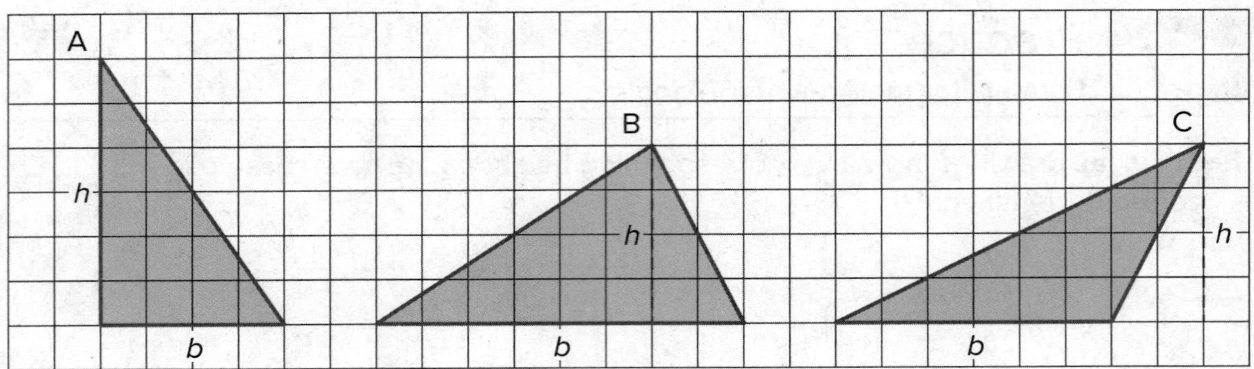

a. Find the area of each triangle.

b. How is the area related to the base and its corresponding height?

3. Here is a right triangle. Name a corresponding height for each base.

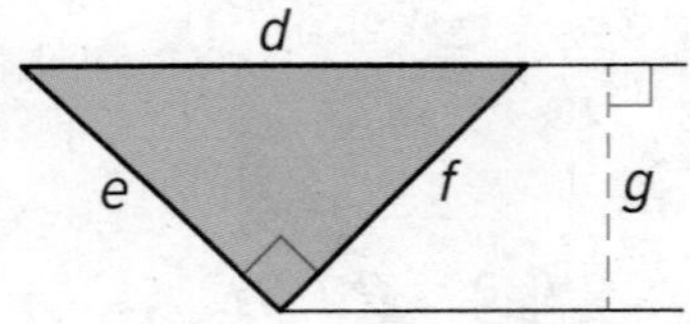

a. side d

b. side e

c. side f

NAME ______________________ DATE ______________ PERIOD ____________

4. Find the area of the shaded triangle. Show your reasoning. **(Lesson 1-8)**

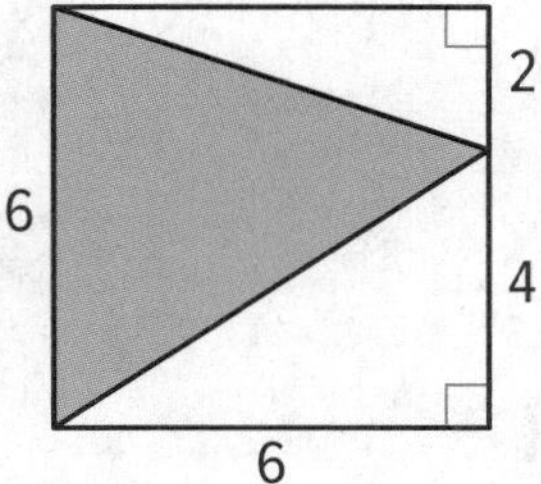

5. Andre drew a line connecting two opposite corners of a parallelogram. Select **all** true statements about the triangles created by the line Andre drew. **(Lesson 1-7)**

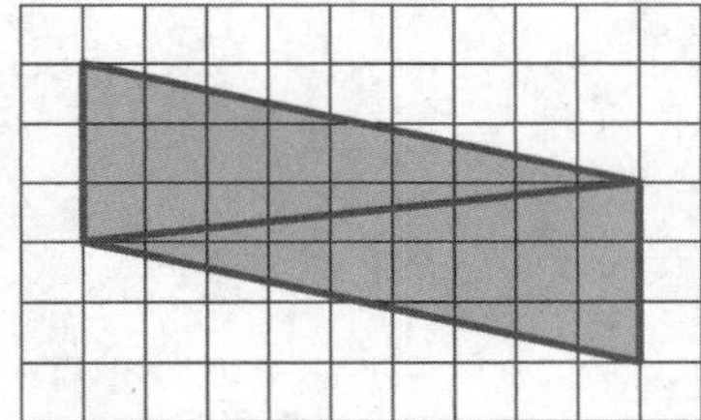

A. Each triangle has two sides that are 3 units long.

B. Each triangle has a side that is the same length as the diagonal line.

C. Each triangle has one side that is 3 units long.

D. When one triangle is placed on top of the other and their sides are aligned, we will see that one triangle is larger than the other.

E. The two triangles have the same area as each other.

6. Here is an octagon. (Note: The diagonal sides of the octagon are *not* 4 inches long.) **(Lesson 1-3)**

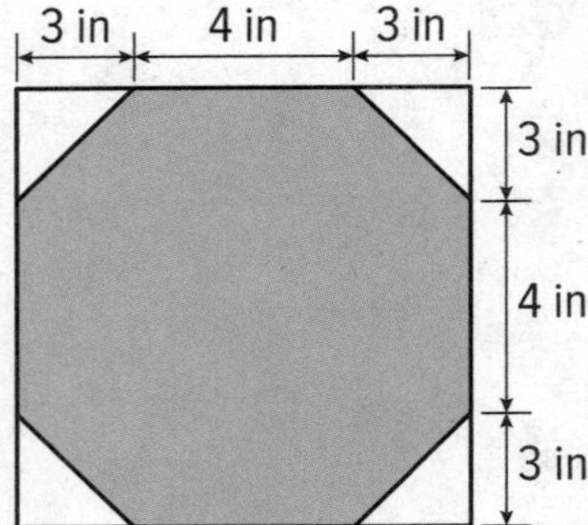

a. While estimating the area of the octagon, Lin reasoned that it must be less than 100 square inches. Do you agree? Explain your reasoning.

b. Find the exact area of the octagon. Show your reasoning.

Lesson 1-10

Bases and Heights of Triangles

NAME ______________________ DATE ____________ PERIOD ____________

Learning Goal Let's use different base-height pairs to find the area of a triangle.

Warm Up

10.1 An Area of 12

On the grid, draw a triangle with an area of 12 square units. Try to draw a non-right triangle. Be prepared to explain how you know the area of your triangle is 12 square units.

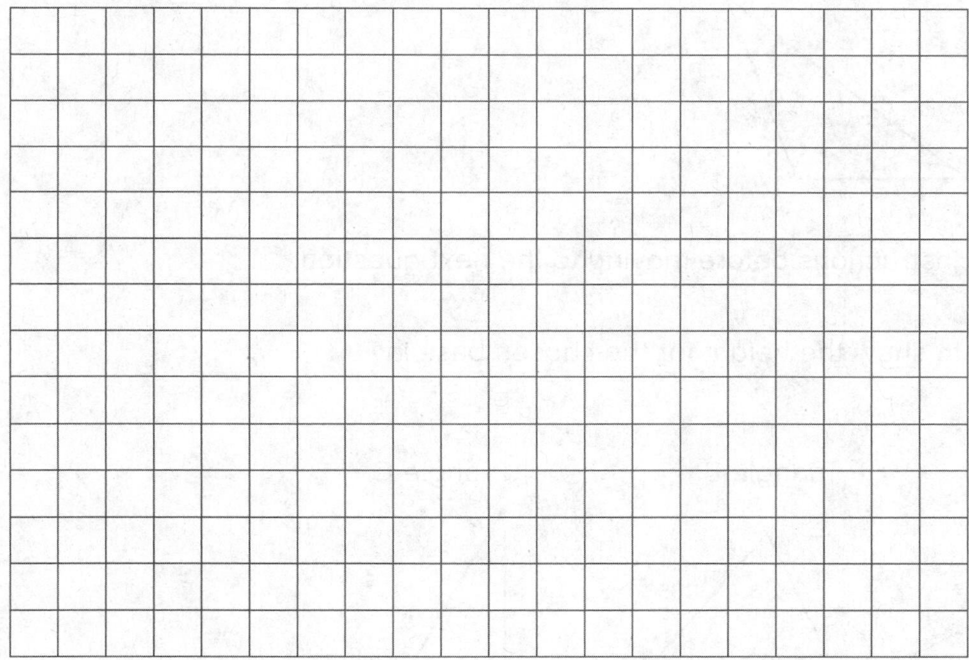

Activity

10.2 Hunting for Heights

1. Here are three copies of the same triangle. The triangle is rotated so that the side chosen as the base is at the bottom and is horizontal. Draw a height that corresponds to each base. Use an index card to help you.

Side a as the base:

Side b as the base:

Side c as the base:

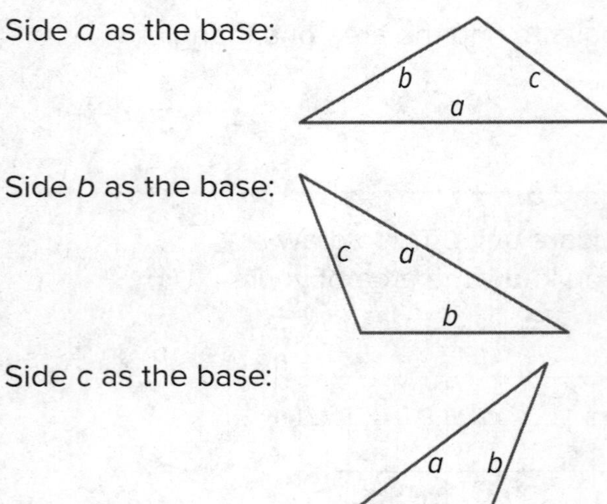

Pause for your teacher's instructions before moving to the next question.

2. Draw a line segment to show the height for the chosen base in each triangle.

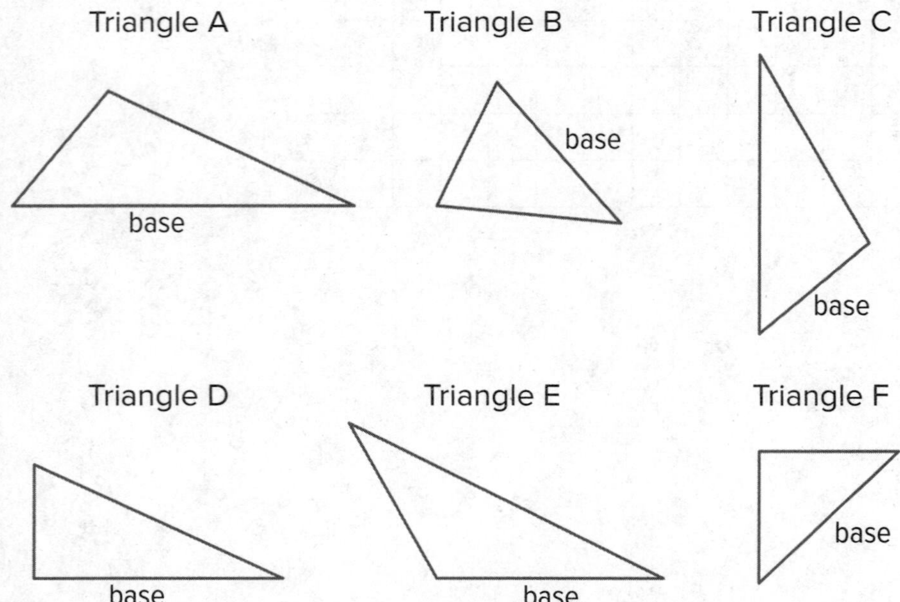

NAME ______________________ DATE ______________ PERIOD ____________

Activity

10.3 Some Bases Are Better Than Others

For each triangle, identify and label a base and height. If needed, draw a line segment to show the height. Then, find the area of the triangle. Show your reasoning. (The side length of each square on the grid is 1 unit.)

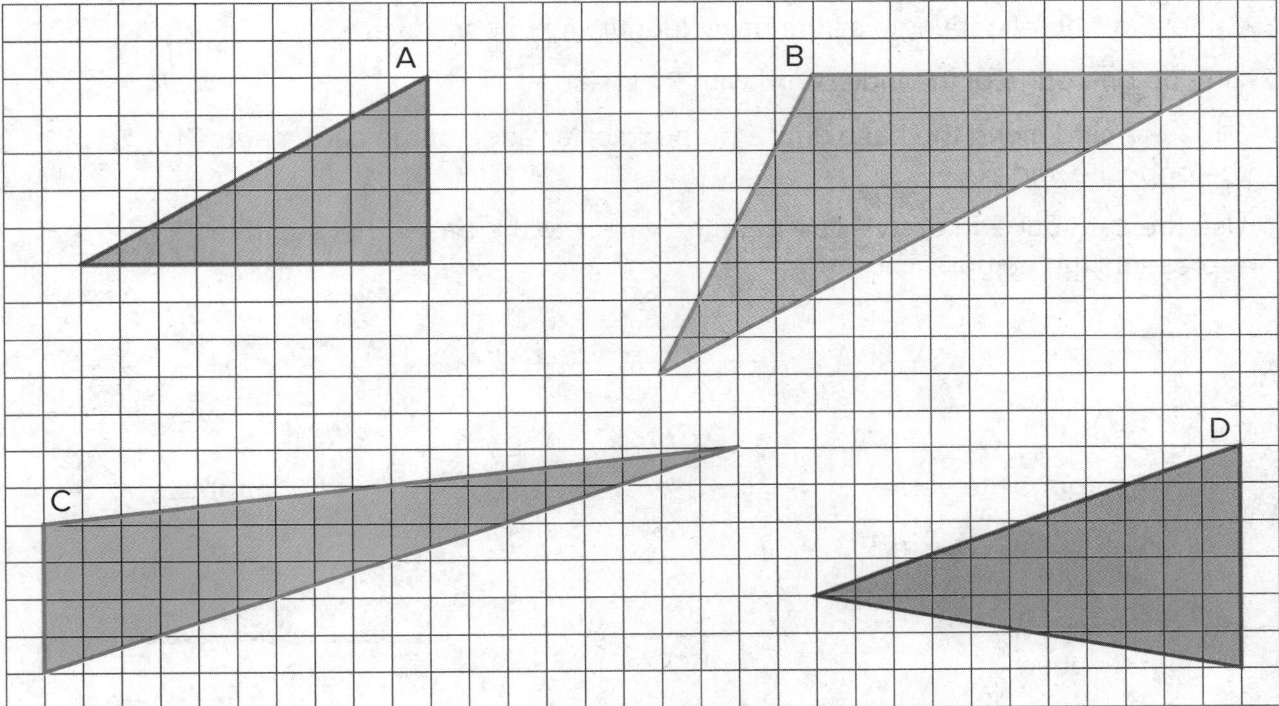

Are you ready for more?

Find the area of this triangle. Show your reasoning.

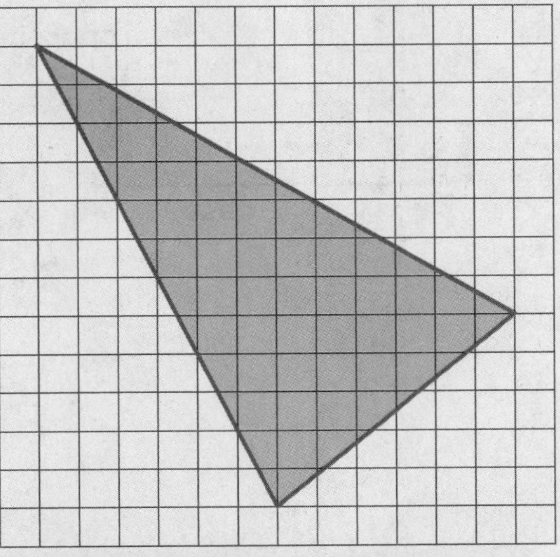

Summary

Bases and Heights of Triangles

A height of a triangle is a perpendicular segment between the side chosen as the base and the opposite vertex. We can use tools with right angles to help us draw height segments.

An index card (or any stiff paper with a right angle) is a handy tool for drawing a line that is perpendicular to another line.

1. Choose a side of a triangle as the base. Identify its opposite vertex.
2. Line up one edge of the index card with that base.
3. Slide the card along the base until a perpendicular edge of the card meets the opposite vertex.
4. Use the card edge to draw a line from the vertex to the base. That segment represents the height.

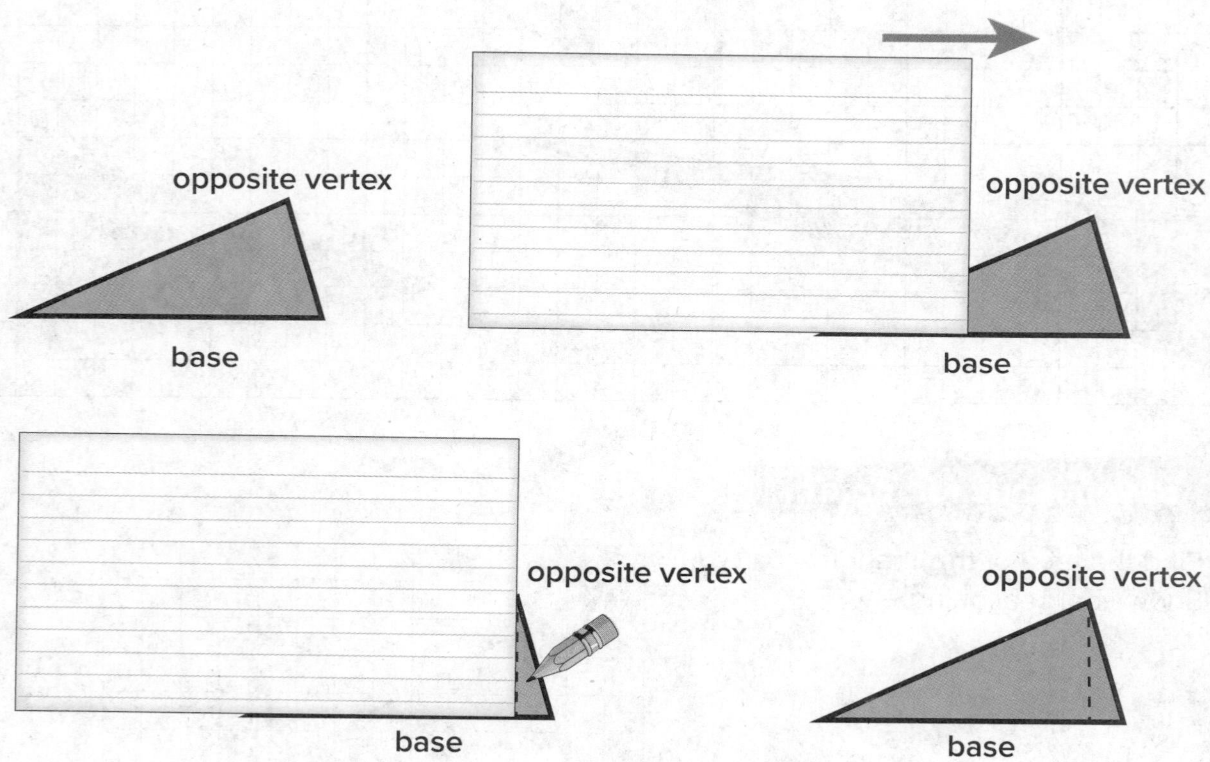

NAME ______________________ DATE ______________ PERIOD ____________

Sometimes we may need to extend the line of the base to identify the height, such as when finding the height of an obtuse triangle, or whenever the opposite vertex is not directly over the base. In these cases, the height segment is typically drawn *outside* of the triangle.

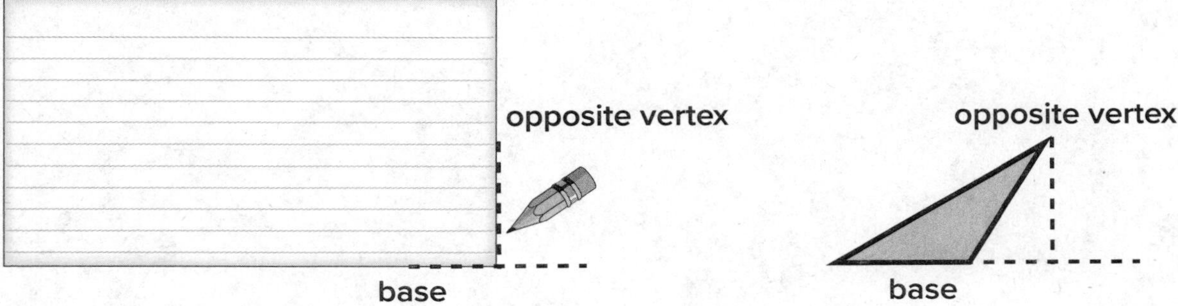

Even though any side of a triangle can be a base, some base-height pairs can be more easily determined than others, so it helps to choose strategically.

For example, when dealing with a right triangle, it often makes sense to use the two sides that make the right angle as the base and the height because one side is already perpendicular to the other.

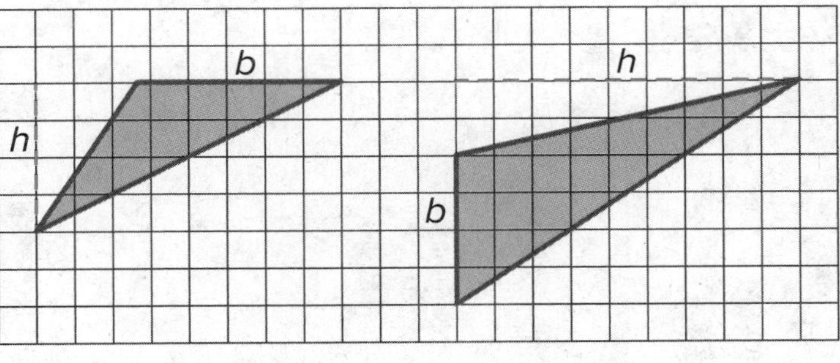

If a triangle is on a grid and has a horizontal or a vertical side, you can use that side as a base and use the grid to find the height, as in these examples.

Glossary

edge

vertex

Practice

Bases and Heights of Triangles

1. For each triangle, a base is labeled *b*. Draw a line segment that shows its corresponding height. Use an index card to help you draw a straight line.

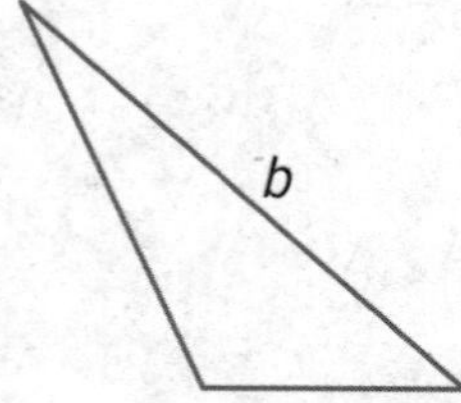

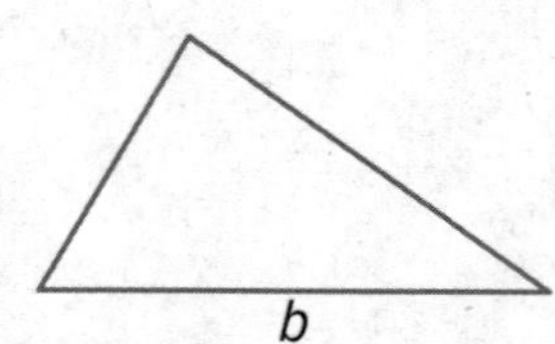

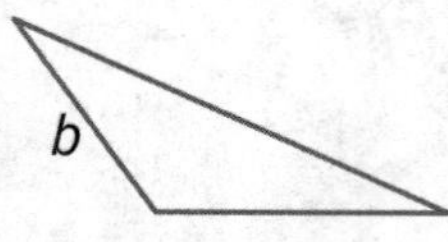

2. Select **all** triangles that have an area of 8 square units. Explain how you know.

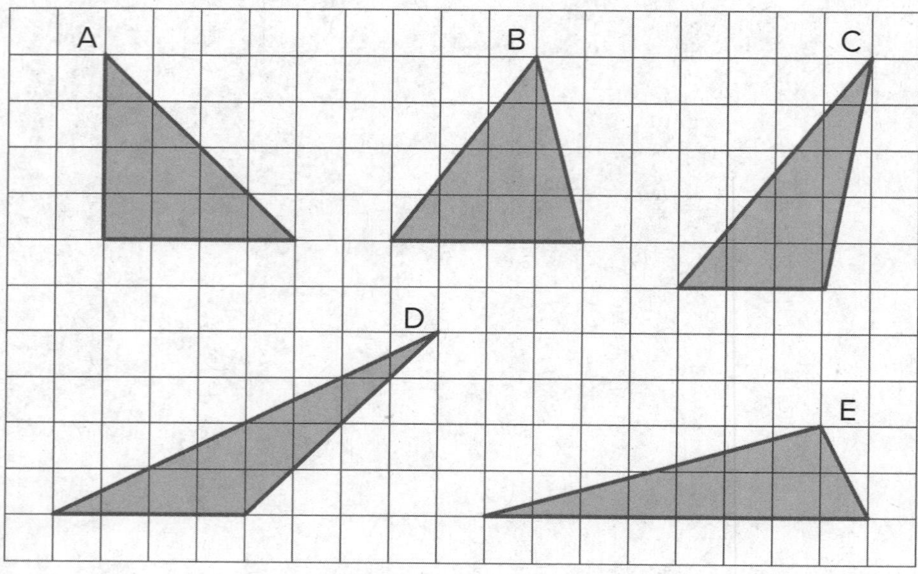

NAME ______________________ DATE ____________ PERIOD ____________

3. Find the area of the triangle. Show your reasoning. If you get stuck, carefully consider which side of the triangle to use as the base.

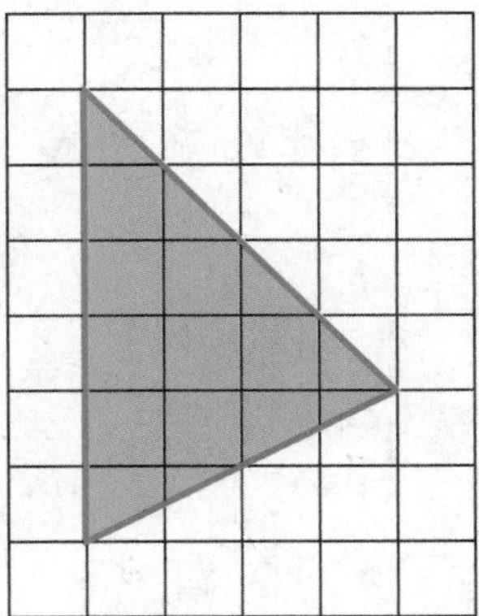

4. Can side *d* be the base for this triangle? If so, which length would be the corresponding height? If not, explain why not.

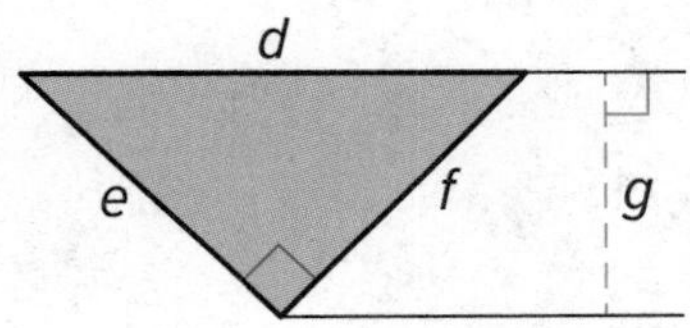

5. Find the area of this shape. Show your reasoning. **(Lesson 1-3)**

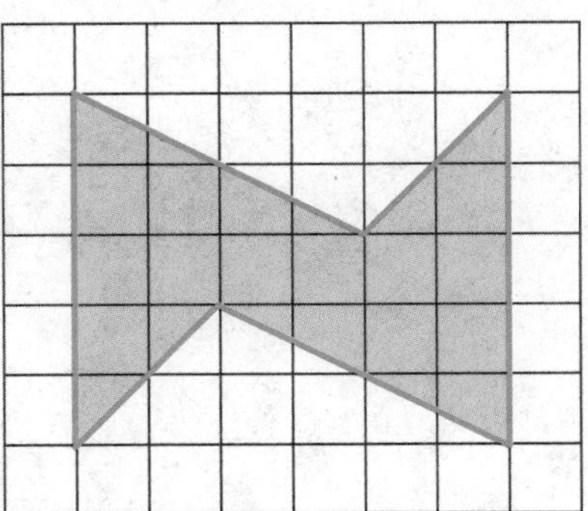

6. On the grid, sketch two different parallelograms that have equal area. Label a base and height of each and explain how you know the areas are the same. **(Lesson 1-6)**

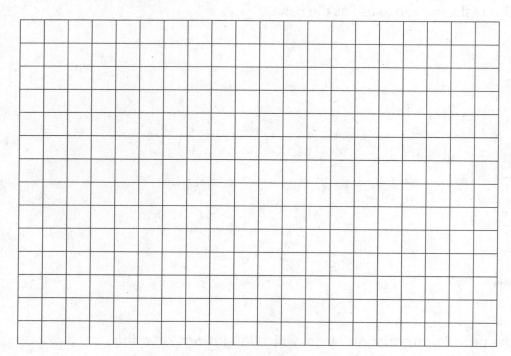

Lesson 1-11

Polygons

NAME ______________________ DATE ______________ PERIOD ____________

Learning Goal Let's investigate polygons and their areas.

Warm Up

11.1 Which One Doesn't Belong: Bases and Heights

Which one doesn't belong?

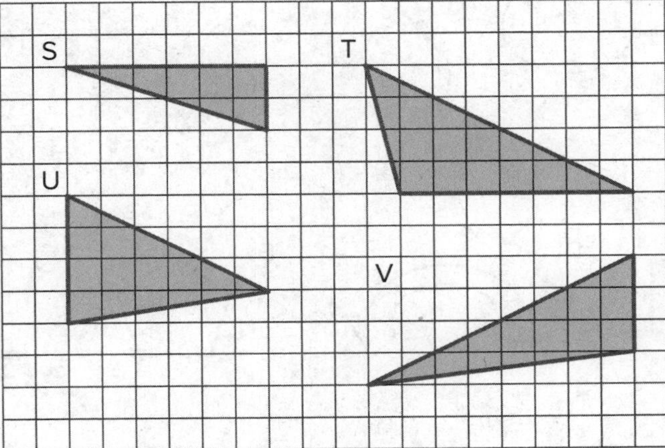

Activity

11.2 What Are Polygons?

Here are five **polygons**.

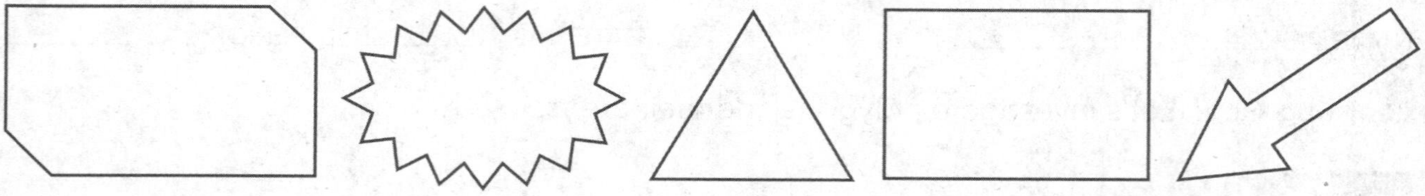

Here are six figures that are *not* polygons.

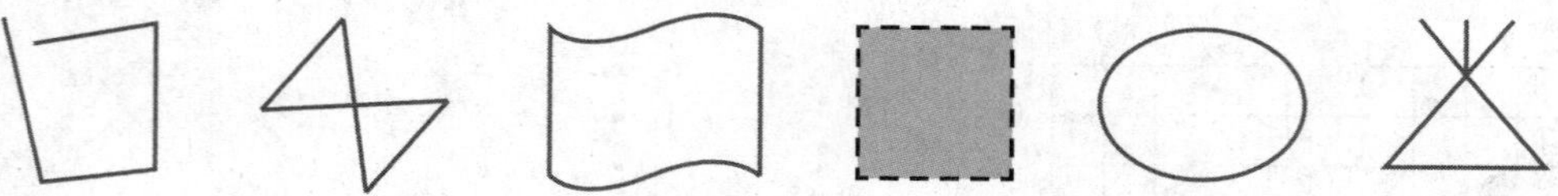

1. Circle the figures that are polygons.

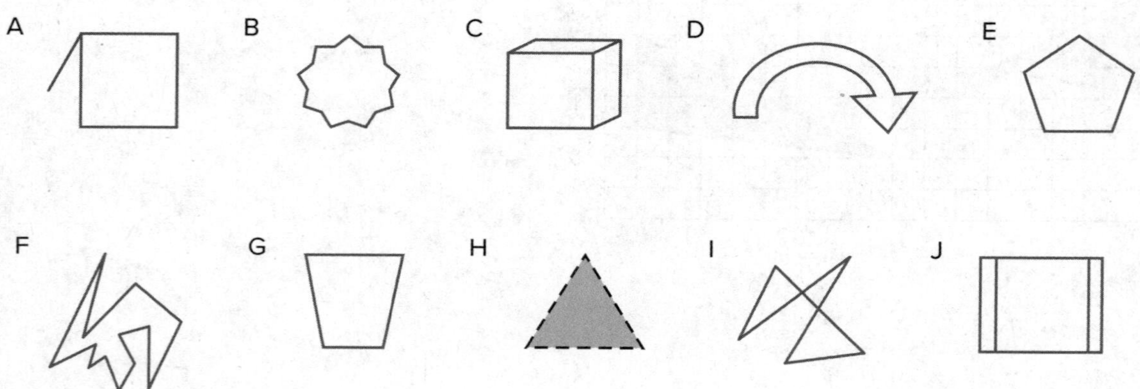

2. What do the figures you circled have in common? What characteristics helped you decide whether a figure was a polygon?

NAME ______________________ DATE ____________ PERIOD ____________

Activity
11.3 Quadrilateral Strategies

Find the area of *two* **quadrilaterals** of your choice. Show your reasoning.

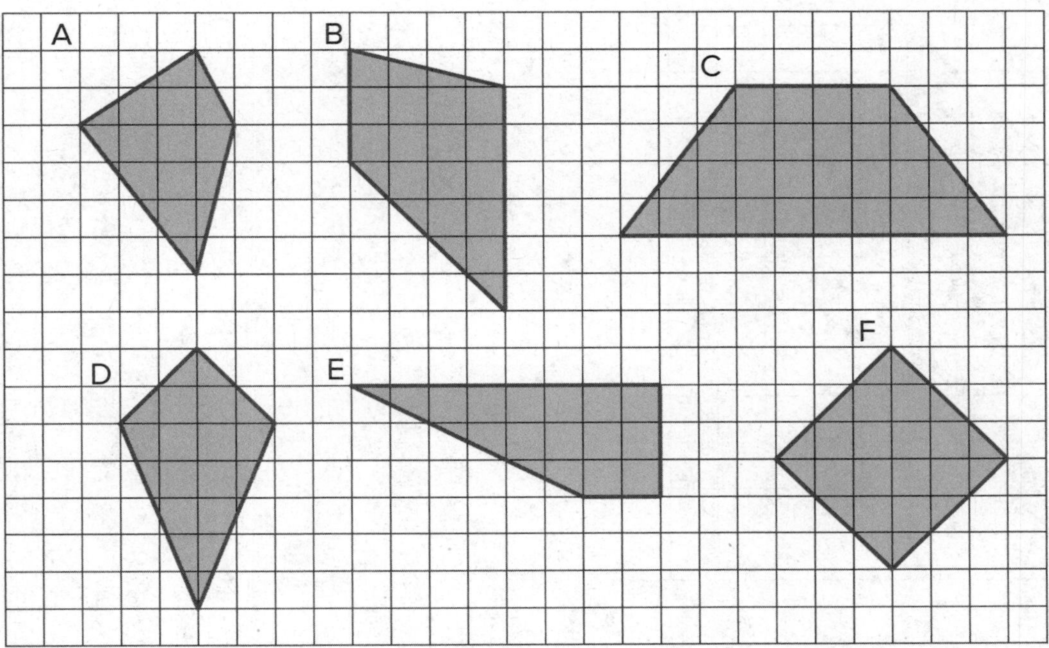

Are you ready for more?

Here is a trapezoid. a and b represent the lengths of its bottom and top sides. The segment labeled h represents its height; it is perpendicular to both the top and bottom sides.

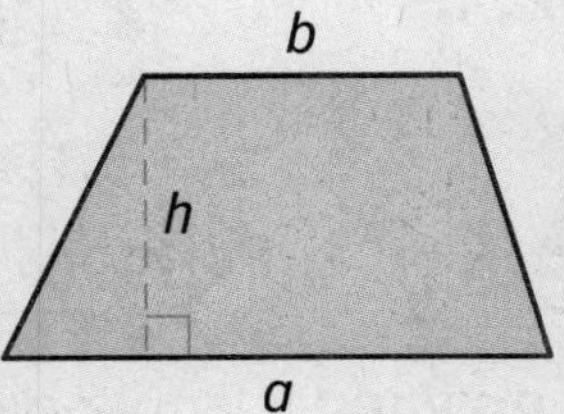

Apply area-reasoning strategies—decomposing, rearranging, duplicating, etc.—on the trapezoid so that you have one or more shapes with areas that you already know how to find.

Use the shapes to help you write a formula for the area of a trapezoid. Show your reasoning.

Activity

11.4 Pinwheel

Find the area of the shaded region in square units. Show your reasoning.

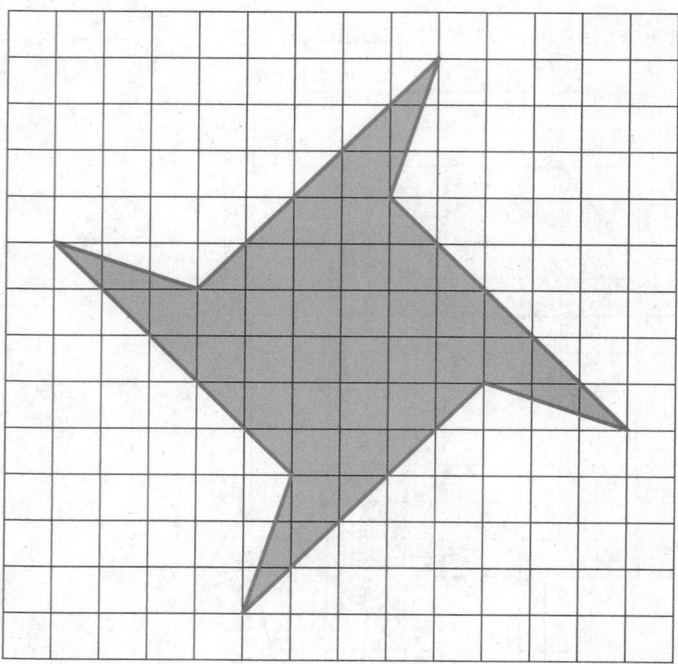

Summary

Polygons

A **polygon** is a two-dimensional figure composed of straight line segments.

- Each end of a line segment connects to one other line segment. The point where two segments connect is a **vertex**. The plural of vertex is vertices.
- The segments are called the **edges** or **sides** of the polygon. The sides never cross each other. There are always an equal number of vertices and sides.

NAME ______________________________ DATE ______________ PERIOD ____________

Here is a polygon with 5 sides. The vertices are labeled A, B, C, D, and E.

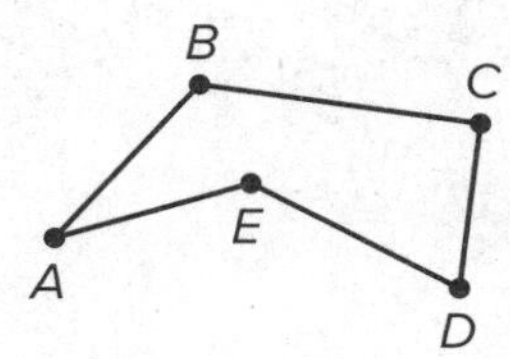

A polygon encloses a region. To find the area of a polygon is to find the area of the region inside it.

We can find the area of a polygon by decomposing the region inside it into triangles and rectangles.

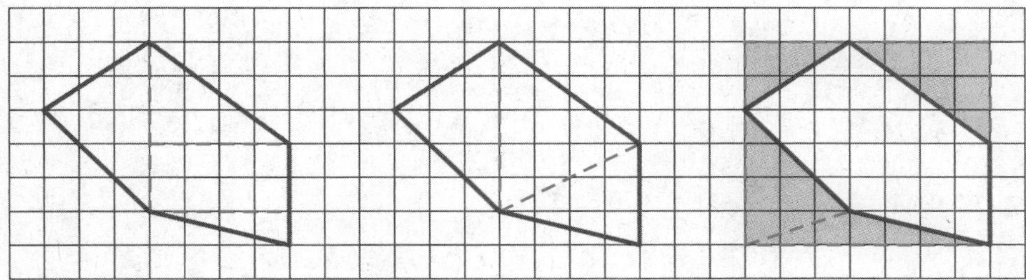

The first two diagrams show the polygon decomposed into triangles and rectangles; the sum of their areas is the area of the polygon. The last diagram shows the polygon enclosed with a rectangle; subtracting the areas of the triangles from the area of the rectangle gives us the area of the polygon.

Glossary

polygon

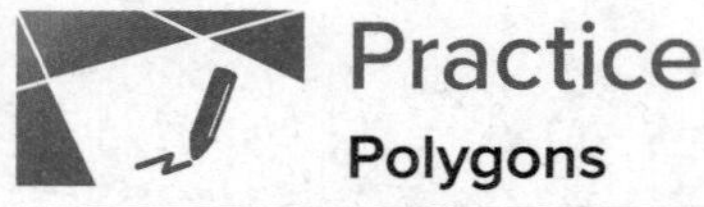

Practice

Polygons

1. Select **all** the polygons.

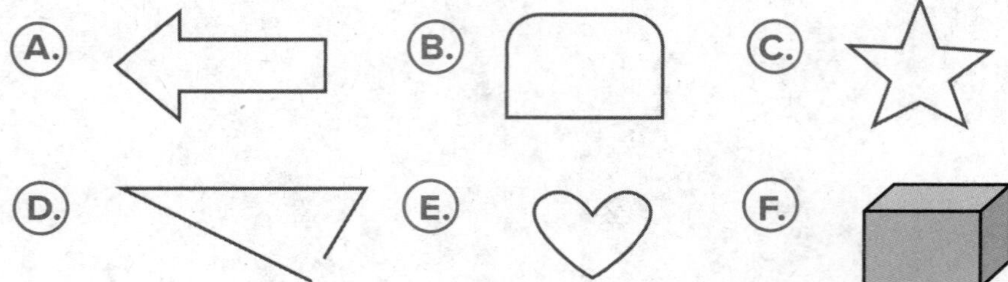

2. Mark each vertex with a large dot. How many edges and vertices does this polygon have?

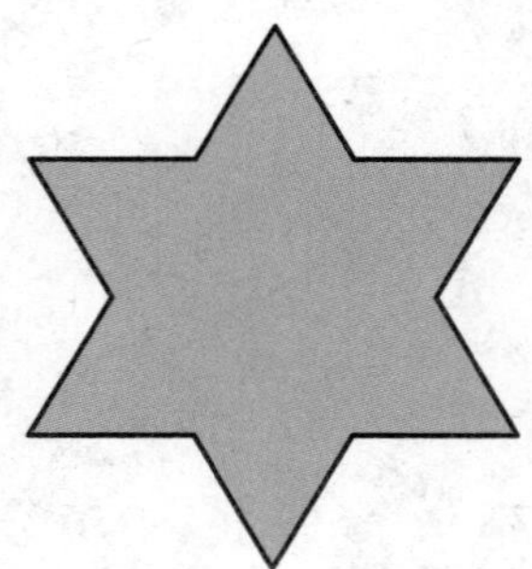

3. Find the area of this trapezoid. Explain or show your strategy.

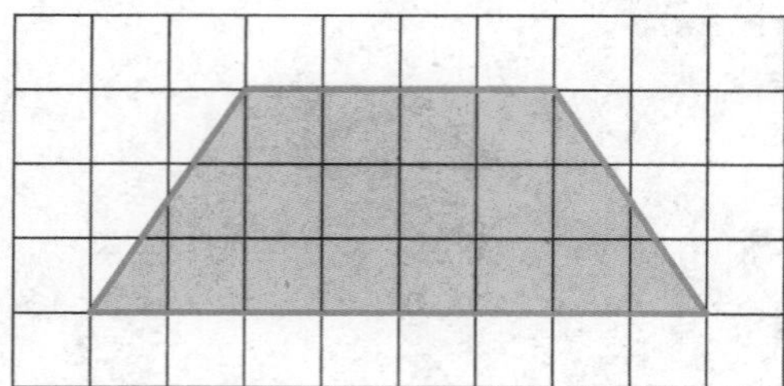

NAME ______________________ DATE ____________ PERIOD ____________

4. Lin and Andre used different methods to find the area of a regular hexagon with 6-inch sides. Lin decomposed the hexagon into six identical, equilateral triangles. Andre decomposed the hexagon into a rectangle and two triangles.

Lin's method

10.4 in
6 in
6 in
6 in

Andre's method

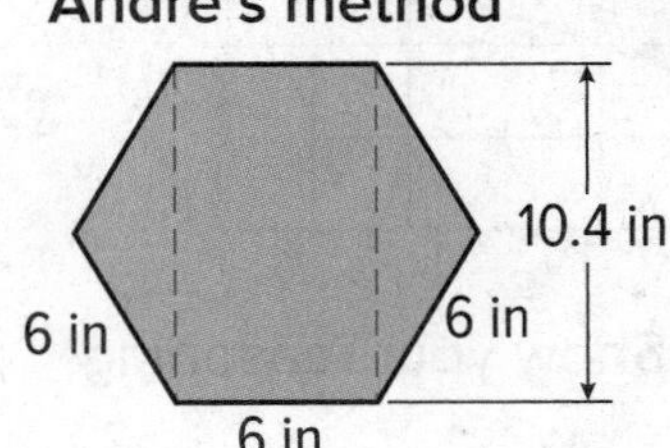

Find the area of the hexagon using each person's method. Show your reasoning.

5. Respond to each of the following. **(Lesson 1-9)**

   a. Identify a base and a corresponding height that can be used to find the area of this triangle. Label the base *b* and the corresponding height *h*.

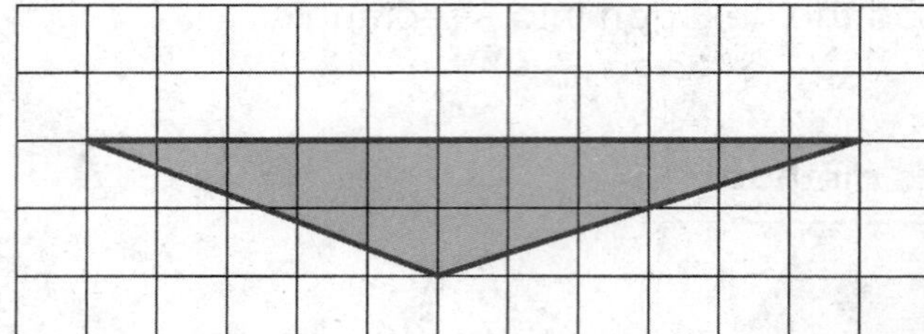

   b. Find the area of the triangle. Show your reasoning.

6. On the grid, draw three different triangles with an area of 8 square units. Label the base and height of each triangle. **(Lesson 1-10)**

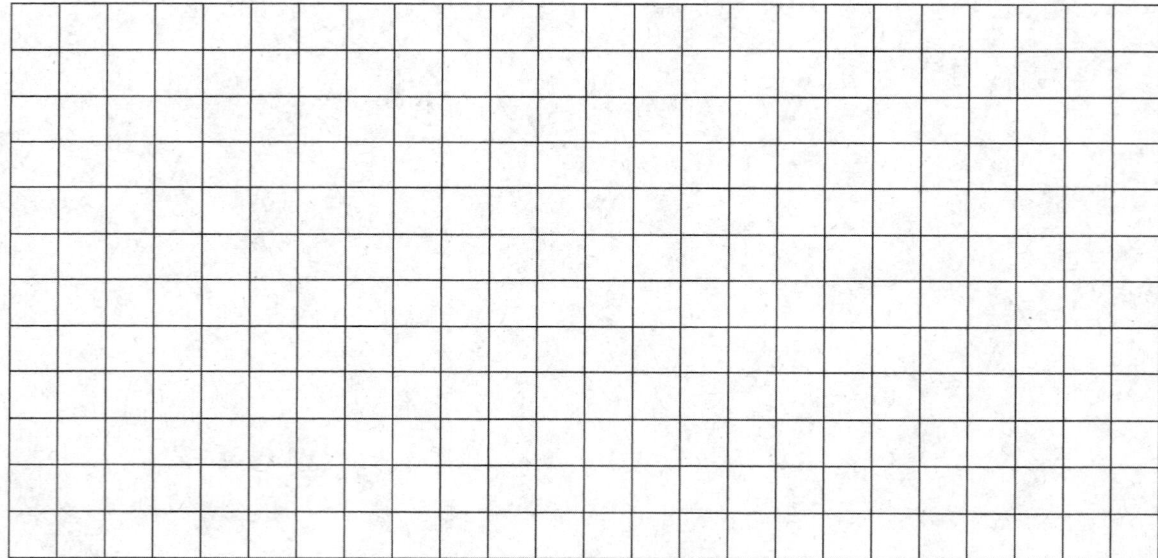

Lesson 1-12

What Is Surface Area?

NAME ______________________ DATE ____________ PERIOD ________

Learning Goal Let's cover the surfaces of some three-dimensional objects.

Warm Up

12.1 Covering the Cabinet (Part 1)

Your teacher will show you a video about a cabinet or some pictures of it.

Estimate an answer to the question: How many sticky notes would it take to cover the cabinet, excluding the bottom?

Activity

12.2 Covering the Cabinet (Part 2)

Earlier, you learned about a cabinet being covered with sticky notes.

1. How could you find the actual number of sticky notes it will take to cover the cabinet, excluding the bottom? What information would you need to know?

2. Use the information you have to find the number of sticky notes to cover the cabinet. Show your reasoning.

Are you ready for more?

How many sticky notes are needed to cover the outside of 2 cabinets pushed together (including the bottom)? What about 3 cabinets? 20 cabinets?

Activity

12.3 Building with Snap Cubes

Here is a sketch of a rectangular prism built from 12 cubes.

It has six **faces**, but you can only see three of them in the sketch. It has a **surface area** of 32 square units.

You have 12 snap cubes from your teacher. Use all of your snap cubes to build a different rectangular prism (with different edge lengths than shown in the prism here).

1. How many faces does your figure have?

2. What is the surface area of your figure in square units?

3. Draw your figure on isometric dot paper. Color each face a different color.

NAME ______________________ DATE ____________ PERIOD ____________

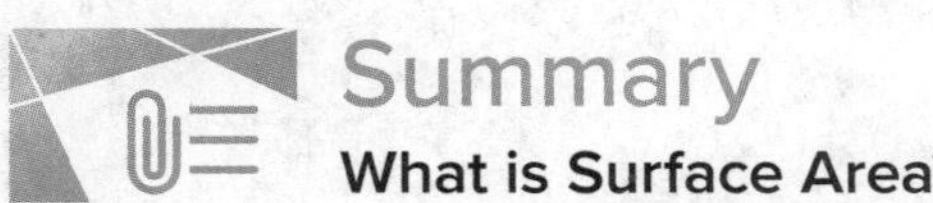

Summary

What is Surface Area?

- The **surface area** of a figure (in square units) is the number of unit squares it takes to cover the entire surface without gaps or overlaps.
- If a three-dimensional figure has flat sides, the sides are called **faces**.
- The surface area is the total of the areas of the faces.

For example, a rectangular prism has six faces. The surface area of the prism is the total of the areas of the six rectangular faces.

So, the surface area of a rectangular prism that has edge lengths 2 cm, 3 cm, and 4 cm has a surface area of...

$(2 \cdot 3) + (2 \cdot 3) + (2 \cdot 4) + (2 \cdot 4) + (3 \cdot 4) + (3 \cdot 4)$ or 52 square centimeters.

Glossary

face

surface area

Practice

What is Surface Area?

1. What is the surface area of this rectangular prism?

   A. 16 square units

   B. 32 square units

   C. 48 square units

   D. 64 square units

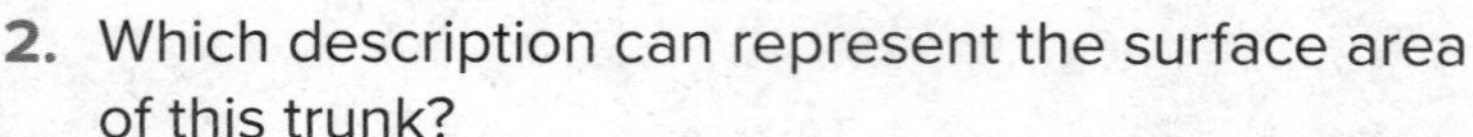

2. Which description can represent the surface area of this trunk?

   A. The number of square inches that cover the top of the trunk.

   B. The number of square feet that cover all the outside faces of the trunk.

   C. The number of square inches of horizontal surface inside the trunk.

   D. The number of cubic feet that can be packed inside the trunk.

NAME _______________ DATE _______________ PERIOD _______________

3. Which figure has a greater surface area?

Figure A

Figure B

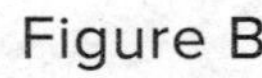

4. A rectangular prism is 4 units high, 2 units wide, and 6 units long. What is its surface area in square units? Explain or show your reasoning.

5. Draw an example of each of these triangles on the grid. **(Lesson 1-9)**

   a. A right triangle with an area of 6 square units.

   b. An acute triangle with an area of 6 square units.

   c. An obtuse triangle with an area of 6 square units.

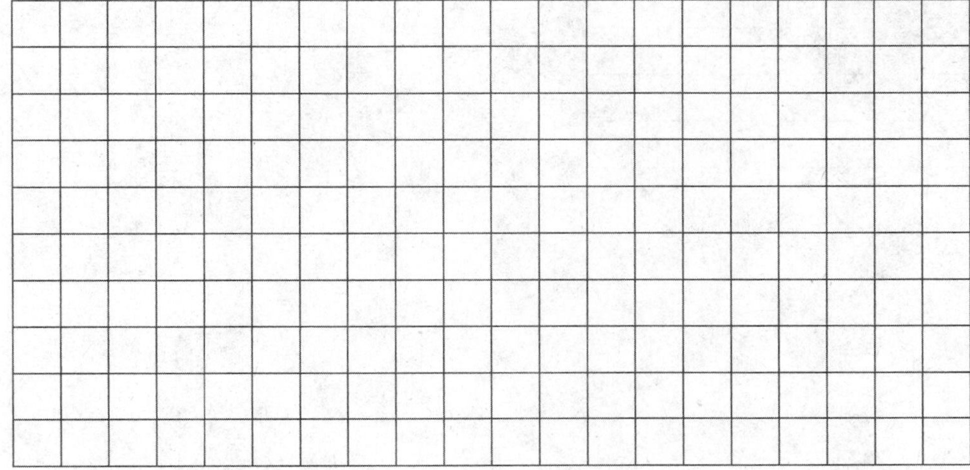

6. Find the area of triangle *MOQ* in square units. Show your reasoning. **(Lesson 1-10)**

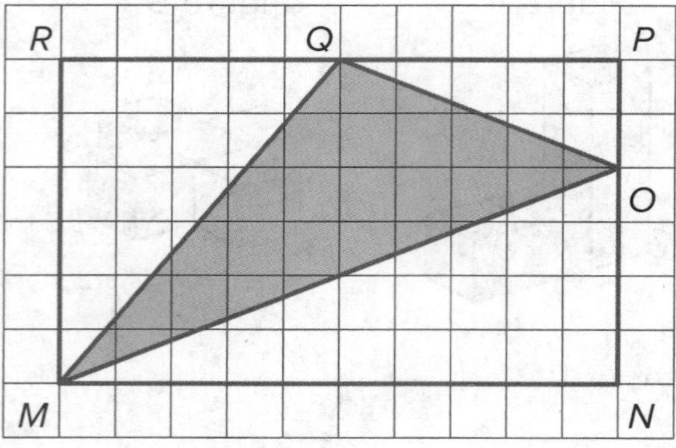

7. Find the area of this shape. Show your reasoning. **(Lesson 1-3)**

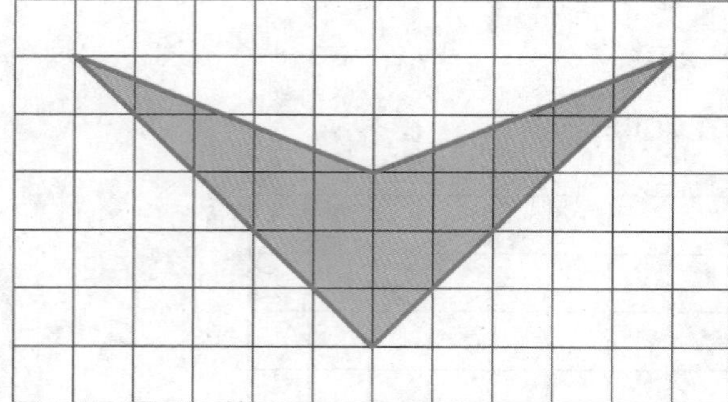

Lesson 1-13

Polyhedra

NAME ______________________ DATE ______________ PERIOD ______________

Learning Goal Let's investigate polyhedra.

Warm Up

13.1 What Are Polyhedra?

Here are pictures that represent **polyhedra**.

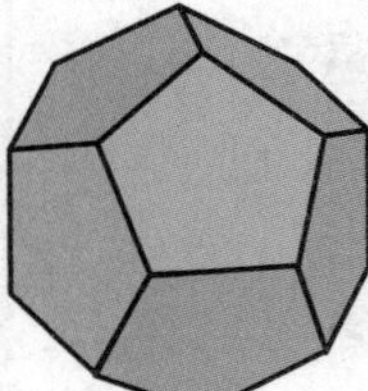
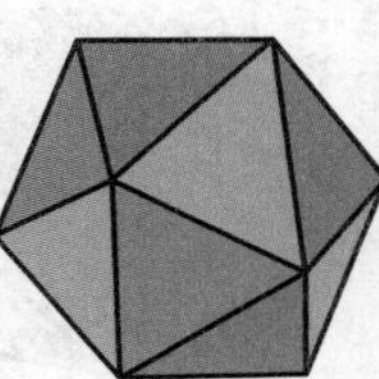

Here are pictures that do *not* represent polyhedra.

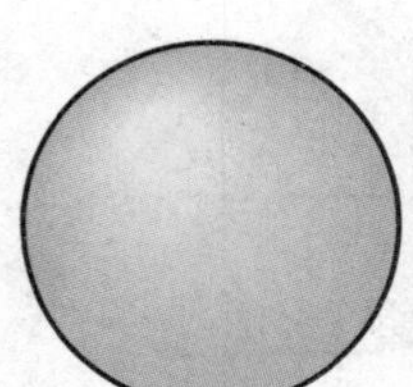
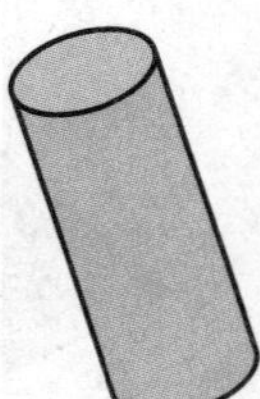
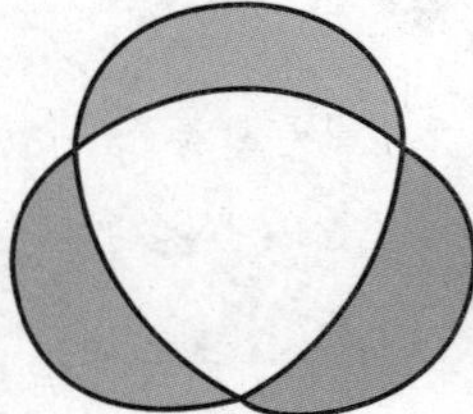
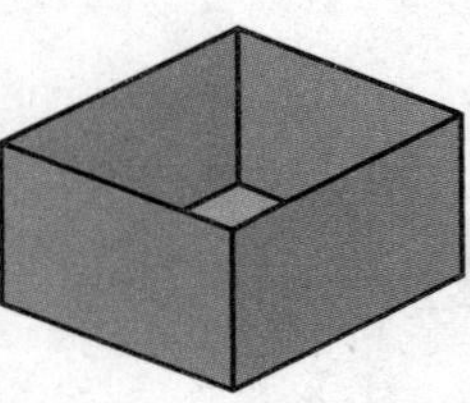

1. Your teacher will give you some figures or objects. Sort them into polyhedra and non-polyhedra.

2. What features helped you distinguish the polyhedra from the other figures?

Activity

13.2 Prisms and Pyramids

1. Here are some polyhedra called **prisms**.

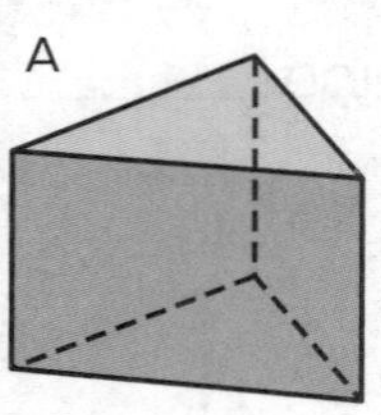

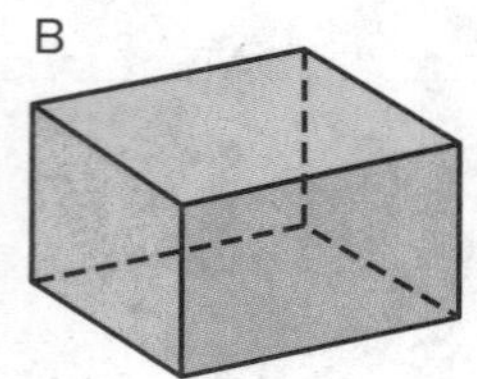

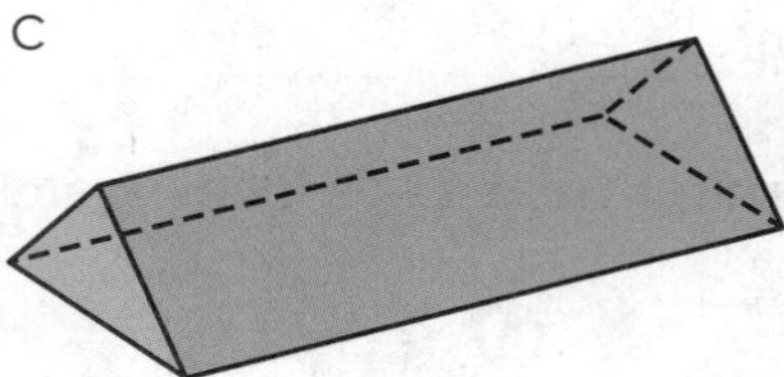

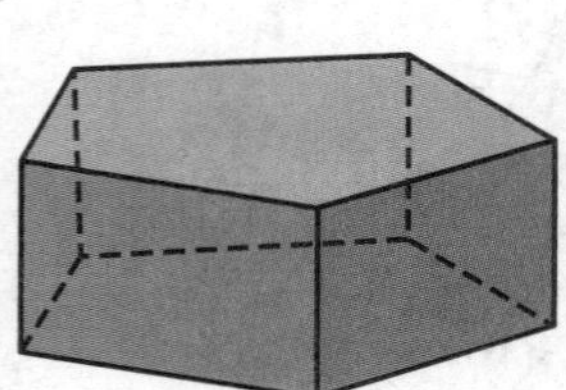

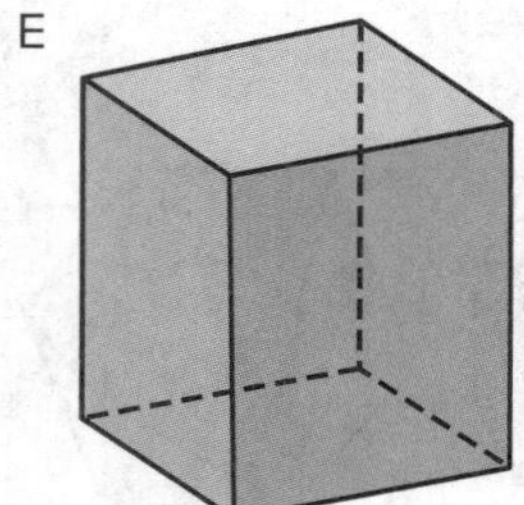

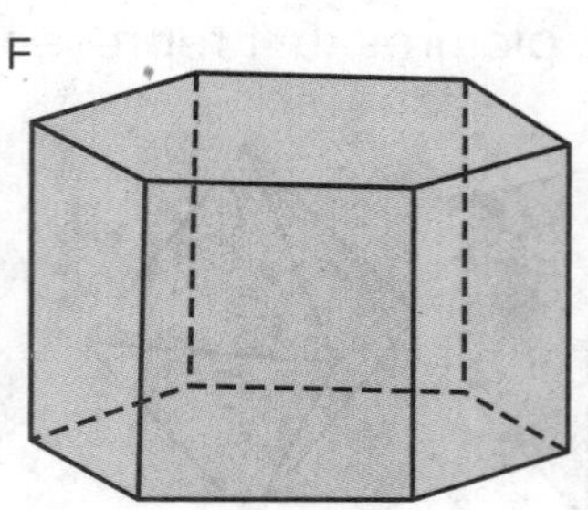

Here are some polyhedra called **pyramids**.

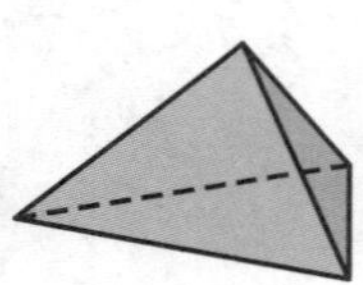

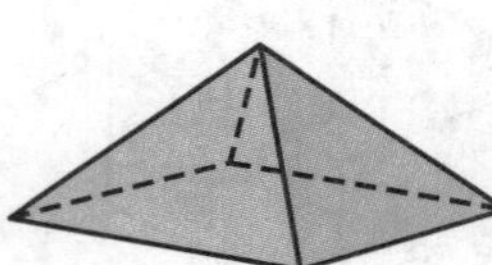

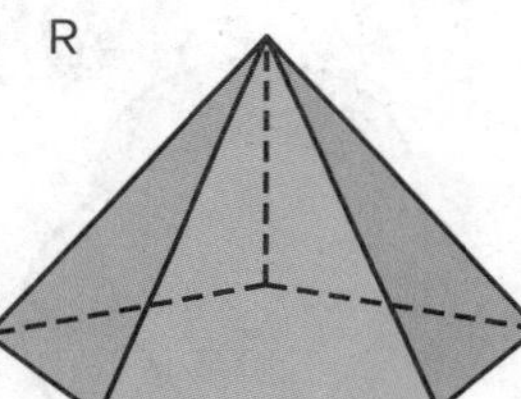

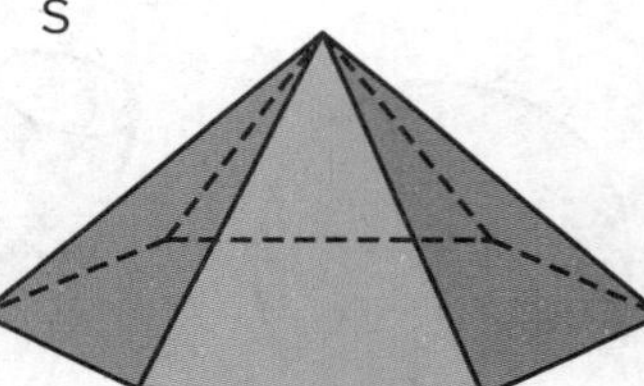

a. Look at the prisms. What are their characteristics or features?

b. Look at the pyramids. What are their characteristics or features?

2. Which of these **nets** can be folded into Pyramid P? Select all that apply.

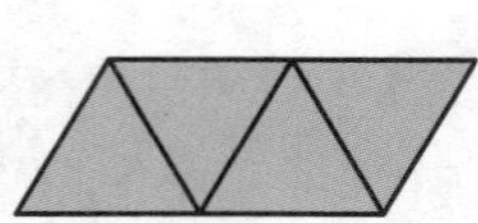

NAME ______________________ DATE ______________ PERIOD ____________

3. Your teacher will give your group a set of polygons and assign a polyhedron.

   a. Decide which polygons are needed to compose your assigned polyhedron. List the polygons and how many of each are needed.

   b. Arrange the cut-outs into a net that, if taped and folded, can be assembled into the polyhedron. Sketch the net. If possible, find more than one way to arrange the polygons (show a different net for the same polyhedron).

Are you ready for more?

What is the smallest number of faces a polyhedron can possibly have? Explain how you know.

Activity

13.3 Assembling Polyhedra

1. Your teacher will give you the net of a polyhedron. Cut out the net and fold it along the edges to assemble a polyhedron. Tape or glue the flaps so that there are no unjoined edges.

2. How many **vertices**, **edges**, and **faces** are in your polyhedron?

Summary

Polyhedra

A **polyhedron** is a three-dimensional figure composed of faces. Each face is a filled-in polygon and meets only one other face along a complete edge. The ends of the edges meet at points that are called vertices.

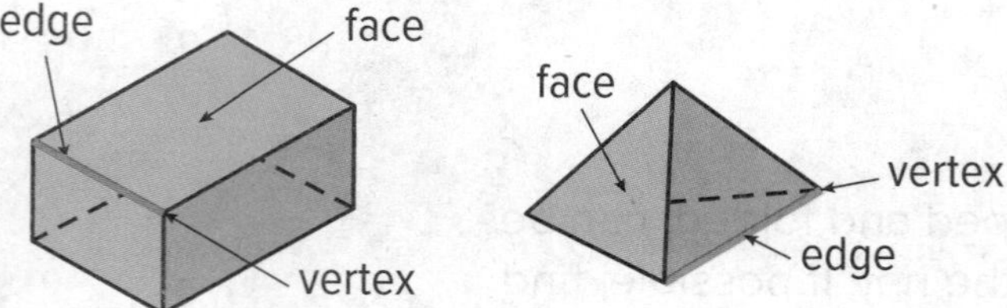

A **polyhedron** always encloses a three-dimensional region.

The plural of polyhedron is **polyhedra**. Here are some drawings of polyhedra.

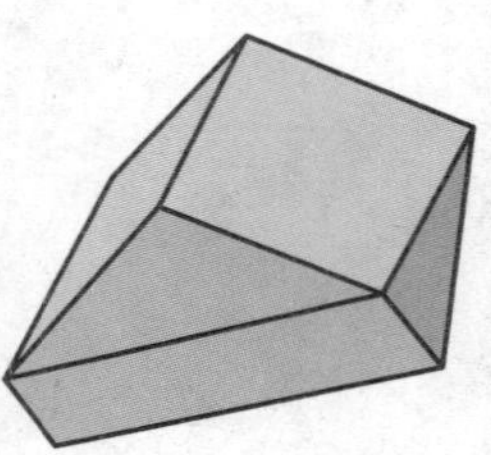

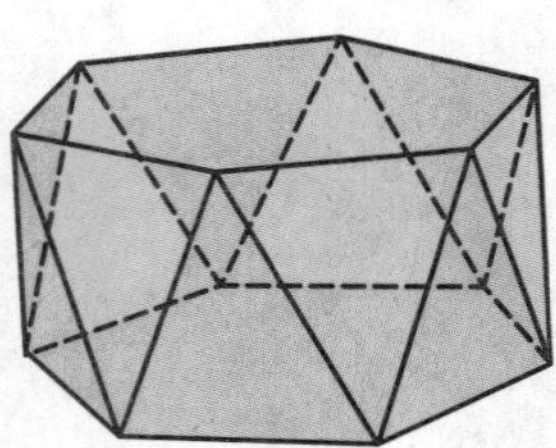

NAME ______________________ DATE ______________ PERIOD ____________

A **prism** is a type of polyhedron with two identical faces that are parallel to each other and that are called **bases**. The bases are connected by a set of rectangles (or sometimes parallelograms).

A prism is named for the shape of its bases. For example, if the base is a pentagon, then it is called a "pentagonal prism."

Triangular Prism

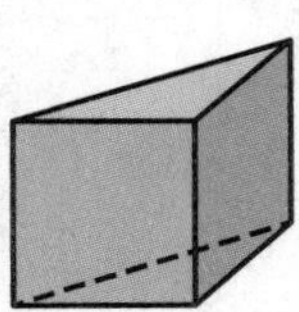

Pentagonal Prism

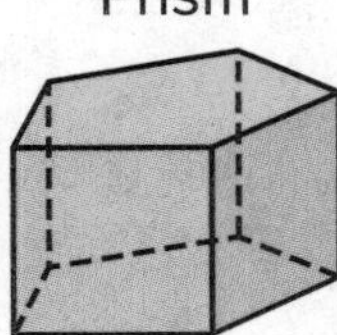

Rectangular Prism

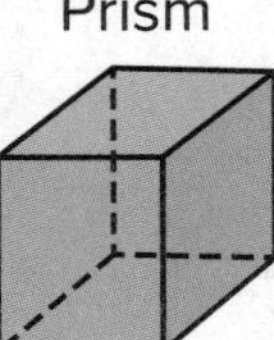

A **pyramid** is a type of polyhedron that has one special face called the base. All of the other faces are triangles that all meet at a single vertex.

A pyramid is named for the shape of its base. For example, if the base is a pentagon, then it is called a "pentagonal pyramid."

Rectangular Pyramid

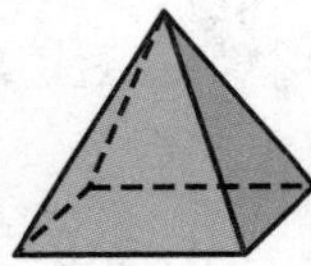

Hexagonal Pyramid

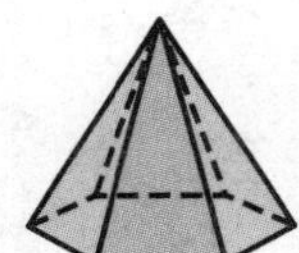

Heptagonal Pyramid

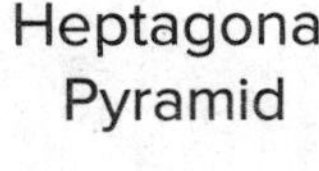

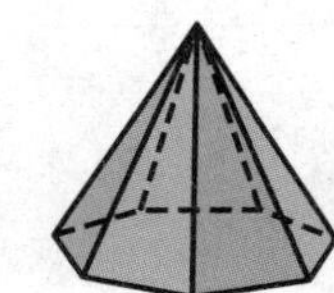

Decagonal Pyramid

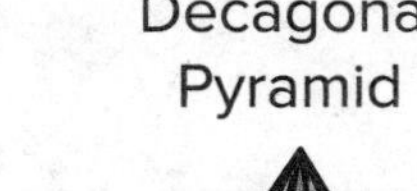

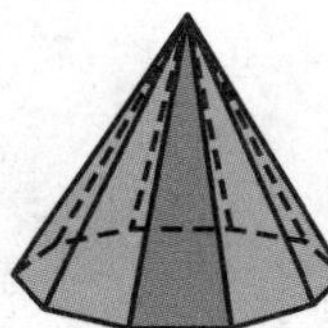

A **net** is a two-dimensional representation of a polyhedron.
It is composed of polygons that form the faces of a polyhedron.

A cube has 6 square faces, so its net is composed of six squares, as shown here.

A net can be cut out and folded to make a model of the polyhedron.

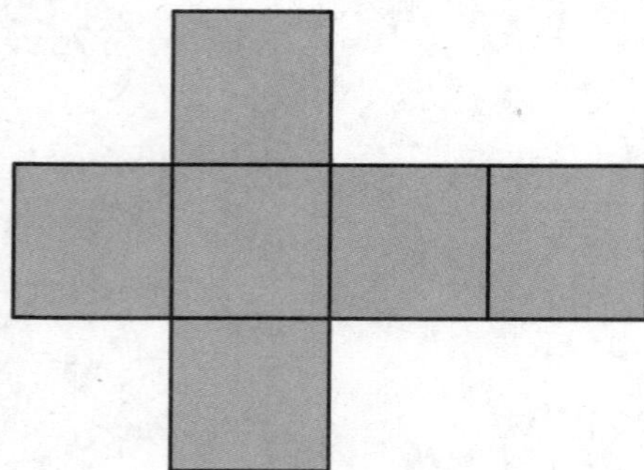

In a cube, every face shares its edges with 4 other squares. In a net of a cube, not all edges of the squares are joined with another edge. When the net is folded, however, each of these open edges will join another edge.

It takes practice to visualize the final polyhedron by just looking at a net.

Glossary

base (of a prism or pyramid)

net

polyhedron

prism

pyramid

NAME ______________________ DATE ____________ PERIOD ________

Practice
Polyhedra

1. Select **all** the polyhedra.

A.
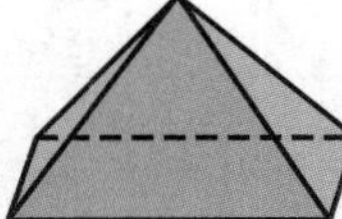

D.
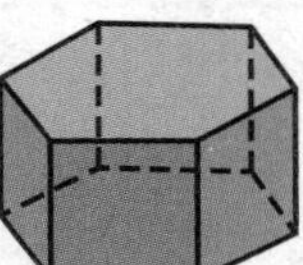

B.
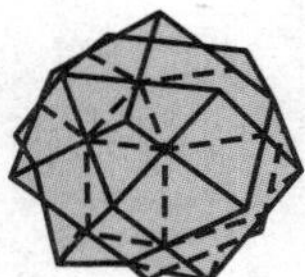

E.
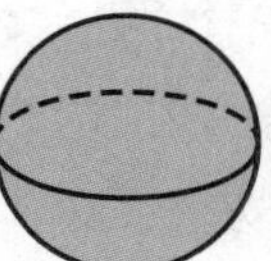

C.
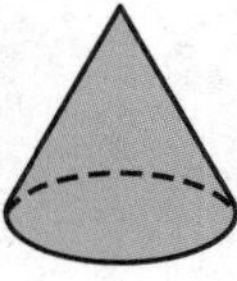

2. Respond to each of the following.

   a. Is this polyhedron a prism, a pyramid, or neither? Explain how you know.

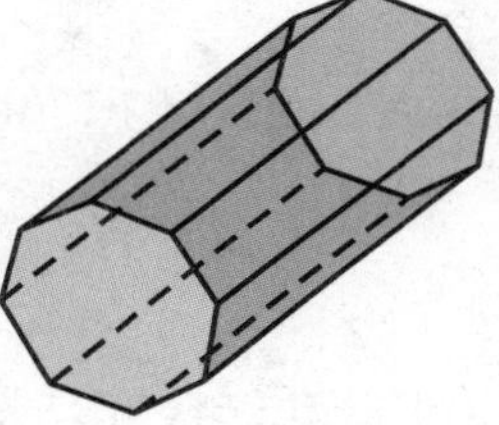

   b. How many faces, edges, and vertices does it have?

3. Tyler said this net cannot be a net for a square prism because not all the faces are squares.

   Do you agree with him? Explain your reasoning.

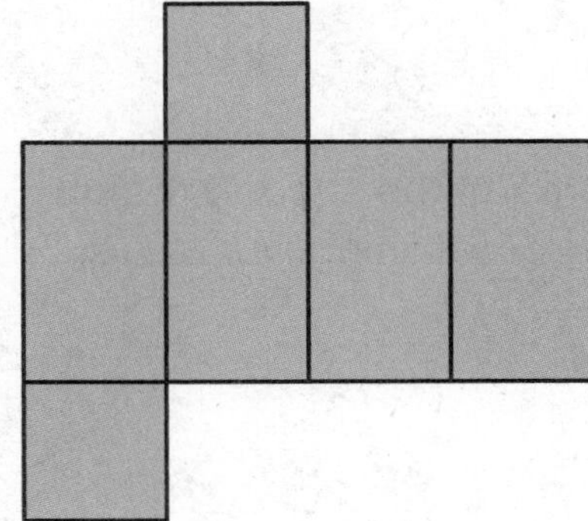

4. Explain why each of these triangles has an area of 9 square units. **(Lesson 1-8)**

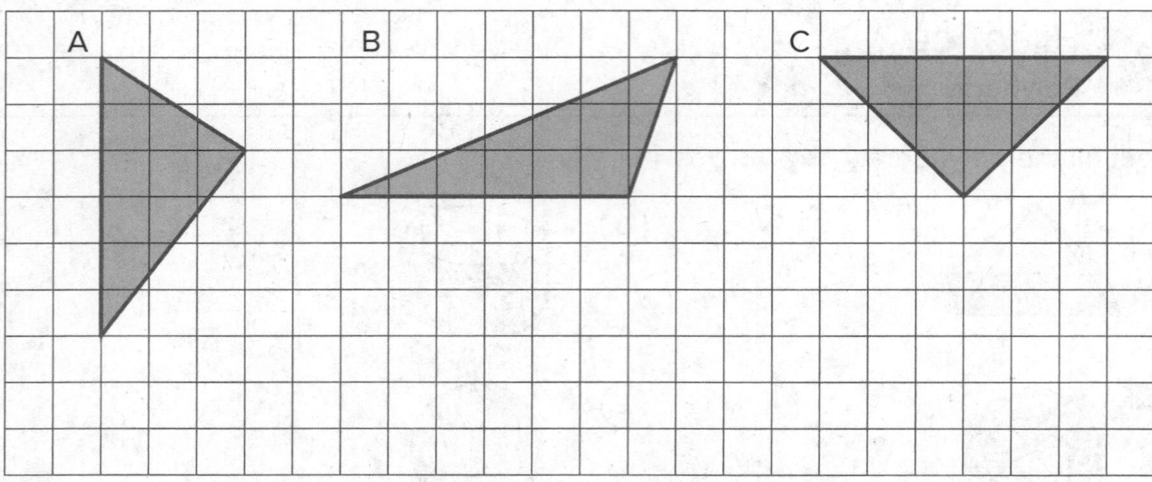

5. Respond to each of the following. **(Lesson 1-9)**

   a. A parallelogram has a base of 12 meters and a height of 1.5 meters. What is its area?

   b. A triangle has a base of 16 inches and a height of $\frac{1}{8}$ inches. What is its area?

   c. A parallelogram has an area of 28 square feet and a height of 4 feet. What is its base?

   d. A triangle has an area of 32 square millimeters and a base of 8 millimeters. What is its height?

6. Find the area of the shaded region. Show or explain your reasoning. **(Lesson 1-3)**

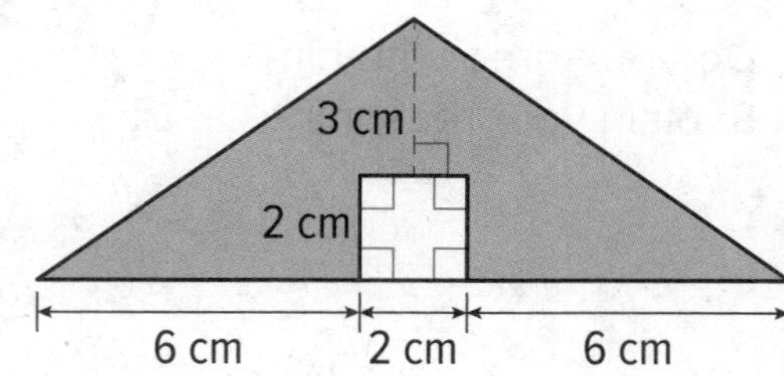

Lesson 1-14

Nets and Surface Area

NAME ______________________ DATE ______________ PERIOD ______________

Learning Goal Let's use nets to find the surface area of polyhedra.

Warm Up

14.1 Matching Nets

Each of the nets can be assembled into a polyhedron. Match each net with its corresponding polyhedron and name the polyhedron. Be prepared to explain how you know the net and polyhedron go together.

Nets

a. 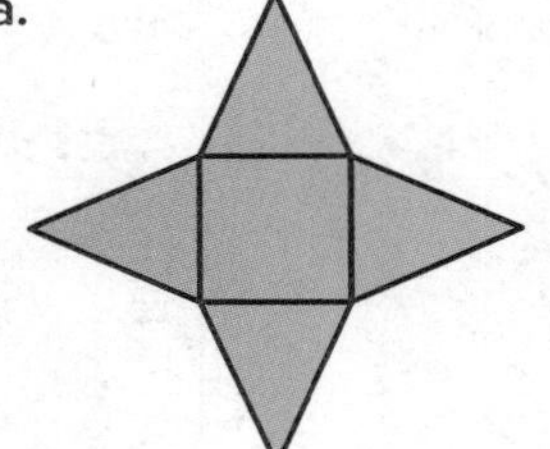

b. 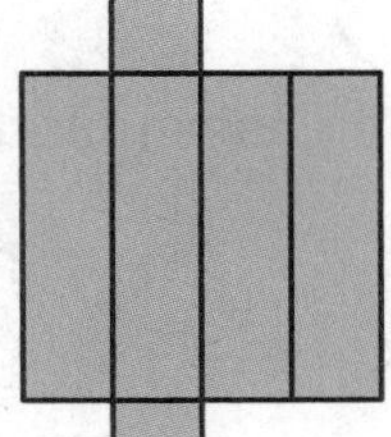

c. 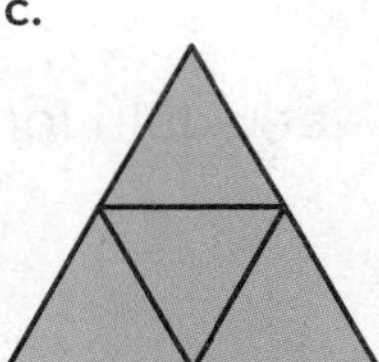

d. 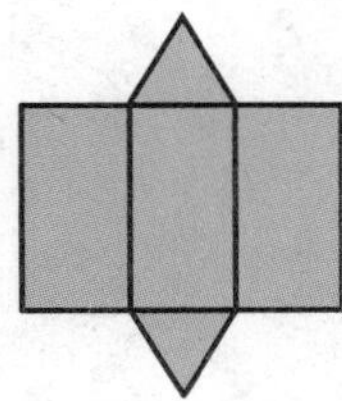

e. 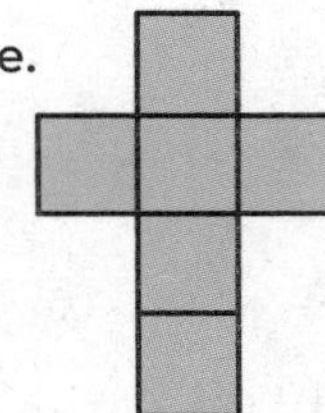

Polyhedra

1 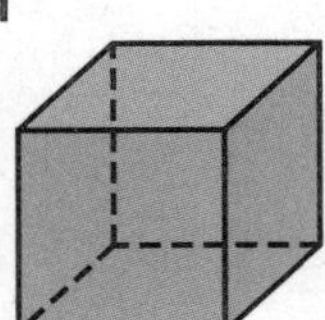

2 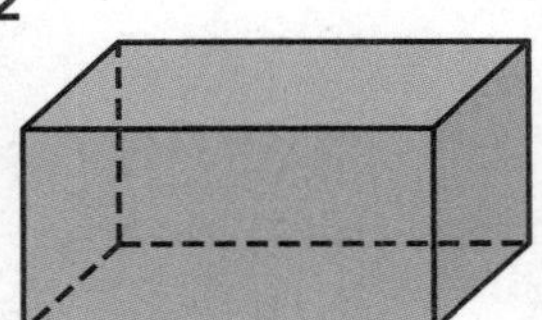

3 

4 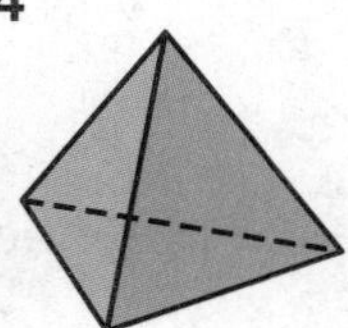

5 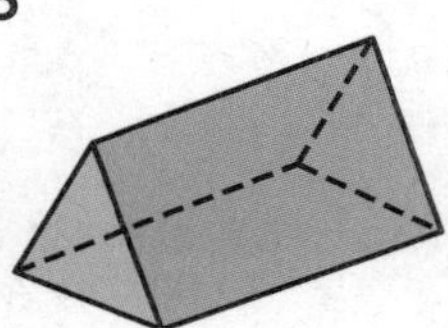

Activity
14.2 Using Nets to Find Surface Area

Refer to the nets shown.

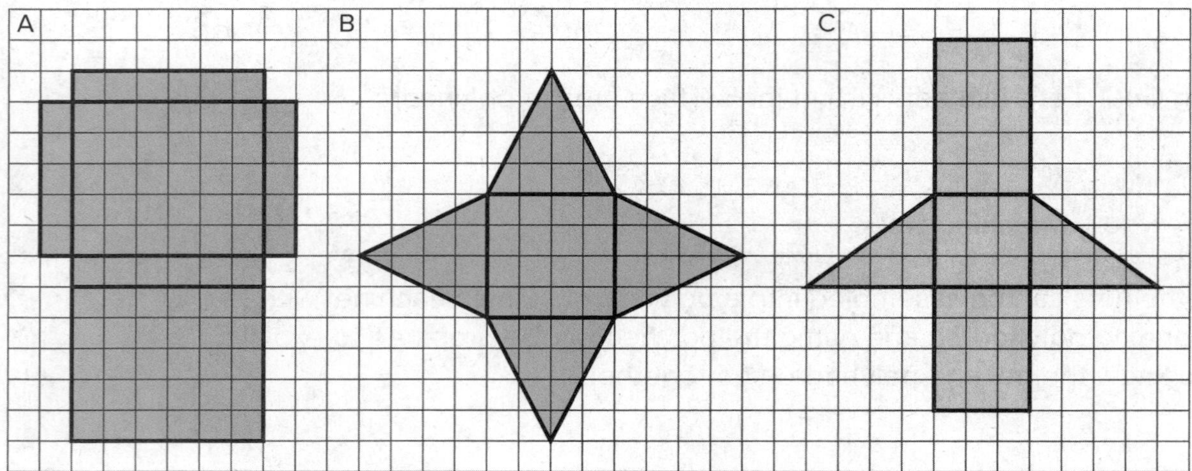

1. Name the polyhedron that each net would form when assembled.

   A:

   B:

   C:

2. Your teacher will give you the nets of three polyhedra. Cut out the nets and assemble the three-dimensional shapes.

3. Find the **surface area** of each polyhedron. Explain your reasoning clearly.

NAME ______________________ DATE ____________ PERIOD ____________

Are you ready for more?

1. For each net, decide if it can be assembled into a rectangular prism.

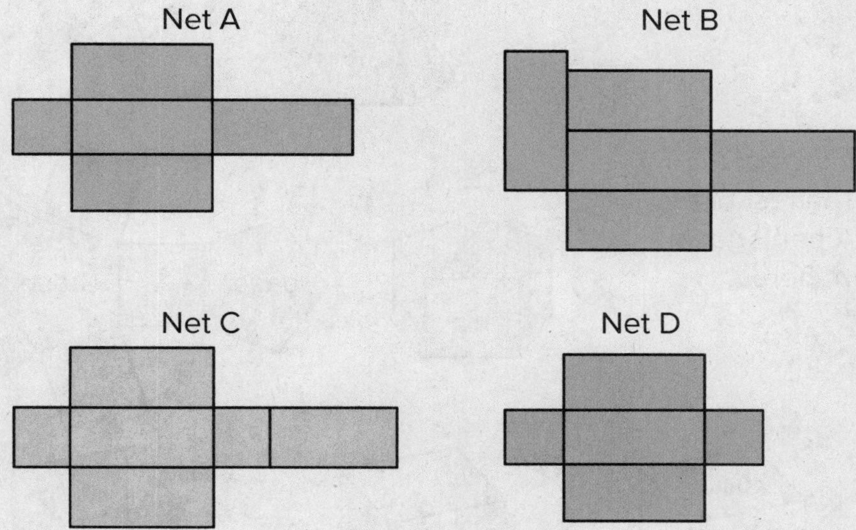

2. For each net, decide if it can be folded into a triangular prism.

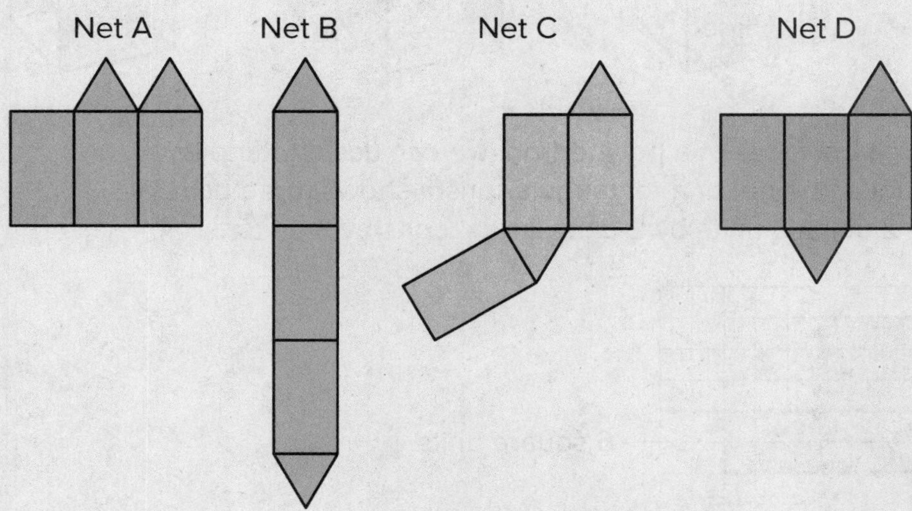

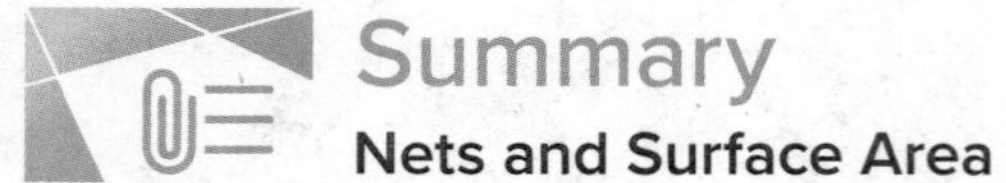

Summary

Nets and Surface Area

A net of a *pyramid* has one polygon that is the base. The rest of the polygons are triangles. A pentagonal pyramid and its net are shown here.

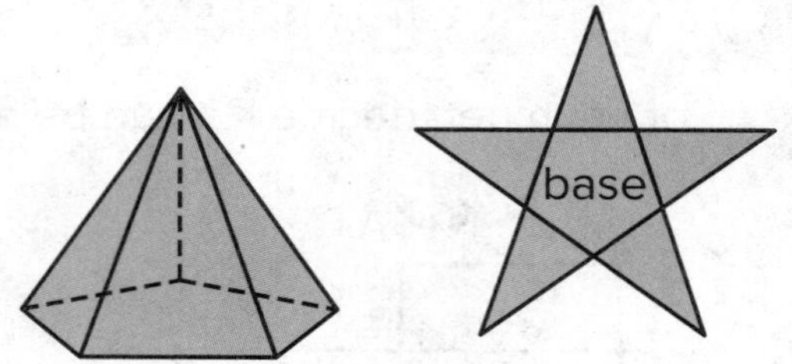

A net of a *prism* has two copies of the polygon that is the base. The rest of the polygons are rectangles. A pentagonal prism and its net are shown here.

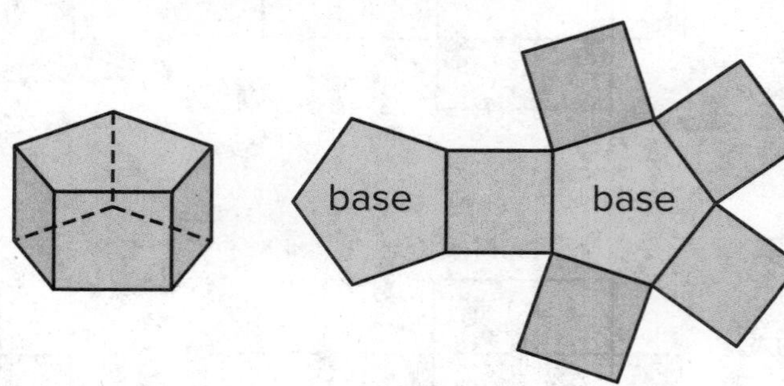

In a rectangular prism, there are three pairs of parallel and identical rectangles. Any pair of these identical rectangles can be the bases.

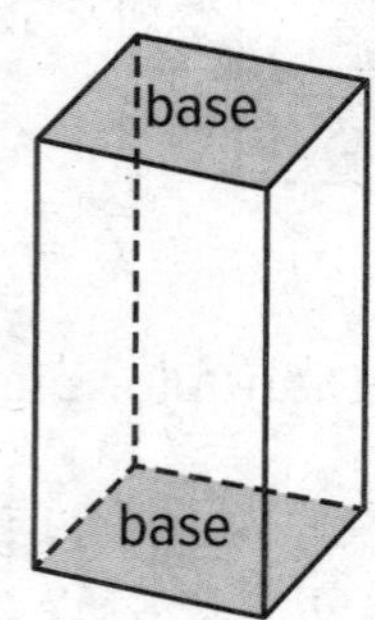

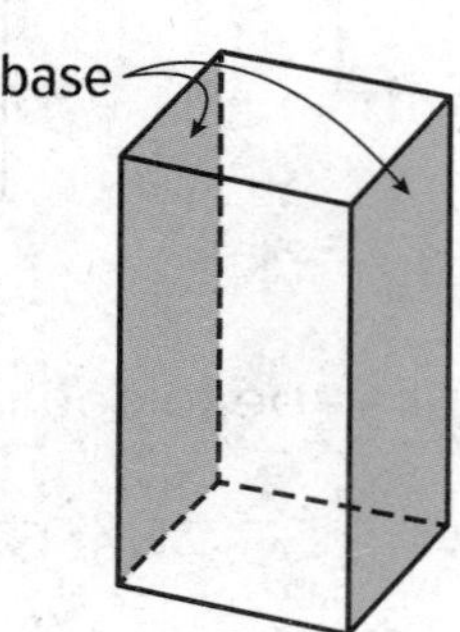

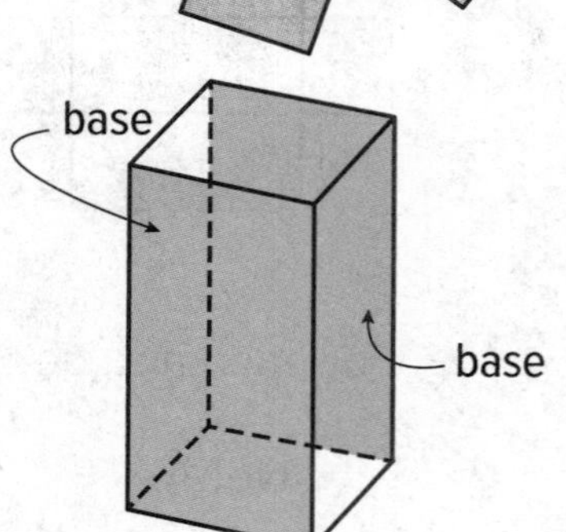

Because a net shows all the faces of a polyhedron, we can use it to find its surface area. For instance, the net of a rectangular prism shows three pairs of rectangles: 4 units by 2 units, 3 units by 2 units, and 4 units by 3 units.

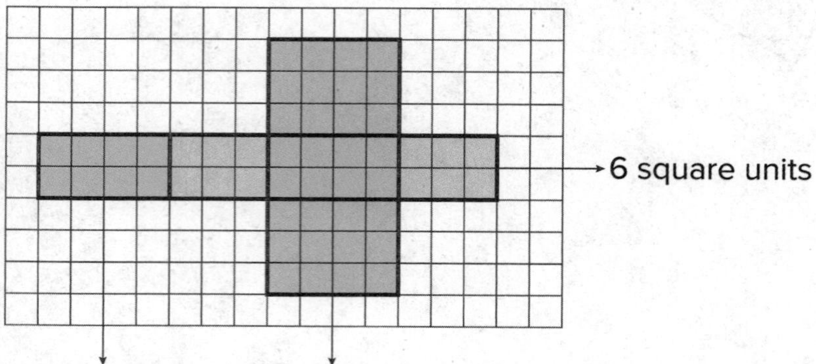

The **surface area** of the rectangular prism is 52 square units because $8 + 8 + 6 + 6 + 12 + 12 = 52$.

NAME ______________________ DATE ______________ PERIOD __________

Practice

Nets and Surface Area

1. Can this net be assembled into a cube? Explain how you know. Label parts of the net with letters or numbers if it helps your explanation.

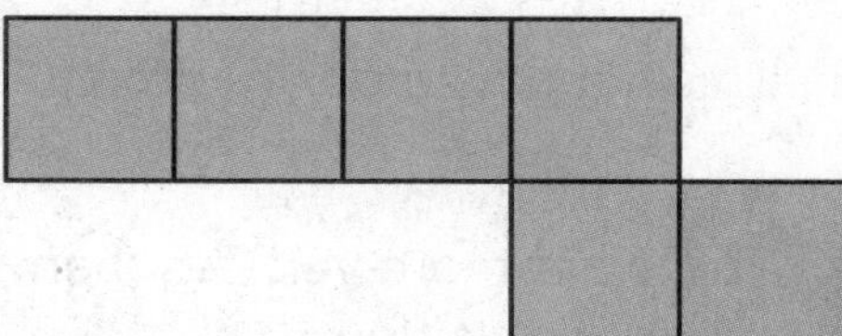

2. a. What polyhedron can be assembled from this net? Explain how you know.

   b. Find the surface area of this polyhedron. Show your reasoning.

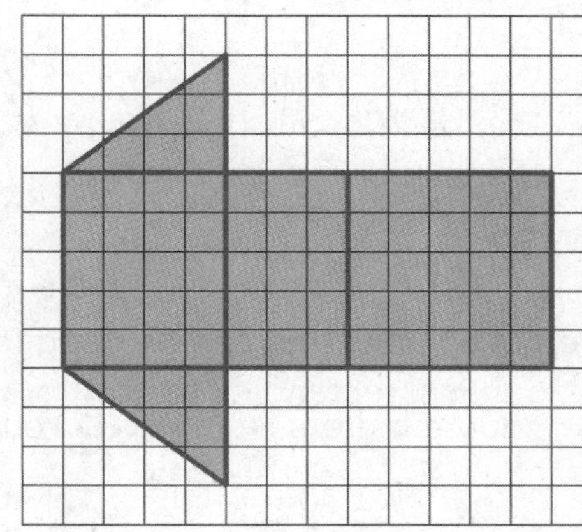

3. Here are two nets. Mai said that both nets can be assembled into the same triangular prism. Do you agree? Explain or show your reasoning.

Net A

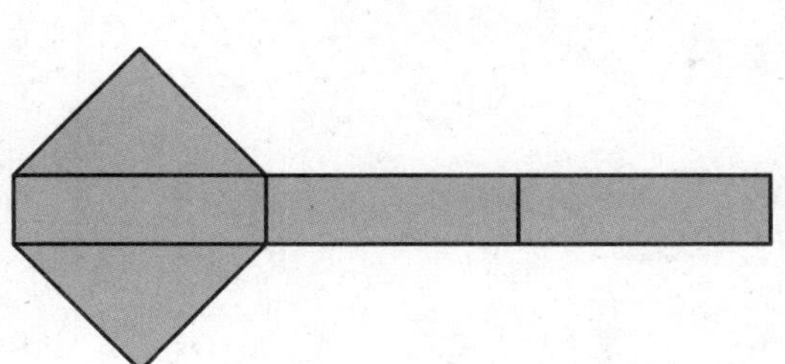

Net B

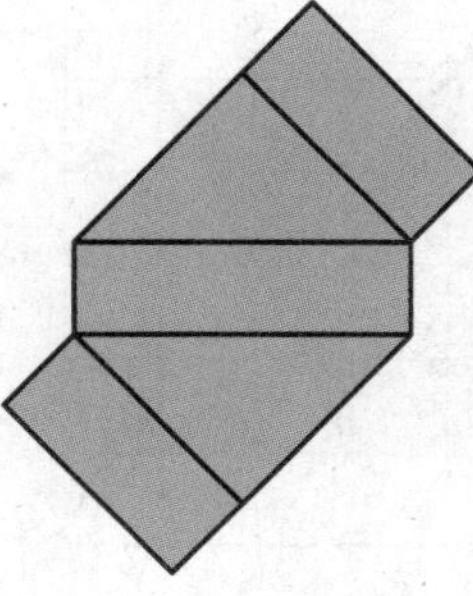

4. Here are two three-dimensional figures. **(Lesson 1-13)**

Tell whether each of the following statements describes Figure A, Figure B, both, or neither.

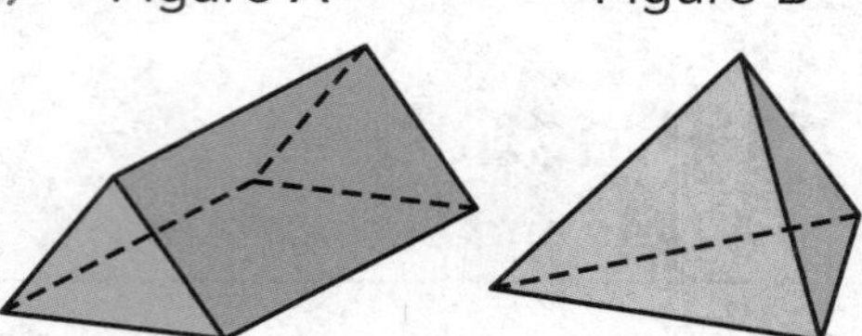

a. This figure is a polyhedron.

b. This figure has triangular faces.

c. There are more vertices than edges in this figure.

d. This figure has rectangular faces.

e. This figure is a pyramid.

f. There is exactly one face that can be the base for this figure.

g. The base of this figure is a triangle.

h. This figure has two identical and parallel faces that can be the base.

5. Select **all** units that can be used for surface area. **(Lesson 1-12)**

A. square meters

B. feet

C. centimeters

D. cubic inches

E. square inches

F. square feet

6. Find the area of this polygon. Show your reasoning. **(Lesson 1-11)**

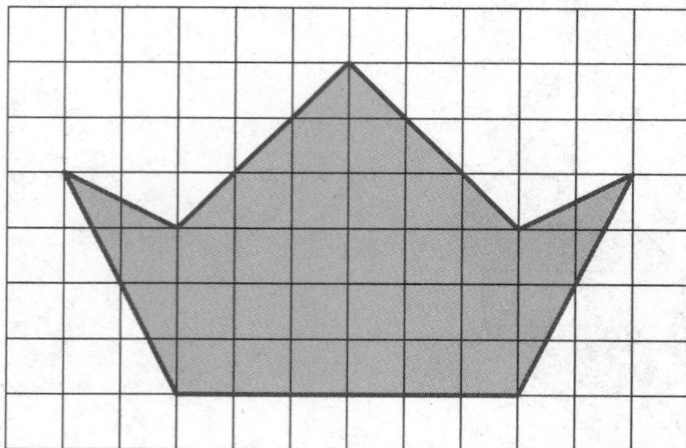

Lesson 1-15

More Nets, More Surface Area

NAME ______________________ DATE ______________ PERIOD ____________

Learning Goal Let's draw nets and find the surface area of polyhedra.

Warm Up

15.1 Notice and Wonder: Wrapping Paper

Kiran is wrapping this box of sports cards as a present for a friend. What do you notice? What do you wonder?

Activity

15.2 Building Prisms and Pyramids

Your teacher will give you a drawing of a polyhedron. You will draw its net and calculate its surface area.

1. What polyhedron do you have?

2. Study your polyhedron. Then, draw its net on graph paper. Use the side length of a grid square as the unit.

3. Label each polygon on the net with a name or number.

4. Find the surface area of your polyhedron. Show your thinking in an organized manner so that it can be followed by others.

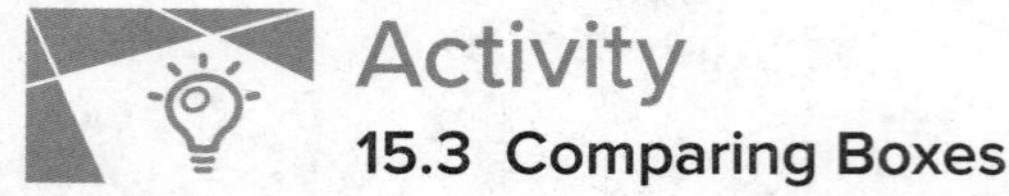

Activity

15.3 Comparing Boxes

Here are the nets of three cardboard boxes that are all rectangular prisms. The boxes will be packed with 1-centimeter cubes. All lengths are in centimeters.

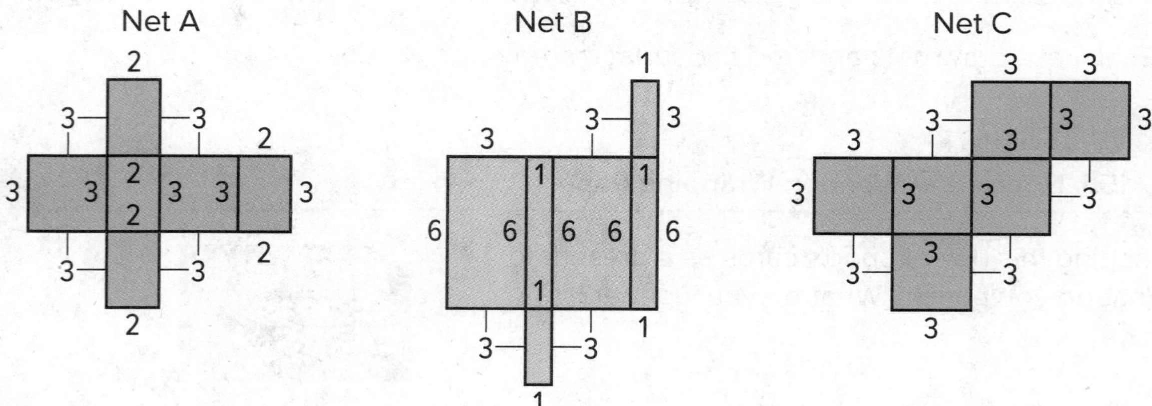

1. Compare the surface areas of the boxes. Which box will use the least cardboard? Show your reasoning.

2. Now compare the volumes of these boxes in cubic centimeters. Which box will hold the most 1-centimeter cubes? Show your reasoning.

NAME ______________________________ DATE ______________ PERIOD ____________

Are you ready for more?

Figure C shows a net of a cube. Draw a different net of a cube. Draw another one. And then another one. How many different nets can be drawn and assembled into a cube?

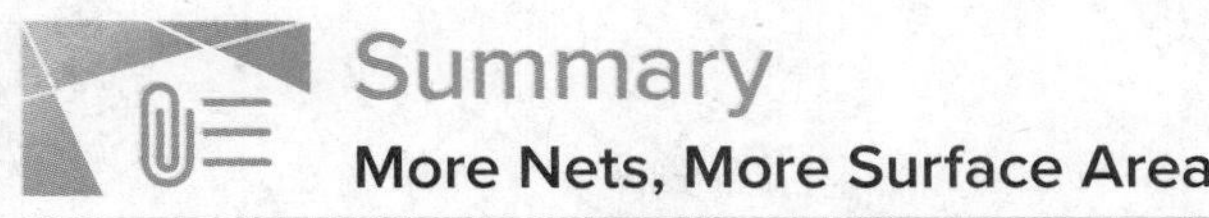

Summary

More Nets, More Surface Area

The **surface area** of a polyhedron is the sum of the areas of all of the faces.

Because a net shows us all faces of a polyhedron at once, it can help us find the surface area. We can find the areas of all polygons in the net and add them.

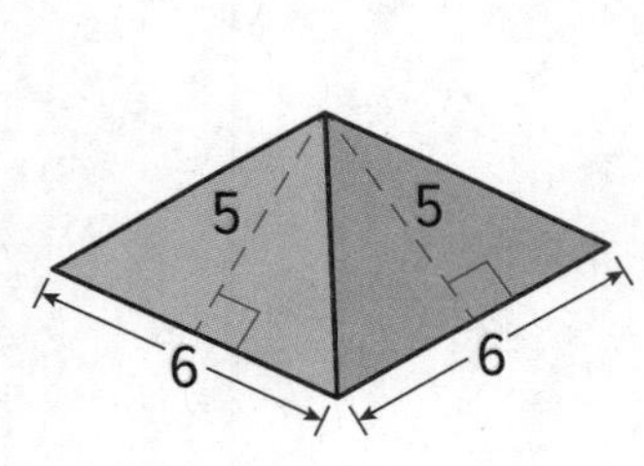

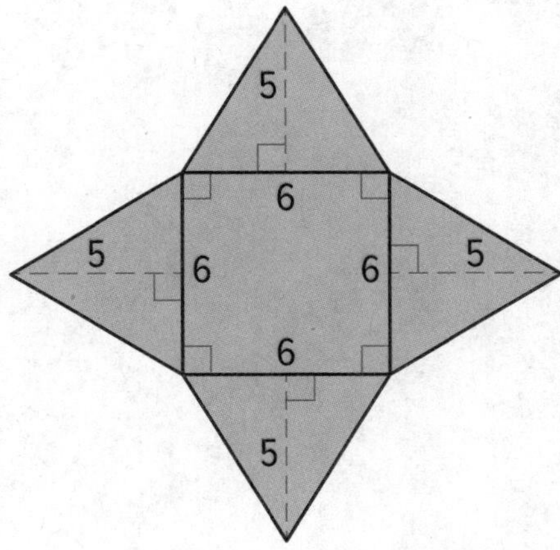

A square pyramid has a square and four triangles for its faces. Its surface area is the sum of the areas of the square base and the four triangular faces:

$$(6 \cdot 6) + 4 \cdot \left(\frac{1}{2} \cdot 5 \cdot 6\right) = 96$$

The surface area of this square pyramid is 96 square units.

Practice

More Nets, More Surface Area

1. Jada drew a net for a polyhedron and calculated its surface area.

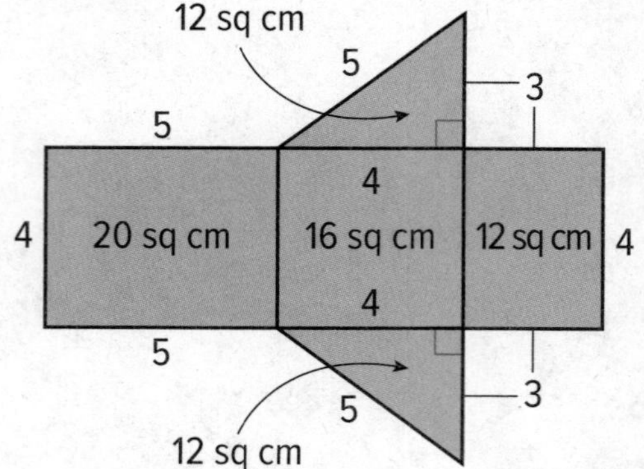

   a. What polyhedron can be assembled from this net?

   b. Jada made some mistakes in her area calculation. What were the mistakes?

   c. Find the surface area of the polyhedron. Show your reasoning.

2. A cereal box is 8 inches by 2 inches by 12 inches. What is its surface area? Show your reasoning. If you get stuck, consider drawing a sketch of the box or its net and labeling the edges with their measurements.

NAME ______________________ DATE ____________ PERIOD ____________

3. Twelve cubes are stacked to make this figure. **(Lesson 1-12)**

a. What is its surface area?

b. How would the surface area change if the top two cubes are removed?

4. Here are two polyhedra and their nets. Label all edges in the net with the correct lengths.

Figure A

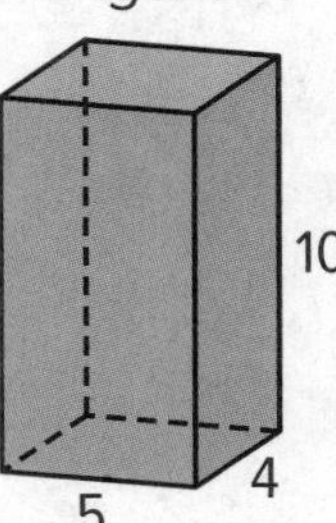

Figure B

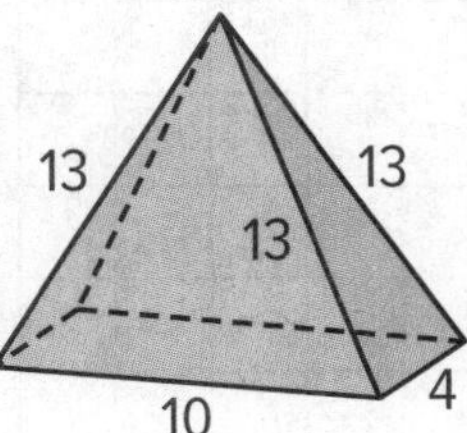

Net of Figure A

Net of Figure B

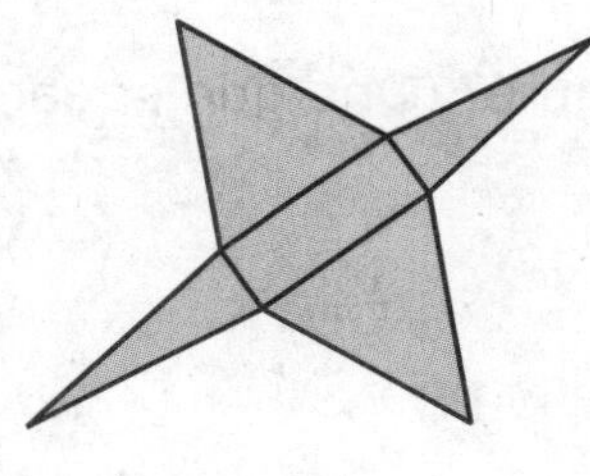

5. Respond to each question. **(Lesson 1-14)**

a. What three-dimensional figure can be assembled from the net?

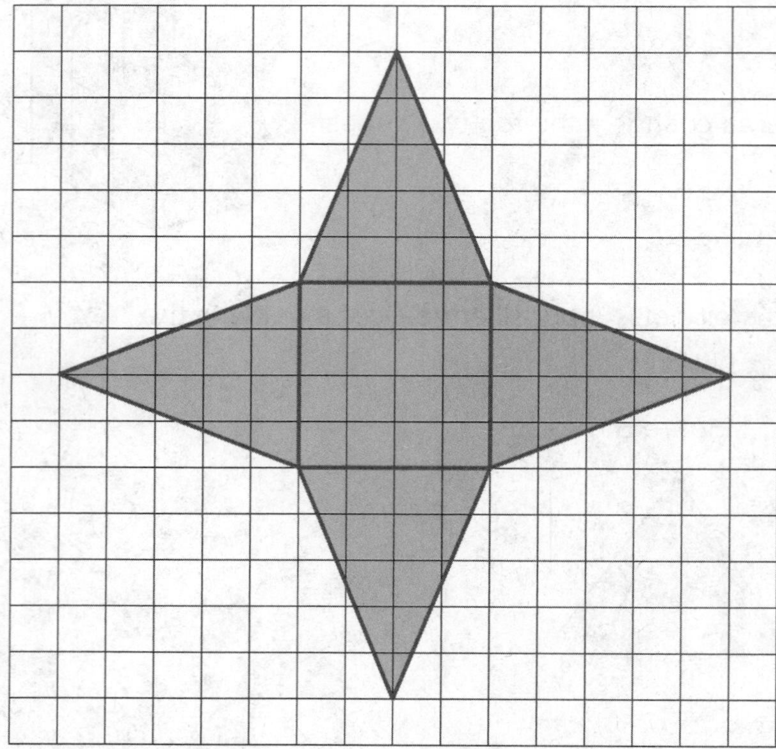

b. What is the surface area of the figure? (One grid square is 1 square unit.)

Lesson 1-16

Distinguishing between Surface Area and Volume

NAME ______________________ DATE ______________ PERIOD ______________

Learning Goal Let's contrast surface area and volume.

Warm Up

16.1 Attributes and Their Measures

For each quantity, choose one or more appropriate units of measurement.

For the last two, think of a quantity that could be appropriately measured with the given units.

Quantities	Units
1. Perimeter of a parking lot:	millimeters (mm)
2. Volume of a semi truck:	feet (ft)
3. Surface area of a refrigerator:	meters (m)
4. Length of an eyelash:	square inches (sq in)
5. Area of a state:	square feet (sq ft)
6. Volume of an ocean:	square miles (sq mi)
7. ______________________: miles	cubic kilometers (cu km)
8. ______________________: cubic meters	cubic yards (cu yd)

Activity

16.2 Building with 8 Cubes

Your teacher will give you 16 cubes. Build two different shapes using 8 cubes for each. For each shape:

1. Give it a name or a label (e.g., Mai's First Shape or Diego's Steps).
2. Determine the volume.
3. Determine the surface area.
4. Record the name, volume, and surface area on a sticky note.

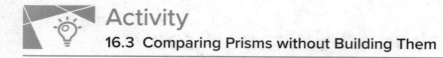

Activity

16.3 Comparing Prisms without Building Them

Three rectangular prisms each have a height of 1 cm.

- Prism A has a base that is 1 cm by 11 cm.
- Prism B has a base that is 2 cm by 7 cm.
- Prism C has a base that is 3 cm by 5 cm.

1. Find the surface area and volume of each prism. Use the dot paper to draw the prisms, if needed.

2. Analyze the volumes and surface areas of the prisms. What do you notice? Write 1–2 observations about them.

Are you ready for more?

Can you find more examples of prisms that have the same surface areas but different volumes? How many can you find?

NAME ________________ DATE ________________ PERIOD ________________

Summary

Distinguishing between Surface Area and Volume

Length is a one-dimensional attribute of a geometric figure. We measure lengths using units like millimeters, centimeters, meters, kilometers, inches, feet, yards, and miles.

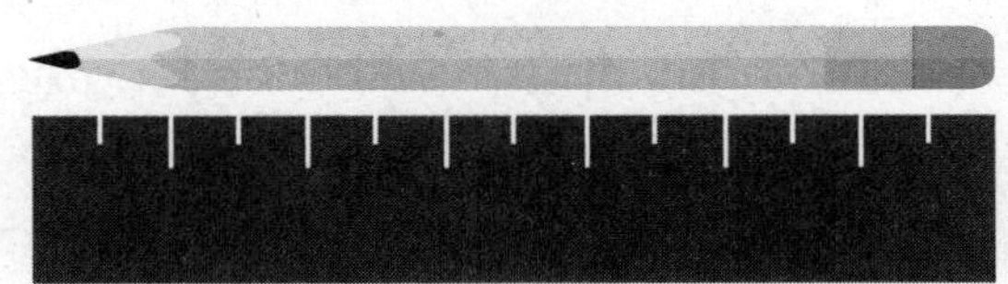

Area is a two-dimensional attribute. We measure area in square units. For example, a square that is 1 centimeter on each side has an area of 1 square centimeter.

Volume is a three-dimensional attribute. We measure volume in cubic units. For example, a cube that is 1 foot on each side has a volume of 1 cubic foot.

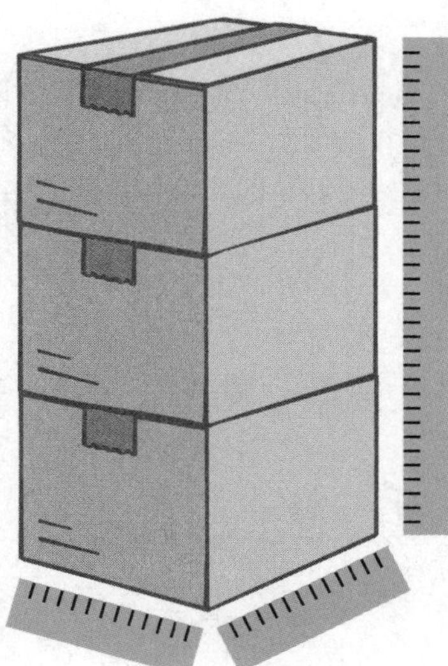

Surface area and volume are different attributes of three-dimensional figures. Surface area is a two-dimensional measure, while volume is a three-dimensional measure.

Two figures can have the same volume but different surface areas. For example:

- A rectangular prism with side lengths of 1 cm, 2 cm, and 2 cm has a volume of 4 cu cm and a surface area of 16 sq cm.
- A rectangular prism with side lengths of 1 cm, 1 cm, and 4 cm has the same volume but a surface area of 18 sq cm.

Similarly, two figures can have the same surface area but different volumes.

- A rectangular prism with side lengths of 1 cm, 1 cm, and 5 cm has a surface area of 22 sq cm and a volume of 5 cu cm.
- A rectangular prism with side lengths of 1 cm, 2 cm, and 3 cm has the same surface area but a volume of 6 cu cm.

Glossary

volume

NAME ______________________ DATE ____________ PERIOD ________

Practice

Distinguishing between Surface Area and Volume

1. Match each quantity with an appropriate unit of measurement.

Quantity	Unit of Measurement
a. The surface area of a tissue box	square meters
b. The amount of soil in a planter box	yards
c. The area of a parking lot	cubic inches
d. The length of a soccer field	cubic feet
e. The volume of a fish tank	square centimeters

2. Here is a figure built from snap cubes.

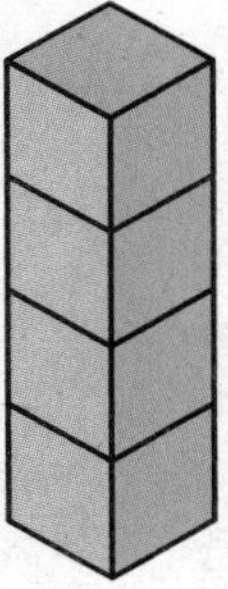

a. Find the volume of the figure in cubic units.

b. Find the surface area of the figure in square units.

c. True or False: If we double the number of cubes being stacked, both the volume and surface area will double. Explain or show how you know.

3. Lin said, “Two figures with the same volume also have the same surface area.”

   a. Which two figures suggest that her statement is true?

   b. Which two figures could show that her statement is *not* true?

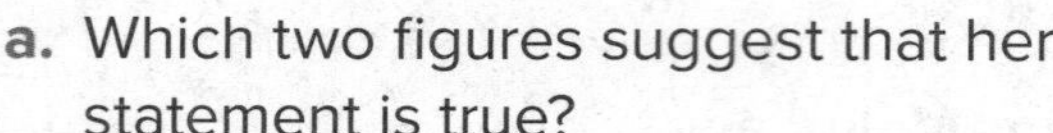

Figure A

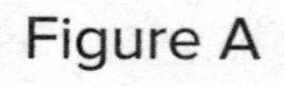

Figure B

Figure C

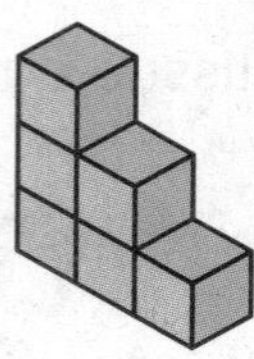

Figure D

Figure E

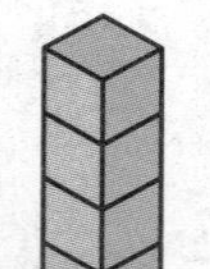

4. Draw a pentagon (five-sided polygon) that has an area of 32 square units. Label all relevant sides or segments with their measurements, and show that the area is 32 square units. **(Lesson 1-11)**

5. Respond to each of the following. **(Lesson 1-15)**

   a. Draw a net for this rectangular prism.

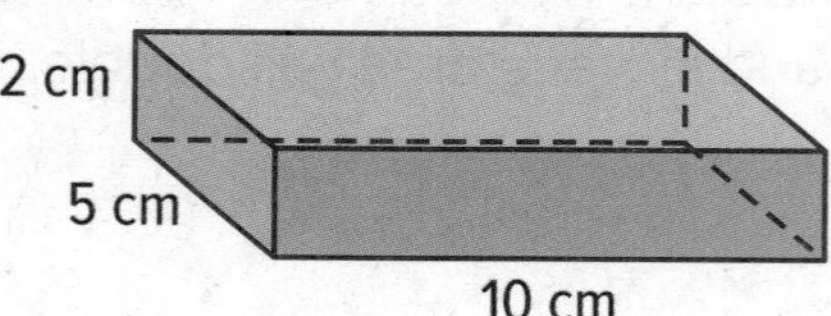

   b. Find the surface area of the rectangular prism.

Lesson 1-17

Squares and Cubes

NAME ______________________ DATE ____________ PERIOD ____________

Learning Goal Let's investigate perfect squares and perfect cubes.

Warm Up

17.1 Perfect Squares

1. The number 9 is a perfect **square**. Find four numbers that are perfect squares and two numbers that are not perfect squares.

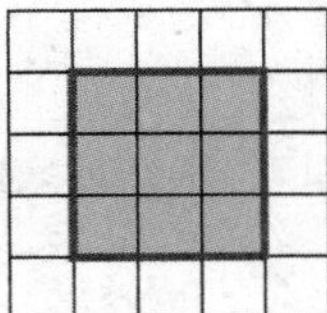

2. A square has side length 7 in. What is its area?

3. The area of a square is 64 sq cm. What is its side length?

Activity

17.2 Building with 32 Cubes

Your teacher will give you 32 snap cubes. Use them to build the largest single cube you can. Each small cube has an edge length of 1 unit.

1. How many snap cubes did you use?

2. What is the edge length of the cube you built?

3. What is the area of each face of the built cube? Be prepared to explain your reasoning.

4. What is the volume of the built cube? Be prepared to explain your reasoning.

Activity
17.3 Perfect Cubes

1. The number 27 is a perfect **cube**. Find four other numbers that are perfect cubes and two numbers that are *not* perfect cubes.

2. A cube has side length 4 cm. What is its volume?

3. A cube has side length 10 inches. What is its volume?

4. A cube has side length *s* units. What is its volume?

Activity
17.4 Introducing Exponents

Make sure to include correct units of measure as part of each answer.

1. A square has side length 10 cm. Use an **exponent** to express its area.

2. The area of a square is 7^2 sq in. What is its side length?

3. The area of a square is 81 m^2. Use an exponent to express this area.

4. A cube has edge length 5 in. Use an exponent to express its volume.

5. The volume of a cube is 6^3 cm^3. What is its edge length?

6. A cube has edge length *s* units. Use an exponent to write an expression for its volume.

NAME ______________________ DATE ______________ PERIOD ____________

Are you ready for more?

The number 15,625 is both a perfect square and a perfect cube. It is a perfect square because it equals 125^2. It is also a perfect cube because it equals 25^3. Find another number that is both a perfect square and a perfect cube. How many of these can you find?

Summary

Squares and Cubes

When we multiply two of the same numbers together, such as $5 \cdot 5$, we say we are **squaring** the number. We can write it like this: 5^2.

Because $5 \cdot 5 = 25$, we write $5^2 = 25$ and we say, "5 squared is 25."

When we multiply three of the same numbers together, such as $4 \cdot 4 \cdot 4$, we say we are **cubing** the number. We can write it like this: 4^3.

Because $4 \cdot 4 \cdot 4 = 64$, we write $4^3 = 64$ and we say, "4 cubed is 64."

We also use this notation for square and cubic units.

- A square with side length 5 inches has area 25 in^2.
- A cube with edge length 4 cm has volume 64 cm^3.

To read 25 in^2, we say "25 square inches," just like before.

The area of a square with side length 7 kilometers is 7^2 km^2.
The volume of a cube with edge length 2 millimeters is 2^3 mm^3.

In general, the area of a square with side length s is s^2, and the volume of a cube with edge length s is s^3.

Glossary

cubed

exponent

squared

Practice

Squares and Cubes

1. What is the volume of this cube?

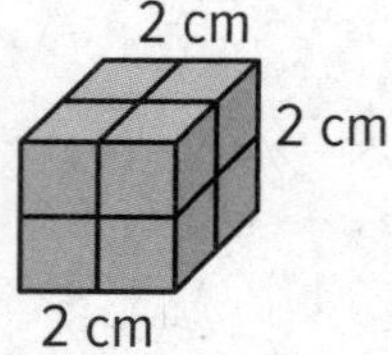

2. Respond to each of the following.

   a. Decide if each number on the list is a perfect square.

   16 20 25 100 125 144 225 10,000

   b. Write a sentence that explains your reasoning.

3. Respond to each of the following.

   a. Decide if each number on the list is a perfect cube.

   1 3 8 9 27 64 100 125

   b. Explain what a perfect cube is.

NAME ______________________ DATE __________ PERIOD __________

4. Respond to each of the following.

a. A square has side length 4 cm. What is its area?

b. The area of a square is 49 m^2. What is its side length?

c. A cube has edge length 3 in. What is its volume?

5. Prism A and Prism B are rectangular prisms. Prism A is 3 inches by 2 inches by 1 inch. Prism B is 1 inch by 1 inch by 6 inches. Select **all** statements that are true about the two prisms. **(Lesson 1-16)**

(A.) They have the same volume.

(B.) They have the same number of faces.

(C.) More inch cubes can be packed into Prism A than into Prism B.

(D.) The two prisms have the same surface area.

(E.) The surface area of Prism B is greater than that of Prism A.

6. Respond to each of the following.

a. What polyhedron can be assembled from this net? **(Lesson 1-14)**

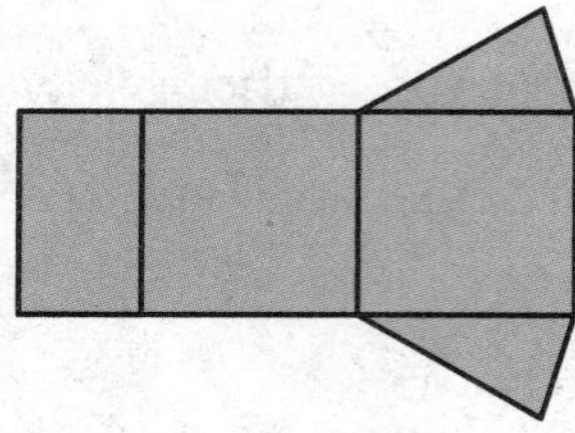

b. What information would you need to find its surface area? Be specific, and label the diagram as needed.

7. Find the surface area of this triangular prism. All measurements are in meters. **(Lesson 1-15)**

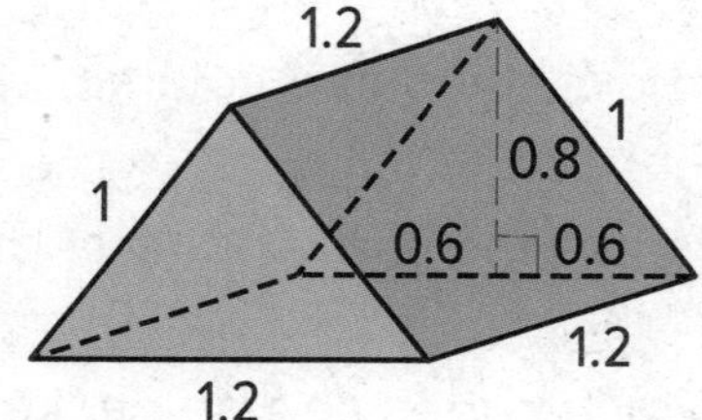

Lesson 1-18

Surface Area of a Cube

NAME ______________________ DATE ____________ PERIOD ____________

Learning Goal Let's write a formula to find the surface area of a cube.

Warm Up

18.1 Exponent Review

Select the greater expression of each pair without calculating the value of each expression. Be prepared to explain your choices.

1. $10 \cdot 3$ or 10^3
2. 13^2 or $12 \cdot 12$
3. $97 + 97 + 97 + 97 + 97 + 97$ or $5 \cdot 97$

Activity

18.2 The Net of a Cube

1. A cube has edge length 5 inches.
   a. In the space at the right, draw a net for this cube and label its sides with measurements.
   b. What is the shape of each face?
   c. What is the area of each face?
   d. What is the surface area of this cube?
   e. What is the volume of this cube?

2. A second cube has edge length 17 units.

   a. In the space at the right, draw a net for this cube and label its sides with measurements.

   b. Explain why the area of each face of this cube is 17^2 square units.

   c. Write an expression for the surface area, in square units.

   d. Write an expression for the volume, in cubic units.

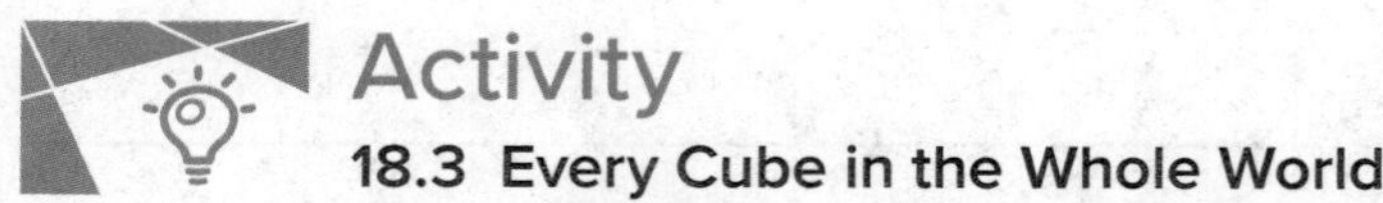

Activity
18.3 Every Cube in the Whole World

A cube has edge length s.

1. In the space at the right, draw a net for the cube.

2. Write an expression for the area of each face. Label each face with its area.

3. Write an expression for the surface area.

4. Write an expression for the volume.

NAME ______________________ DATE ______________ PERIOD ____________

Summary

Surface Area of a Cube

The volume of a cube with edge length s is s^3.

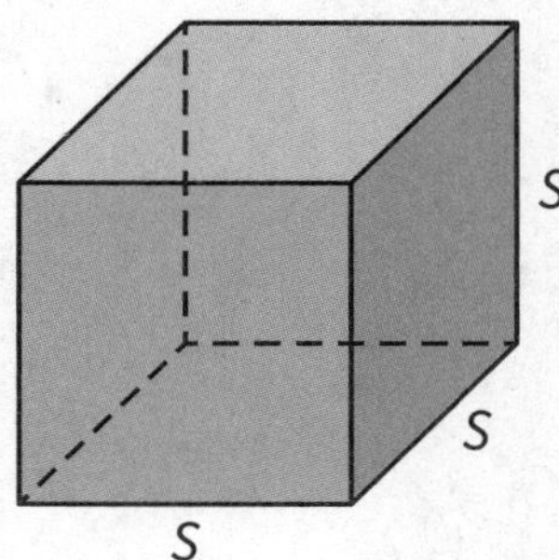

A cube has 6 faces that are all identical squares. The surface area of a cube with edge length s is $6 \cdot s^2$.

Practice

Surface Area of a Cube

1. Respond to each of the following.

   a. What is the volume of a cube with edge length 8 in.?

   b. What is the volume of a cube with edge length $\frac{1}{3}$ cm?

   c. A cube has a volume of 8 ft^3. What is its edge length?

2. a. What three-dimensional figure can be assembled from this net?

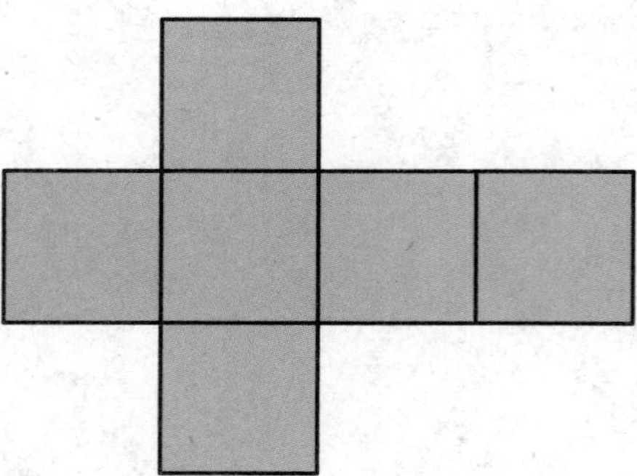

   b. If each square has a side length of 61 cm, write an expression for the surface area and another for the volume of the figure.

3. Respond to each of the following.

   a. Draw a net for a cube with edge length x cm.

   b. What is the surface area of this cube?

   c. What is the volume of this cube?

NAME ______________________ DATE ____________ PERIOD ________

4. Here is a net for a rectangular prism that was not drawn accurately. **(Lesson 1-14)**

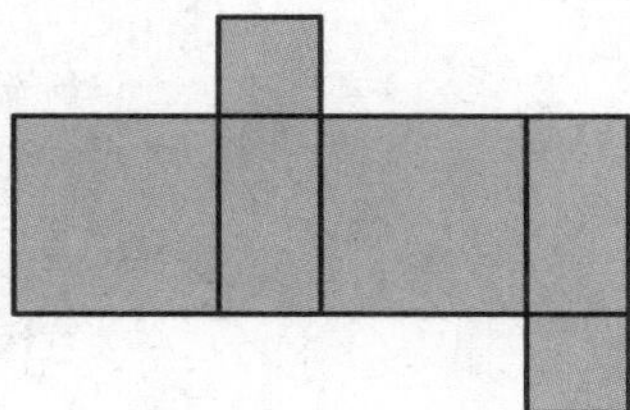

a. Explain what is wrong with the net.

b. Draw a net that can be assembled into a rectangular prism.

c. Create another net for the same prism.

5. State whether each figure is a polyhedron. Explain how you know. **(Lesson 1-13)**

Figure A

Figure B

6. Here is Elena's work for finding the surface area of a rectangular prism that is 1 foot by 1 foot by 2 feet. **(Lesson 1-12)**

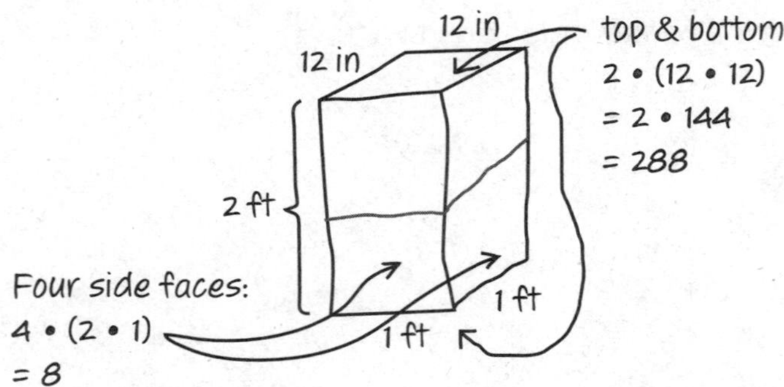

She concluded that the surface area of the prism is 296 square feet. Do you agree with her? Explain your reasoning.

Lesson 1-19

Designing a Tent

NAME ______________________ DATE ______________ PERIOD ____________

Learning Goal Let's design some tents.

Activity

19.1 Tent Design—Part 1

Have you ever been camping?

You might know that sleeping bags are all about the same size, but tents come in a variety of shapes and sizes.

Your task is to design a tent to accommodate up to four people and estimate the amount of fabric needed to make your tent. Your design and estimate must be based on the information given and have mathematical justification.

First, look at these examples of tents, the average specifications of a camping tent, and standard sleeping bag measurements. Talk to a partner about:

- Similarities and differences among the tents
- Information that will be important in your designing process
- The pros and cons of the various designs

Tent Styles

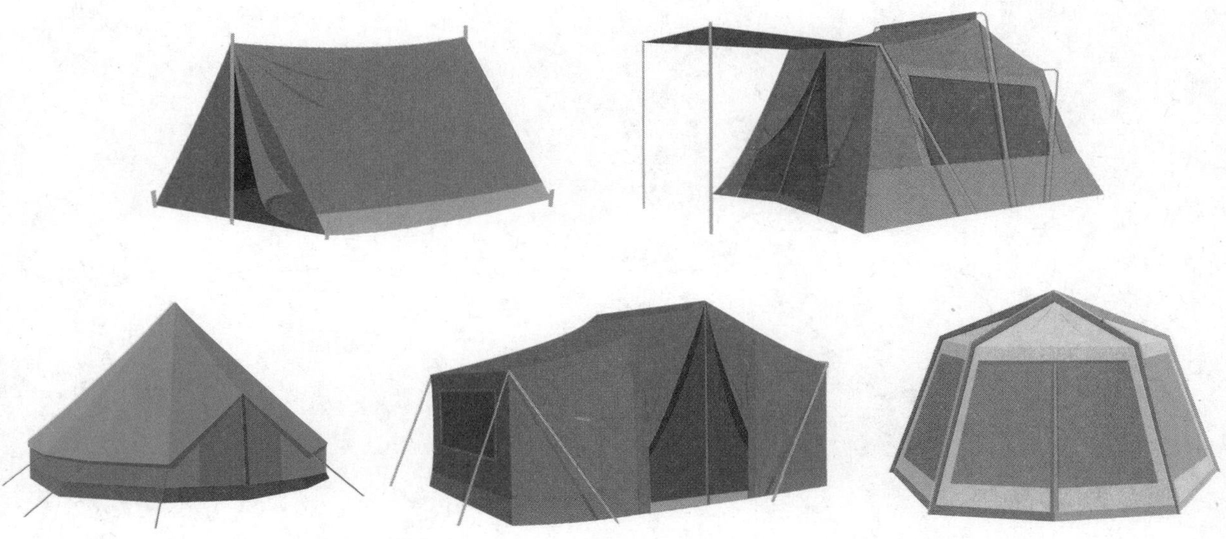

Tent Height Specifications

Height Description	Height of Tent	Notes
Sitting Height	3 feet	Campers are able to sit, lie, or crawl inside the tent.
Kneeling Height	4 feet	Campers are able to kneel inside the tent. Found mainly in 3- to 4-person tents.
Stooping Height	5 feet	Campers are able to move around on their feet inside the tent, but most campers will not be able to stand upright.
Standing Height	6 feet	Most adult campers are able to stand upright inside the tent.
Roaming Height	7 feet	Adult campers are able to stand upright and walk around inside the tent.

Sleeping Bag Measurements

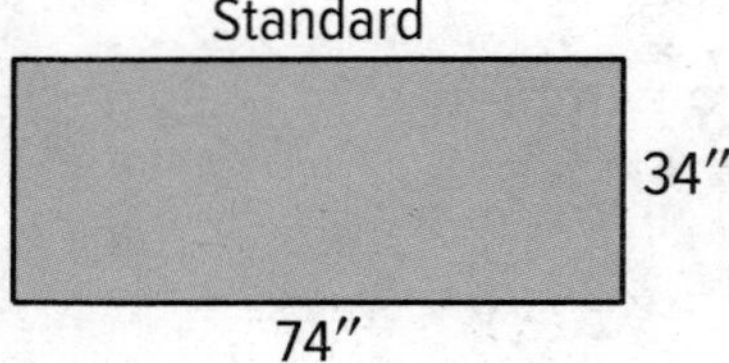

1. Create and sketch your tent design. The tent must include a floor.

NAME ______________________ DATE ____________ PERIOD ________

2. What decisions were important when choosing your tent design?

3. How much fabric do you estimate will be necessary to make your tent? Show your reasoning and provide mathematical justification.

Activity
19.2 Tent Design — Part 2

1. Explain your tent design and fabric estimate to your partner or partners. Be sure to explain why you chose this design and how you found your fabric estimate.

2. Compare the estimated fabric necessary for each tent in your group. Discuss the following questions:

   - Which tent design used the least fabric? Why?

   - Which tent design used the most fabric? Why?

   - Which change in design most impacted the amount of fabric needed for the tent? Why?

Learning Targets

Lesson	Learning Target(s)
1-1 Tiling the Plane	• I can explain the meaning of area.
1-2 Finding Area by Decomposing and Rearranging	• I can explain how to find the area of a figure that is composed of other shapes. • I know how to find the area of a figure by decomposing it and rearranging the parts. • I know what it means for two figures to have the same area.
1-3 Reasoning to Find Area	• I can use different reasoning strategies to find the area of shapes.
1-4 Parallelograms	• I can use reasoning strategies and what I know about the area of a rectangle to find the area of a parallelogram. • I know how to describe the features of a parallelogram using mathematical vocabulary.

(continued on the next page)

(continued from the previous page)

Lesson	Learning Target(s)
1-5 Bases and Heights of Parallelograms	• I can identify pairs of base and height of a parallelogram. • I can write and explain the formula for the area of a parallelogram. • I know what the terms "base" and "height" refer to in a parallelogram.
1-6 Area of Parallelograms	• I can use the area formula to find the area of any parallelogram.
1-7 From Parallelograms to Triangles	• I can explain the special relationship between a pair of identical triangles and a parallelogram.
1-8 Area of Triangles	• I can use what I know about parallelograms to reason about the area of triangles.

Lesson	Learning Target(s)
1-9 Formula for the Area of a Triangle	• I can use the area formula to find the area of any triangle. • I can write and explain the formula for the area of a triangle. • I know what the terms "base" and "height" refer to in a triangle.
1-10 Bases and Heights of Triangles	• I can identify pairs of base and corresponding height of any triangle. • When given information about a base of a triangle, I can identify and draw a corresponding height.
1-11 Polygons	• I can describe the characteristics of a polygon using mathematical vocabulary. • I can reason about the area of any polygon by decomposing and rearranging it, and by using what I know about rectangles and triangles.
1-12 What Is Surface Area?	• I know what the surface area of a three-dimensional object means.

(continued on the next page)

(continued from the previous page)

Lesson	Learning Target(s)
1-13 Polyhedra	• I can describe the features of a polyhedron using mathematical vocabulary. • I can explain the difference between prisms and pyramids. • I understand the relationship between a polyhedron and its net.
1-14 Nets and Surface Area	• I can match polyhedra to their nets and explain how I know. • When given a net of a prism or a pyramid, I can calculate its surface area.
1-15 More Nets, More Surface Area	• I can calculate the surface area of prisms and pyramids. • I can draw the nets of prisms and pyramids.
1-16 Distinguishing between Surface Area and Volume	• I can explain how it is possible for two polyhedra to have the same surface area but different volumes, or to have different surface areas but the same volume. • I know how one-, two-, and three-dimensional measurements and units are different.

Lesson	Learning Target(s)
1-17 Squares and Cubes	• I can write and explain the formula for the volume of a cube, including the meaning of the exponent. • When I know the edge length of a cube, I can find the volume and express it using appropriate units.
1-18 Surface Area of a Cube	• I can write and explain the formula for the surface area of a cube. • When I know the edge length of a cube, I can find its surface area and express it using appropriate units.
1-19 Designing a Tent	• I can apply what I know about the area of polygons to find the surface area of three-dimensional objects. • I can use surface area to reason about real-world objects.

(continued on the next page)

(continued from the previous page)

Notes:

Unit 2

Introducing Ratios

Burcu Atalay Tankut/Moment/Getty Images

To make a smoothie, you can follow a recipe. To double or triple the recipe, you can use ratios. You'll learn more about ratios in this unit.

Topics

- What Are Ratios?
- Equivalent Ratios
- Representing Equivalent Ratios
- Solving Ratio and Rate Problems
- Part-Part-Whole Ratios
- Let's Put It to Work

Unit 2

Introducing Ratios

What Are Ratios?

Equivalent Ratios

Representing Equivalent Ratios

Solving Ratio and Rate Problems

Part-Part-Whole Ratios

Let's Put It to Work

Lesson 2-1

Introducing Ratios and Ratio Language

NAME ______________________ DATE ____________ PERIOD ________

Learning Goal Let's describe two quantities at the same time.

Warm Up

1.1 What Kind and How Many?

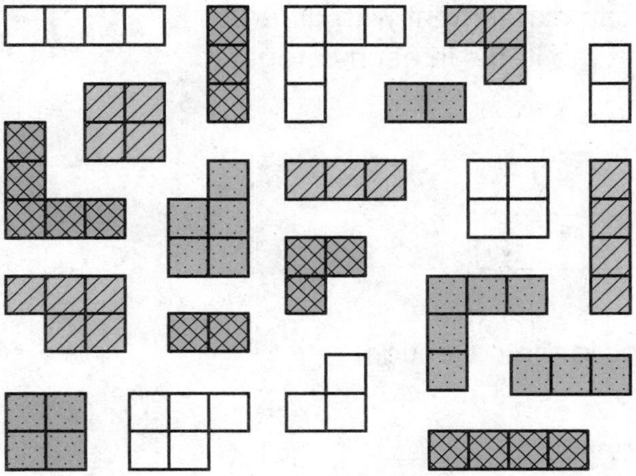

Think of ways you could sort these figures. What categories would you use? How many groups would you have?

Activity

1.2 The Teacher's Collection

1. Think of a way to sort your teacher's collection into two or three categories. Count the items in each category and record the information in the table.

Category Name			
Category Amount			

Pause here so your teacher can review your work.

2. Write at least two sentences that describe **ratios** in the collection. Remember, there are many ways to write a ratio.

- The ratio of *one category* to *another category* is __________ to __________.
- The ratio of *one category* to *another category* is __________ : __________.
- There are __________ of *one category* for every __________ of *another category*.

Activity

1.3 The Student's Collection

1. Sort your collection into three categories. You can experiment with different ways of arranging these categories. Then, count the items in each category, and record the information in the table.

Category Name			
Category Amount			

2. Write at least two sentences that describe **ratios** in the collection. Remember, there are many ways to write a ratio.

- The ratio of *one category* to *another category* is __________ to __________.
- The ratio of *one category* to *another category* is __________ : __________.
- There are __________ of *one category* for every __________ of *another category*.

Pause here so your teacher can review your sentences.

3. Make a visual display of your items that clearly shows one of your statements. Be prepared to share your display with the class.

NAME ______________________ DATE ____________ PERIOD ____________

Are you ready for more?

1. Use two colors to shade the rectangle so there are 2 square units of one color for every 1 square unit of the other color.

2. The rectangle you just colored has an area of 24 square units. Draw a different shape that does *not* have an area of 24 square units, but that can also be shaded with two colors in a 2 : 1 ratio. Shade your new shape using two colors.

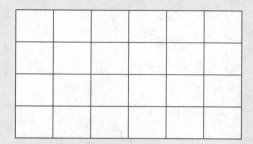

Summary

Introducing Ratios and Ratio Language

A **ratio** is an association between two or more quantities. There are many ways to describe a situation in terms of ratios.

For example, look at this collection.

Here are some correct ways to describe the collection:

- The ratio of squares to circles is 6 : 3.
- The ratio of circles to squares is 3 to 6.

Notice that the shapes can be arranged in equal groups, which allow us to describe the shapes using other numbers.

- There are 2 squares for every 1 circle.
- There is 1 circle for every 2 squares.

Glossary

ratio

Practice

Introducing Ratios and Ratio Language

1. In a fruit basket there are 9 bananas, 4 apples, and 3 plums.

   a. The ratio of bananas to apples is ________ : ________.

   b. The ratio of plums to apples is ________ to ________.

   c. For every ________ apples, there are ________ plums.

   d. For every 3 bananas there is one ________.

2. Complete the sentences to describe this picture.

   a. The ratio of dogs to cats is ________.

   b. For every ________ dogs, there are ________ cats.

3. Write two different sentences that use ratios to describe the number of eyes and legs in this picture.

NAME ______________________ DATE ______________ PERIOD ____________

4. Choose an appropriate unit of measurement for each quantity. **(Lesson 1-17)**

Quantity	**Unit of Measurement**
a. area of a rectangle	cm
b. volume of a prism	cm^3
c. side of a square	cm^2
d. area of a square	
e. volume of a cube	

5. Find the volume and surface area of each prism. **(Lesson 1-16)**

a. Prism A: 3 cm by 3 cm by 3 cm

b. Prism B: 5 cm by 5 cm by 1 cm

c. Compare the volumes of the prisms and then their surface areas. Does the prism with the greater volume also have the greater surface area?

6. Which figure is a triangular prism? Select **all** that apply. **(Lesson 1-13)**

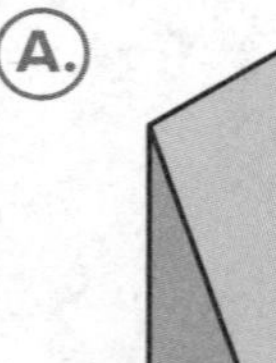

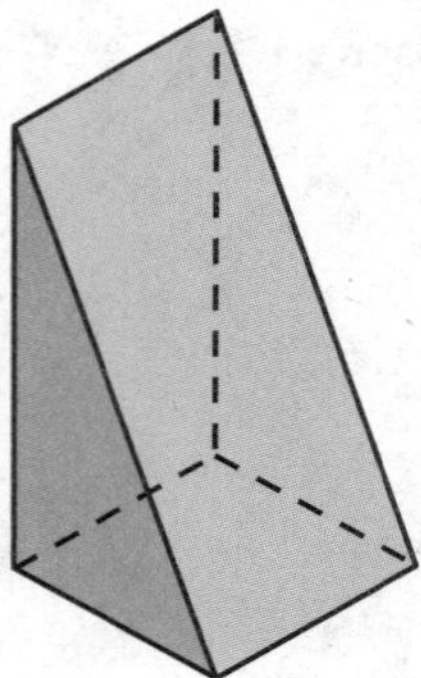

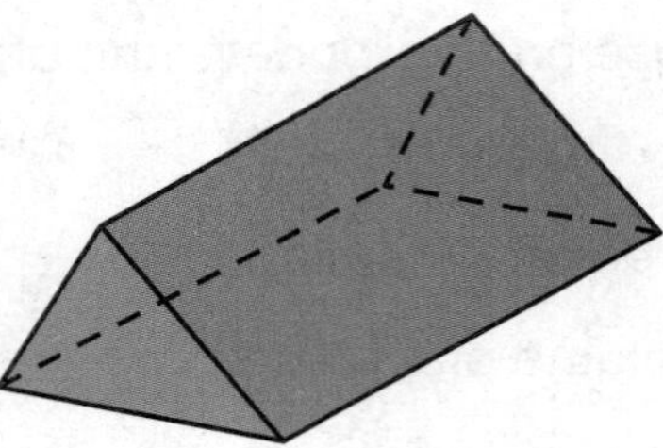

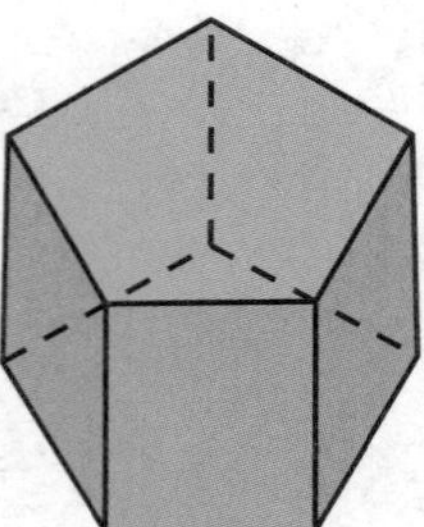

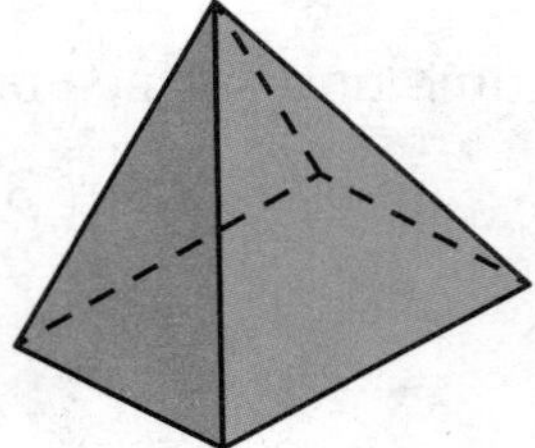

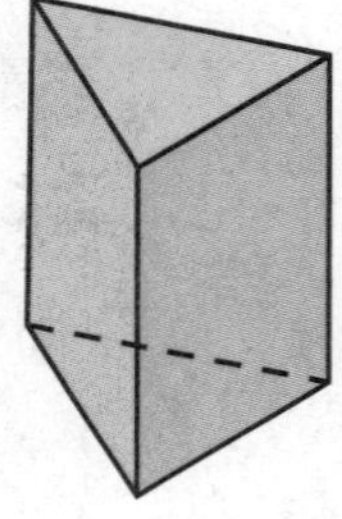

Lesson 2-2

Representing Ratios with Diagrams

NAME _______________ DATE _______________ PERIOD _______________

Learning Goal Let's use diagrams to represent ratios.

Warm Up

2.1 Number Talk: Dividing by 4 and Multiplying by $\frac{1}{4}$

Find the value of each expression mentally.

1. $24 \div 4$ **2.** $\frac{1}{4} \cdot 24$ **3.** $24 \cdot \frac{1}{4}$ **4.** $5 \div 4$

Activity

2.2 A Collection of Snap Cubes

Here is a collection of snap cubes.

1. Choose two of the colors in the image, and draw a diagram showing the number of snap cubes for these two colors.

2. Trade papers with a partner. On their paper, write a sentence to describe a ratio shown in their diagram. Your partner will do the same for your diagram.

3. Return your partner's paper. Read the sentence written on your paper. If you disagree, explain your thinking.

Activity
2.3 Blue Paint and Art Paste

Elena mixed 2 cups of white paint with 6 tablespoons of blue paint.

Here is a diagram that represents this situation.

White Paint (cups)

Blue Paint (tablespoons)

1. Discuss each statement and circle all those that correctly describe this situation. Make sure that both you and your partner agree with each circled answer.
   a. The ratio of cups of white paint to tablespoons of blue paint is 2 : 6.
   b. For every cup of white paint, there are 2 tablespoons of blue paint.
   c. There is 1 cup of white paint for every 3 tablespoons of blue paint.
   d. There are 3 tablespoons of blue paint for every cup of white paint.
   e. For each tablespoon of blue paint, there are 3 cups of white paint.
   f. For every 6 tablespoons of blue paint, there are 2 cups of white paint.
   g. The ratio of tablespoons of blue paint to cups of white paint is 6 to 2.

2. Jada mixed 8 cups of flour with 2 pints of water to make paste for an art project.
   a. Draw a diagram that represents the situation.

   b. Write at least two sentences describing the ratio of flour and water.

NAME ______________________ DATE ____________ PERIOD ____________

Activity

2.4 Card Sort: Spaghetti Sauce

Your teacher will give you cards describing different recipes for spaghetti sauce. In the diagrams:

- a circle represents a cup of tomato sauce
- a square represents a tablespoon of oil
- a triangle represents a teaspoon of oregano

1. Take turns with your partner to match a sentence with a diagram.
   a. For each match that you find, explain to your partner how you know it's a match.
   b. For each match that your partner finds, listen carefully to their explanation. If you disagree, discuss your thinking and work to reach an agreement.
2. After you and your partner have agreed on all of the matches, check your answers with the answer key. If there are any errors, discuss why and revise your matches.
3. There were two diagrams that each matched with two different sentences. Which were they?
   - Diagram ________ matched with both sentences _______ and _______.
   - Diagram ________ matched with both sentences _______ and _______.
4. Select one of the other diagrams and invent another sentence that could describe the ratio shown in the diagram.

Are you ready for more?

Create a diagram that represents any of the ratios in a recipe of your choice. Is it possible to include more than 2 ingredients in your diagram?

Summary

Representing Ratios with Diagrams

Ratios can be represented using diagrams. The diagrams do not need to include realistic details. For example, a recipe for lemonade says, "Mix 2 scoops of lemonade powder with 6 cups of water."

Instead of this:

We can draw something like this:

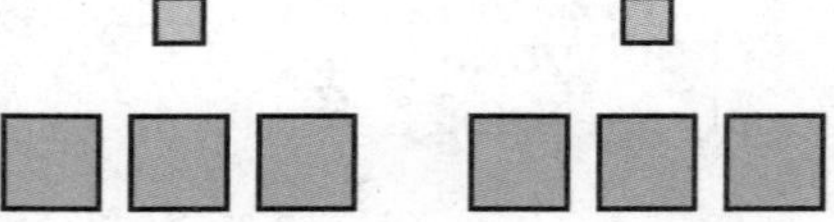

This diagram shows that the ratio of cups of water to scoops of lemonade powder is 6 to 2. We can also see that for every scoop of lemonade powder, there are 3 cups of water.

NAME ______________________ DATE ____________ PERIOD ____________

Practice
Representing Ratios with Diagrams

1. Here is a diagram that describes the cups of green and white paint in a mixture.

Green Paint (cups) ■■ ■■
White Paint (cups) □ □

Select **all** the statements that correctly describe this diagram.

A. The ratio of cups of white paint to cups of green paint is 2 to 4.

B. For every cup of green paint, there are two cups of white paint.

C. The ratio of cups of green paint to cups of white paint is 4 : 2.

D. For every cup of white paint, there are two cups of green paint.

E. The ratio of cups of green paint to cups of white paint is 2 : 4.

2. To make a snack mix, combine 2 cups of raisins with 4 cups of pretzels and 6 cups of almonds.

   a. Create a diagram to represent the quantities of each ingredient in this recipe.

   b. Use your diagram to complete each sentence.

   - The ratio of ____________ to ____________ to ____________ is _____ : _____ : _____.
   - There are _____ cups of pretzels for every cup of raisins.
   - There are _____ cups of almonds for every cup of raisins.

3. Respond to the following questions. **(Lesson 1-17)**

   a. A square is 3 inches by 3 inches. What is its area?

   b. A square has a side length of 5 feet. What is its area?

   c. The area of a square is 36 square centimeters. What is the length of each side of the square?

4. Find the area of this quadrilateral. Explain or show your strategy. **(Lesson 1-11)**

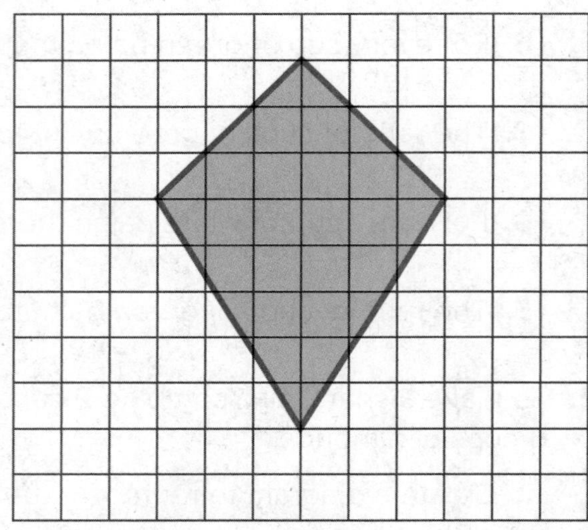

5. Complete each equation with a number that makes it true. **(Lesson 2-1)**

   a. $\frac{1}{8} \cdot 8 =$ ________

   b. $\frac{3}{8} \cdot 8 =$ ________

   c. $\frac{1}{8} \cdot 7 =$ ________

   d. $\frac{3}{8} \cdot 7 =$ ________

Lesson 2-3

Recipes

NAME _______________ DATE _______________ PERIOD _______________

Learning Goal Let's explore how ratios affect the way a recipe tastes.

Warm Up

3.1 Flower Pattern

This flower is made up of yellow hexagons, red trapezoids, and green triangles.

1. Write sentences to describe the ratios of the shapes that make up this pattern.

2. How many of each shape would be in two copies of this flower pattern?

Activity

3.2 Powdered Drink Mix

Here are diagrams representing three mixtures of powdered drink mix and water.

Diagram A | Diagram B | Diagram C

Key: □ = 1 teaspoon drink mix ■ = 1 cup of water

1. How would the taste of Mixture A compare to the taste of Mixture B?
2. Use the diagrams to complete each statement:
   a. Mixture B uses __________ cups of water and __________ teaspoons of drink mix. The ratio of cups of water to teaspoons of drink mix in Mixture B is __________.
   b. Mixture C uses __________ cups of water and __________ teaspoons of drink mix. The ratio of cups of water to teaspoons of drink mix in Mixture C is __________.
3. How would the taste of Mixture B compare to the taste of Mixture C?

Are you ready for more?

Sports drinks use sodium (better known as salt) to help people replenish electrolytes. Here are the nutrition labels of two sports drinks.

Label A

Nutrition Facts	
Serving Size 8 fl oz (240 mL)	
Servings per container 4	
Amount per serving	
Calories 50	
	% Daily Value*
Total Fat 0 g	0%
Sodium 110 mg	5%
Potassium 30 mg	1%
Total Carbohydrate 14 g	5%

Label B

Nutrition Facts	
Serving Size 12 fl oz (355 mL)	
Servings per container 2.5	
Amount per serving	
Calories 80	
	% Daily Value*
Total Fat 0 g	0%
Sodium 150 mg	6%
Potassium 35 mg	1%
Total Carbohydrate 21 g	7%

1. Which of these drinks is saltier? Explain how you know.

2. If you wanted to make sure a sports drink was less salty than both of the ones given, what ratio of sodium to water would you use?

NAME ______________________ DATE ____________ PERIOD ________

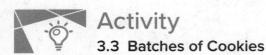

Activity

3.3 Batches of Cookies

A recipe for one batch of cookies calls for 5 cups of flour and 2 teaspoons of vanilla.

1. Draw a diagram that shows the amount of flour and vanilla needed for *two* batches of cookies.

2. How many batches can you make with 15 cups of flour and 6 teaspoons of vanilla? Show the additional batches by adding more ingredients to your diagram.

3. How much flour and vanilla would you need for 5 batches of cookies?

4. Whether the ratio of cups of flour to teaspoons of vanilla is 5 : 2, 10 : 4, or 15 : 6, the recipes would make cookies that taste the same. We call these equivalent ratios.

   a. Find another ratio of cups of flour to teaspoons of vanilla that is equivalent to these ratios.

   b. How many batches can you make using this new ratio of ingredients?

Summary

Recipes

A recipe for fizzy juice says, "Mix 5 cups of cranberry juice with 2 cups of soda water."

To double this recipe, we would use 10 cups of cranberry juice with 4 cups of soda water. To triple this recipe, we would use 15 cups of cranberry juice with 6 cups of soda water.

This diagram shows a single batch of the recipe, a double batch, and a triple batch.

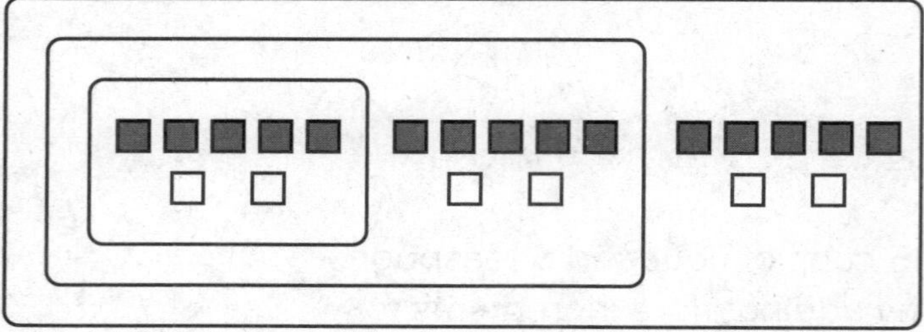

We say that the ratios 5 : 2, 10 : 4, and 15 : 6 are equivalent ratios. Even though the amounts of each ingredient within a single, double, or triple batch are not the same, they would make fizzy juice that tastes the same.

NAME ______________________ DATE ____________ PERIOD ____________

Practice
Recipes

1. A recipe for 1 batch of spice mix says, "Combine 3 teaspoons of mustard seeds, 5 teaspoons of chili powder, and 1 teaspoon of salt." How many batches are represented by the diagram? Explain or show your reasoning.

Mustard Seeds (tsp)	■■■ ■■■ ■■■
Chili Powder (tsp)	■■■■■■■■■■■■■■■
Salt (tsp)	■ ■ ■

2. Priya makes chocolate milk by mixing 2 cups of milk and 5 tablespoons of cocoa powder. Draw a diagram that clearly represents two batches of her chocolate milk.

3. In a recipe for fizzy grape juice, the ratio of cups of sparkling water to cups of grape juice concentrate is 3 to 1.

   a. Find two more ratios of cups of sparkling water to cups of juice concentrate that would make a mixture that tastes the same as this recipe.

   b. Describe another mixture of sparkling water and grape juice that would taste different than this recipe.

4. Write the missing number under each tick mark on the number line. **(Lesson 2-1)**

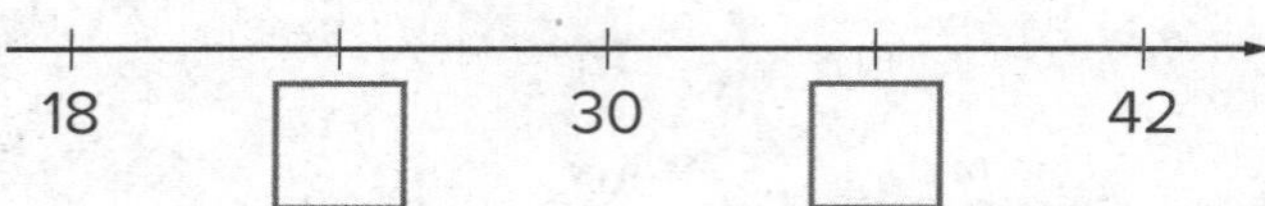

5. At the kennel, there are 6 dogs for every 5 cats. **(Lesson 2-1)**

   a. The ratio of dogs to cats is __________ to __________.

   b. The ratio of cats to dogs is __________ to __________.

   c. For every __________ dogs there are __________ cats.

   d. The ratio of cats to dogs is __________ : __________.

6. Elena has 80 unit cubes. What is the volume of the largest cube she can build with them? **(Lesson 1-17)**

7. Fill in the blanks to make each equation true. **(Lesson 2-1)**

   a. $3 \cdot \frac{1}{3} =$ ______________

   b. $10 \cdot \frac{1}{10} =$ ______________

   c. $19 \cdot \frac{1}{19} =$ ______________

   d. $a \cdot \frac{1}{a} =$ ______________ (As long as a does not equal 0)

   e. $5 \cdot$ ______________ $= 1$

   f. $17 \cdot$ ______________ $= 1$

   g. $b \cdot$ ______________ $= 1$

Lesson 2-4

Color Mixtures

NAME ____________________ DATE __________ PERIOD __________

Learning Goal Let's see what color-mixing has to do with ratios.

Warm Up

4.1 Number Talk: Adjusting a Factor

Find the value of each product mentally.

1. $6 \cdot 15$
2. $12 \cdot 15$
3. $6 \cdot 45$
4. $13 \cdot 45$

Activity

4.2 Turning Green

Your teacher mixed milliliters of blue water and milliliters of yellow water in the ratio 5 : 15.

1. Doubling the original recipe:
   a. Draw a diagram to represent the amount of each color that you will combine to double your teacher's recipe.

   b. Use a marker to label an empty cup with the ratio of blue water to yellow water in this double batch.

c. Predict whether these amounts of blue and yellow will make the same shade of green as your teacher's mixture. Next, check your prediction by measuring those amounts and mixing them in the cup.

d. Is the ratio in your mixture equivalent to the ratio in your teacher's mixture? Explain your reasoning.

2. Tripling the original recipe:

a. Draw a diagram to represent triple your teacher's recipe.

b. Label an empty cup with the ratio of blue water to yellow water.

c. Predict whether these amounts will make the same shade of green. Next, check your prediction by mixing those amounts.

d. Is the ratio in your new mixture equivalent to the ratio in your teacher's mixture? Explain your reasoning.

NAME ______________________ DATE ____________ PERIOD ____________

3. Next, invent your own recipe for a *bluer* shade of green water.

a. Draw a diagram to represent the amount of each color you will combine.

b. Label the final empty cup with the ratio of blue water to yellow water in this recipe.

c. Test your recipe by mixing a batch in the cup. Does the mixture yield a bluer shade of green?

d. Is the ratio you used in this recipe equivalent to the ratio in your teacher's mixture? Explain your reasoning.

Are you ready for more?

Someone has made a shade of green by using 17 ml of blue and 13 ml of yellow. They are sure it cannot be turned into the original shade of green by adding more blue or yellow. Either explain how more can be added to create the original green shade or explain why this is impossible.

Activity

4.3 Perfect Purple Water

The recipe for Perfect Purple Water says, "Mix 8 ml of blue water with 3 ml of red water."

Jada mixes 24 ml of blue water with 9 ml of red water. Andre mixes 16 ml of blue water with 9 ml of red water.

1. Which person will get a color mixture that is the same shade as Perfect Purple Water? Explain or show your reasoning.

2. Find another combination of blue water and red water that will also result in the same shade as Perfect Purple Water. Explain or show your reasoning.

NAME ______________________ DATE ____________ PERIOD ____________

Summary
Color Mixtures

When mixing colors, doubling or tripling the amount of each color will create the same shade of the mixed color. In fact, you can always multiply the amount of *each* color by *the same number* to create a different amount of the same mixed color.

For example, a batch of dark orange paint uses 4 ml of red paint and 2 ml of yellow paint.

- To make two batches of dark orange paint, we can mix 8 ml of red paint with 4 ml of yellow paint.
- To make three batches of dark orange paint, we can mix 12 ml of red paint with 6 ml of yellow paint.

Here is a diagram that represents 1, 2, and 3 batches of this recipe.

Red Paint (ml)

Yellow Paint (ml)

1 Batch Orange

2 Batches Orange

3 Batches Orange

We say that the ratios 4 : 2, 8 : 4, and 12 : 6 are *equivalent* because they describe the same color mixture in different numbers of batches, and they make the same shade of orange.

Practice

Color Mixtures

1. Here is a diagram showing a mixture of red paint and green paint needed for 1 batch of a particular shade of brown. Add to the diagram so that it shows 3 batches of the same shade of brown paint.

Red Paint (cups)

Green Paint (cups)

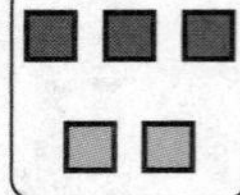

2. Diego makes green paint by mixing 10 tablespoons of yellow paint and 2 tablespoons of blue paint. Which of these mixtures produce the same shade of green paint as Diego's mixture? Select **all** that apply.

   A. For every 5 tablespoons of blue paint, mix in 1 tablespoon of yellow paint.

   B. Mix tablespoons of blue paint and yellow paint in the ratio 1 : 5.

   C. Mix tablespoons of yellow paint and blue paint in the ratio 15 to 3.

   D. Mix 11 tablespoons of yellow paint and 3 tablespoons of blue paint.

   E. For every tablespoon of blue paint, mix in 5 tablespoons of yellow paint.

NAME _______________ DATE _______________ PERIOD _______________

3. To make 1 batch of sky blue paint, Clare mixes 2 cups of blue paint with 1 gallon of white paint.

   a. Explain how Clare can make 2 batches of sky blue paint.

   b. Explain how to make a mixture that is a darker shade of blue than the sky blue.

   c. Explain how to make a mixture that is a lighter shade of blue than the sky blue.

4. A smoothie recipe calls for 3 cups of milk, 2 frozen bananas, and 1 tablespoon of chocolate syrup. **(Lesson 2-2)**

   a. Create a diagram to represent the quantities of each ingredient in the recipe.

   b. Write 3 different sentences that use a ratio to describe the recipe.

5. Write the missing number under each tick mark on the number line. **(Lesson 2-1)**

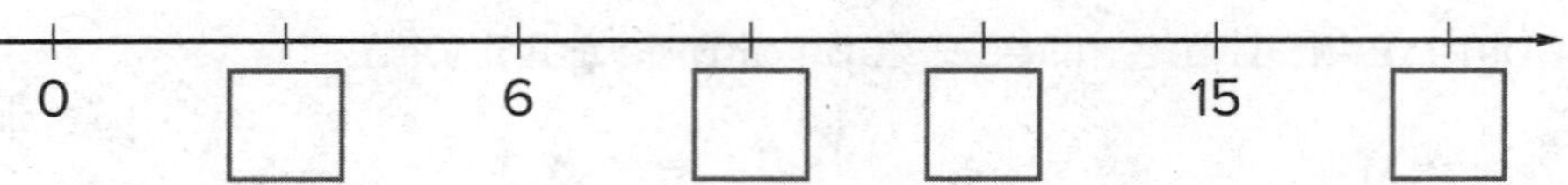

6. Find the area of the parallelogram. Show your reasoning. **(Lesson 1-4)**

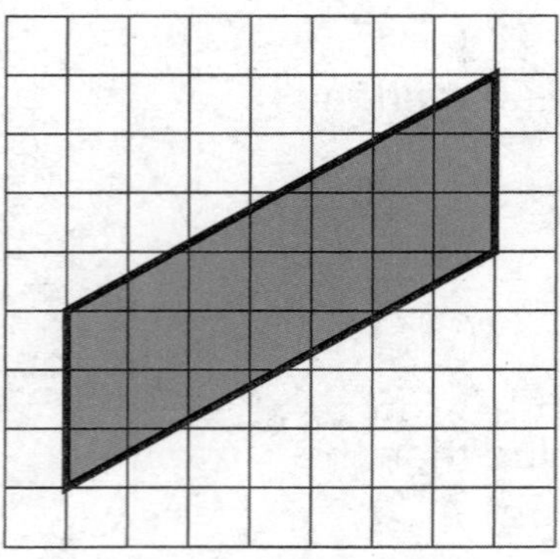

7. Complete each equation with a number that makes it true. **(Lesson 2-1)**

a. $11 \cdot \frac{1}{4} =$ ________________

b. $7 \cdot \frac{1}{4} =$ ________________

c. $13 \cdot \frac{1}{27} =$ ________________

d. $13 \cdot \frac{1}{99} =$ ________________

e. $x \cdot \frac{1}{y} =$ ________________
(As long as y does not equal 0.)

Lesson 2-5

Defining Equivalent Ratios

NAME ______________________ DATE ____________ PERIOD ____________

Learning Goal Let's investigate equivalent ratios some more.

Warm Up

5.1 Dots and Half Dots

Dot Pattern 1

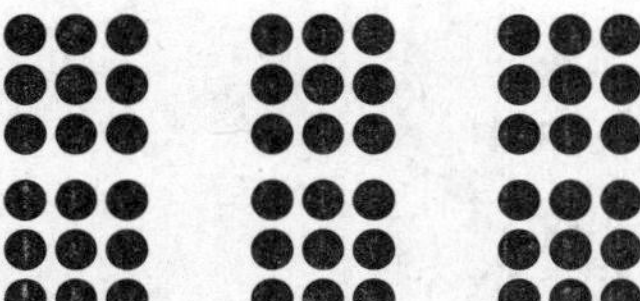

Dot Pattern 2

Activity

5.2 Tuna Casserole

Here is a recipe for tuna casserole.

rotofrank/iStockphoto/Getty Images

Ingredients

- 3 cups cooked elbow-shaped pasta
- 6-ounce can tuna, drained
- 10-ounce can cream of chicken soup
- 1 cup shredded cheddar cheese
- $1\frac{1}{2}$ cups French fried onions

Instructions

Combine the pasta, tuna, soup, and half of the cheese. Transfer into a 9-inch by 18-inch baking dish. Put the remaining cheese on top. Bake 30 minutes at 350 degrees. During the last 5 minutes, add the French fried onions. Let sit for 10 minutes before serving.

1. What is the ratio of the ounces of soup to the cups of shredded cheese to the cups of pasta in one batch of casserole?

2. How much of each of these 3 ingredients would be needed to make:
   a. twice the amount of casserole?
   b. half the amount of casserole?
   c. five times the amount of casserole?
   d. one-fifth the amount of casserole?

3. What is the ratio of cups of pasta to ounces of tuna in one batch of casserole?

4. How many batches of casserole would you make if you used the following amounts of ingredients?
   a. 9 cups of pasta and 18 ounces of tuna?
   b. 36 ounces of tuna and 18 cups of pasta?
   c. 1 cup of pasta and 2 ounces of tuna?

NAME ______________________ DATE ____________ PERIOD ____________

Are you ready for more?

1. The recipe says to use a 9 inch by 18 inch baking dish. Determine the length and width of a baking dish with the same height that could hold:
   a. twice the amount of casserole
   b. half the amount of casserole
   c. five times the amount of casserole
   d. one-fifth the amount of casserole

Activity
5.3 What Are Equivalent Ratios?

The ratios 5 : 3 and 10 : 6 are **equivalent ratios**.

1. Is the ratio 15 : 12 equivalent to these? Explain your reasoning.

2. Is the ratio 30 : 18 equivalent to these? Explain your reasoning.

3. Give two more examples of ratios that are equivalent to 5 : 3.

4. How do you know when ratios are equivalent and when they are *not* equivalent?

5. Write a definition of *equivalent ratios*.

Pause here so your teacher can review your work and assign you a ratio to use for your visual display.

6. Create a visual display that includes:
   - the title "Equivalent Ratios"
   - your best definition of *equivalent ratios*
   - the ratio your teacher assigned to you
   - at least two examples of ratios that are equivalent to your assigned ratio
   - an explanation of how you know these examples are equivalent
   - at least one example of a ratio that is *not* equivalent to your assigned ratio
   - an explanation of how you know this example is *not* equivalent

   Be prepared to share your display with the class.

Summary
5.3 Defining Equivalent Ratios

All ratios that are **equivalent** to *a*:*b* can be made by multiplying both *a* and *b* by the same number.

For example, the ratio 18 : 12 is equivalent to 9 : 6 because both 9 and 6 are multiplied by the same number: 2.

9 : 6

• 2 • 2

18 : 12

3:2 is also equivalent to 9 : 6, because both 9 and 6 are multiplied by the same number: $\frac{1}{3}$.

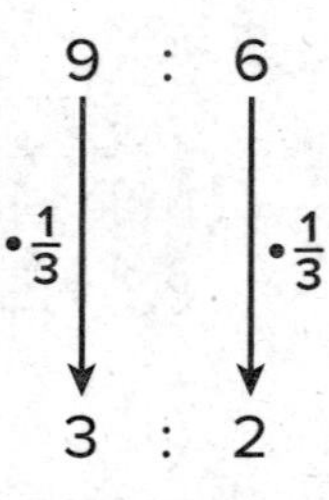

Is 18:15 equivalent to 9 : 6?

No, because 18 is 9 • 2, but 15 is *not* 6 • 2.

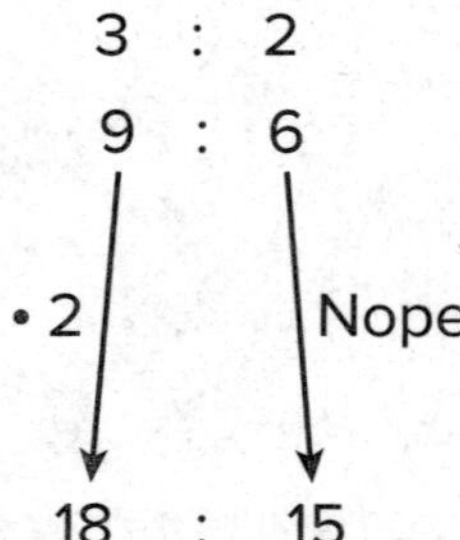

Glossary

equivalent ratios

NAME ______________________ DATE ____________ PERIOD ____________

Practice

Defining Equivalent Ratios

1. Each of these is a pair of equivalent ratios. For each pair, explain why they are equivalent ratios or draw a diagram that shows why they are equivalent ratios.

   a. 4 : 5 and 8 : 10

   b. 18 : 3 and 6 : 1

   c. 2 : 7 and 10,000 : 35,000

2. Explain why 6:4 and 18:8 are not equivalent ratios.

3. Are the ratios 3 : 6 and 6 : 3 equivalent? Why or why not?

4. This diagram represents 3 batches of light yellow paint. Draw a diagram that represents 1 batch of the same shade of light yellow paint. **(Lesson 2-4)**

White Paint (cups) □□□□□□□□□□

Yellow Paint (cups) ■■■■■■■■■■■■■■■□□

5. In the fruit bowl there are 6 bananas, 4 apples, and 3 oranges. **(Lesson 2-1)**

   a. For every 4 ______________, there are 3 ______________.

   b. The ratio of ______________ to ______________ is 6 : 3.

   c. The ratio of ______________ to ______________ is 4 to 6.

   d. For every 1 orange, there are ______________ bananas.

6. Write fractions for points *A* and *B* on the number line. **(Lesson 2-1)**

0 *A* *B* 1

Lesson 2-6

Introducing Double Number Line Diagrams

NAME ______________________ DATE ______________ PERIOD ______________

Learning Goal Let's use number lines to represent equivalent ratios.

Warm Up

6.1 Number Talk: Adjusting Another Factor

Find the value of each product mentally.

1. $(4.5) \cdot 4$ **2.** $(4.5) \cdot 8$ **3.** $\frac{1}{10} \cdot 65$ **4.** $\frac{2}{10} \cdot 65$

Activity

6.2 Drink Mix on a Double Number Line

The other day, we made drink mixtures by mixing 4 teaspoons of powdered drink mix for every cup of water. Here are two ways to represent multiple batches of this recipe.

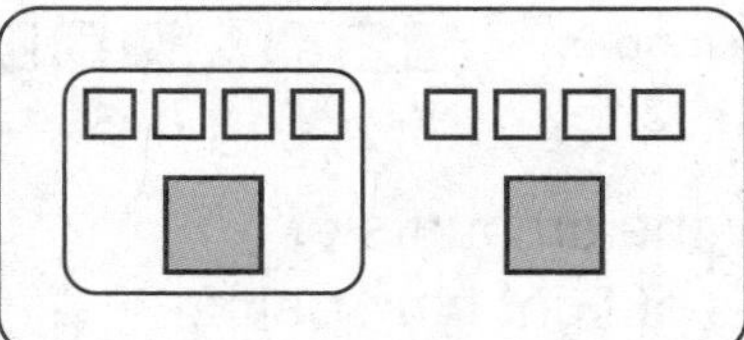

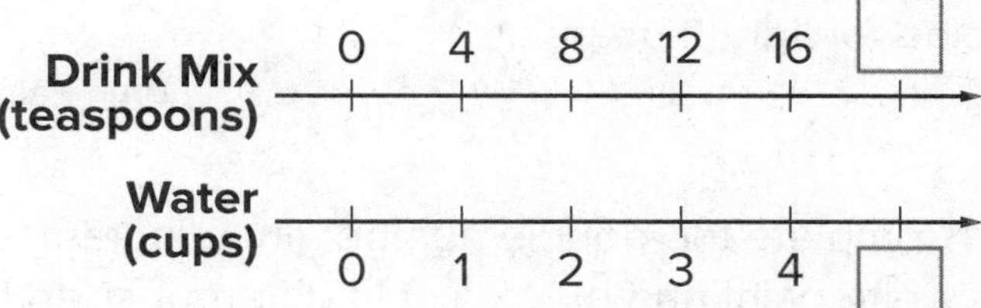

1. How can we tell that 4 : 1 and 12 : 3 are equivalent ratios?

2. How are these representations the same? How are these representations different?

3. How many teaspoons of drink mix should be used with 3 cups of water?

4. How many cups of water should be used with 16 teaspoons of drink mix?

5. What numbers should go in the empty boxes on the **double number line diagram**? What do these numbers mean?

Are you ready for more?

Recall that a *perfect square* is a number of objects that can be arranged into a square. For example, 9 is a perfect square because 9 objects can be arranged into 3 rows of 3. 16 is also a perfect square, because 16 objects can be arranged into 4 rows of 4. In contrast, 12 is not a perfect square because you can't arrange 12 objects into a square.

1. How many whole numbers starting with 1 and ending with 100 are perfect squares?

2. What about whole numbers starting with 1 and ending with 1,000?

Activity

6.3 Blue Paint on a Double Number Line

Here is a diagram showing Elena's recipe for light blue paint.

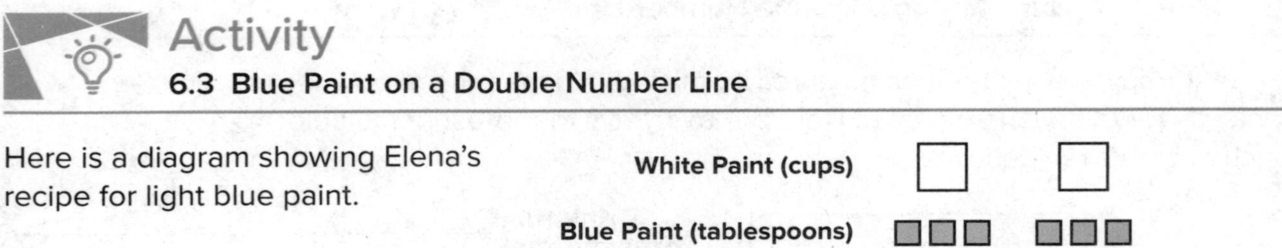

1. Complete the double number line diagram to show the amounts of white paint and blue paint in different-sized batches of light blue paint.

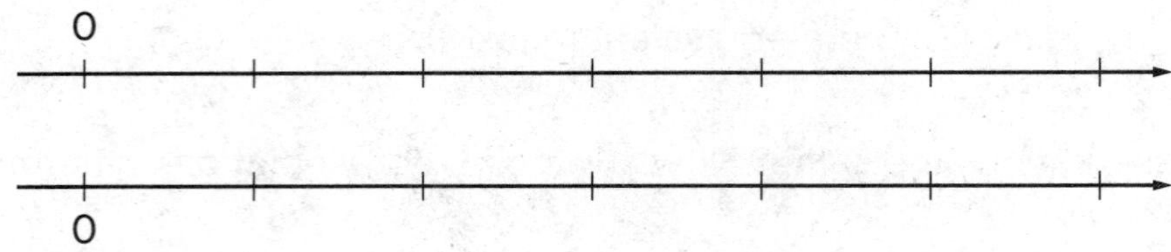

2. Compare your double number line diagram with your partner's. Discuss your thinking. If needed, revise your diagram.

3. How many cups of white paint should Elena mix with 12 tablespoons of blue paint? How many batches would this make?

NAME ______________________ DATE ____________ PERIOD ____________

4. How many tablespoons of blue paint should Elena mix with 6 cups of white paint? How many batches would this make?

5. Use your double number line diagram to find another amount of white paint and blue paint that would make the same shade of light blue paint.

6. How do you know that these mixtures would make the same shade of light blue paint?

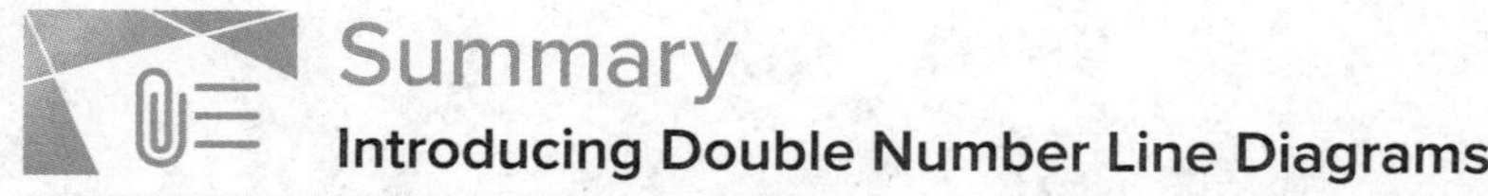

Summary

Introducing Double Number Line Diagrams

You can use a **double number line diagram** to find many equivalent ratios. For example, a recipe for fizzy juice says, "Mix 5 cups of cranberry juice with 2 cups of soda water." The ratio of cranberry juice to soda water is 5 : 2. Multiplying both ingredients by the same number creates equivalent ratios.

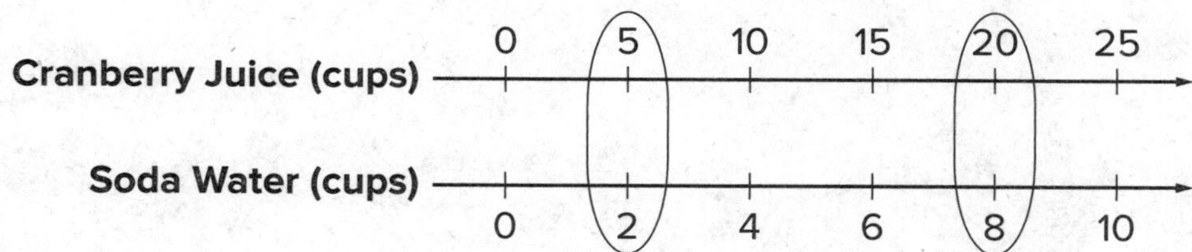

This double number line shows that the ratio 20 : 8 is equivalent to 5 : 2. If you mix 20 cups of cranberry juice with 8 cups of soda water, it makes 4 times as much fizzy juice that tastes the same as the original recipe.

Glossary

double number line diagram

Practice

Introducing Double Number Line Diagrams

1. A particular shade of orange paint has 2 cups of yellow paint for every 3 cups of red paint. On the double number line, circle the numbers of cups of yellow and red paint needed for 3 batches of orange paint.

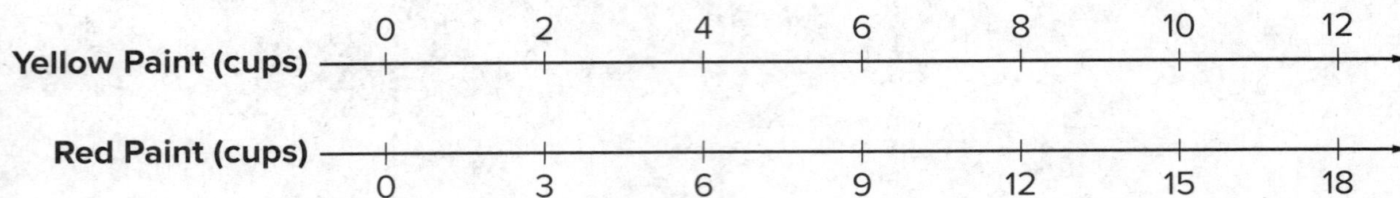

2. This double number line diagram shows the amount of flour and eggs needed for 1 batch of cookies.

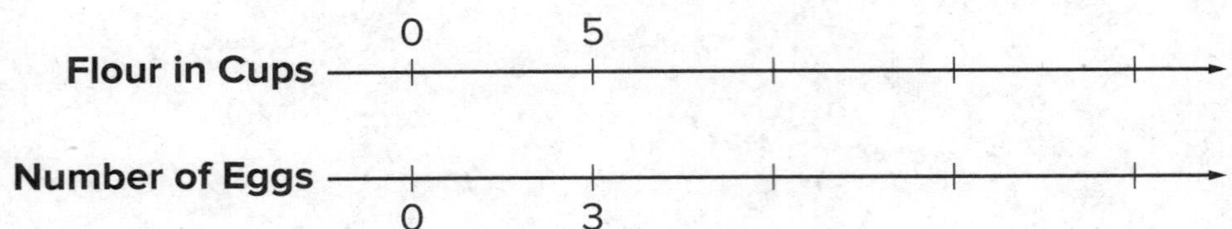

a. Complete the diagram to show the amount of flour and eggs needed for 2, 3, and 4 batches of cookies.

b. What is the ratio of cups of flour to eggs?

c. How much flour and how many eggs are used in 4 batches of cookies?

d. How much flour is used with 6 eggs?

e. How many eggs are used with 15 cups of flour?

NAME _______________ DATE _______________ PERIOD _______________

3. Here is a representation showing the amounts of red and blue paint that make 2 batches of purple paint.

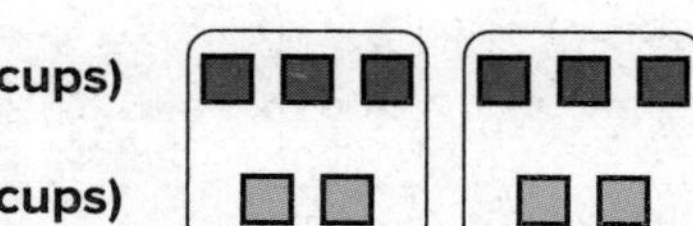

a. On the double number line, label the tick marks to represent amounts of red and blue paint used to make batches of this shade of purple paint.

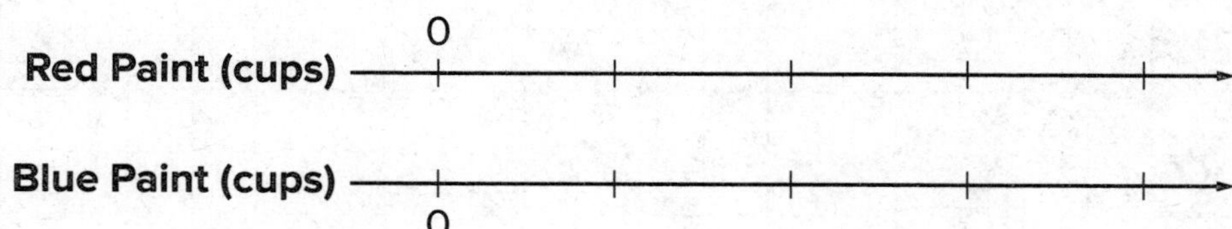

b. How many batches are made with 12 cups of red paint?

c. How many batches are made with 6 cups of blue paint?

4. Diego estimates that there will need to be 3 pizzas for every 7 kids at his party. Select **all** the statements that express this ratio. **(Lesson 2-1)**

A. The ratio of kids to pizzas is 7 : 3.

B. The ratio of pizzas to kids is 3 to 7.

C. The ratio of kids to pizzas is 3 : 7.

D. The ratio of pizzas to kids is 7 to 3.

E. For every 7 kids, there need to be 3 pizzas.

5. a. Draw a parallelogram that is not a rectangle that has an area of 24 square units. Explain or show how you know the area is 24 square units.

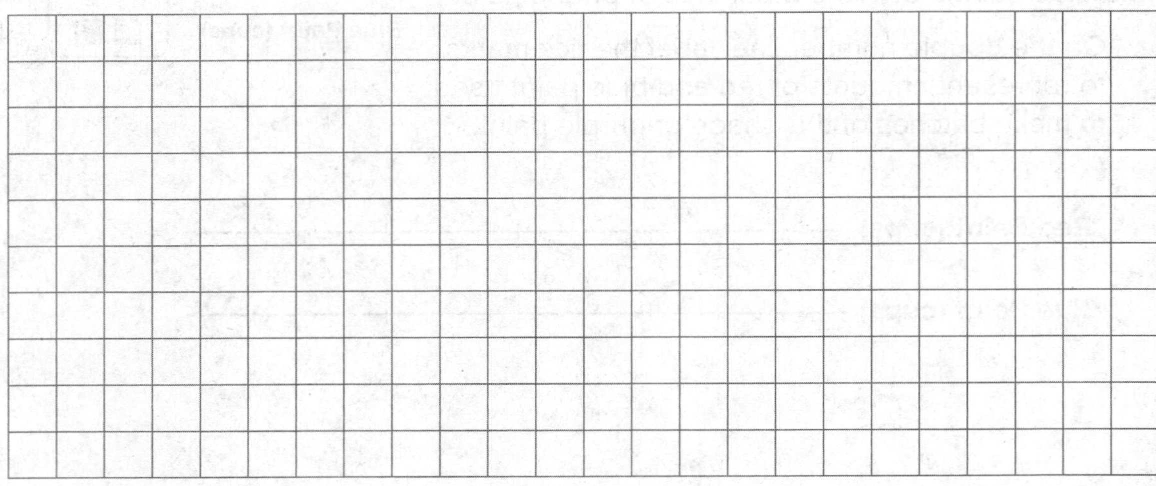

b. Draw a triangle that has an area of 24 square units. Explain or show how you know the area is 24 square units. **(Lesson 1-6)**

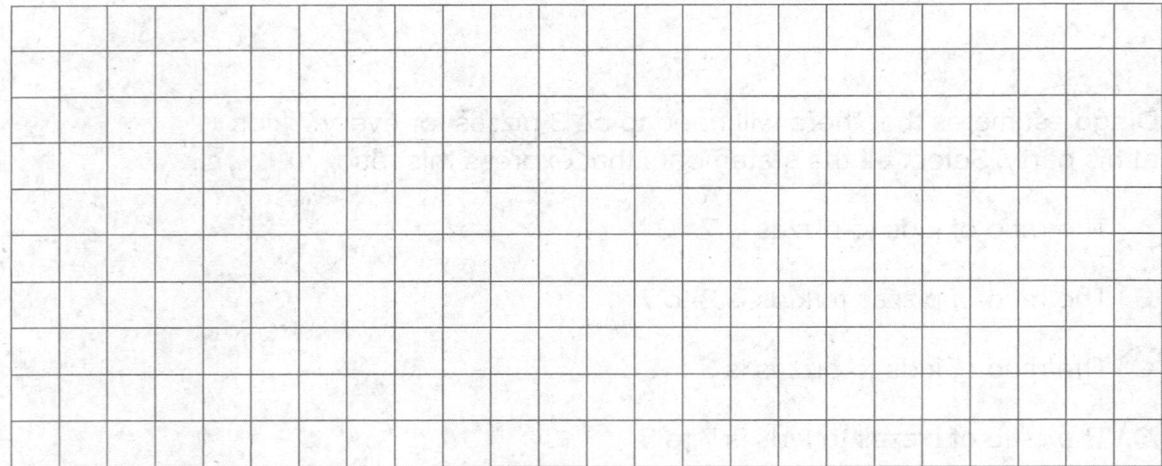

Lesson 2-7

Creating Double Number Line Diagrams

NAME _______________ DATE _______________ PERIOD _______________

Learning Goal Let's draw double number line diagrams like a pro.

Warm Up

7.1 Ordering on a Number Line

1. Locate and label the following numbers on the number line:

$\frac{1}{2}$ $\frac{1}{4}$ $1\frac{3}{4}$ 1.5 1.75

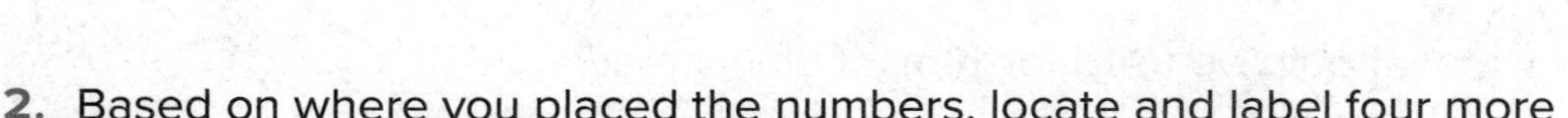

2. Based on where you placed the numbers, locate and label four more fractions or decimals on the number line.

Activity

7.2 Just a Little Green

The other day, we made green water by mixing 5 ml of blue water with 15 ml of yellow water. We want to make a very small batch of the same shade of green water. We need to know how much yellow water to mix with only 1 ml of blue water.

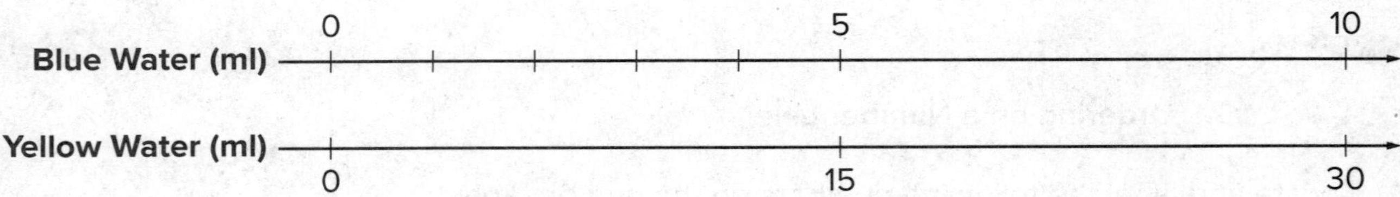

1. On the number line for blue water, label the four tick marks shown.
2. On the number line for yellow water, draw and label tick marks to show the amount of yellow water needed for each amount of blue water.
3. How much yellow water should be used for 1 ml of blue water? Circle where you can see this on the double number line.
4. How much yellow water should be used for 11 ml of blue water?
5. How much yellow water should be used for 8 ml of blue water?
6. Why is it useful to know how much yellow water should be used with 1 ml of blue water?

NAME ______________________ DATE ______________ PERIOD ____________

Activity

7.3 Art Paste on a Double Number Line

A recipe for art paste says "For every 2 pints of water, mix in 8 cups of flour."

1. Follow the instructions to draw a double number line diagram at the right representing the recipe for art paste.
   a. Use a ruler to draw two parallel lines.
   b. Label the first line "pints of water." Label the second line "cups of flour."
   c. Draw at least 6 equally spaced tick marks that line up on both lines.
   d. Along the water line, label the tick marks with the amount of water in 0, 1, 2, 3, 4, and 5 batches of art paste.
   e. Along the flour line, label the tick marks with the amount of flour in 0, 1, 2, 3, 4, and 5 batches of art paste.
2. Compare your double number line diagram with your partner's. Discuss your thinking. If needed, revise your diagram.
3. Next, use your double number line to answer these questions.
   a. How much flour should be used with 10 pints of water?
   b. How much water should be used with 24 cups of flour?
   c. How much flour **per** pint of water does this recipe use?

Are you ready for more?

A square with a side of 10 units overlaps a square with a side of 8 units in such a way that its corner *B* is placed exactly at the center of the smaller square. As a result of the overlapping, the two sides of the large square intersect the two sides of the small square exactly at points *C* and *E*, as shown. The length of *CD* is 6 units. What is the area of the overlapping region *CDEB*?

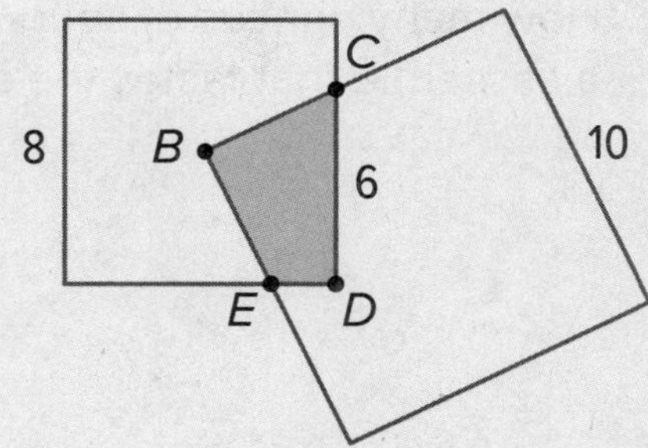

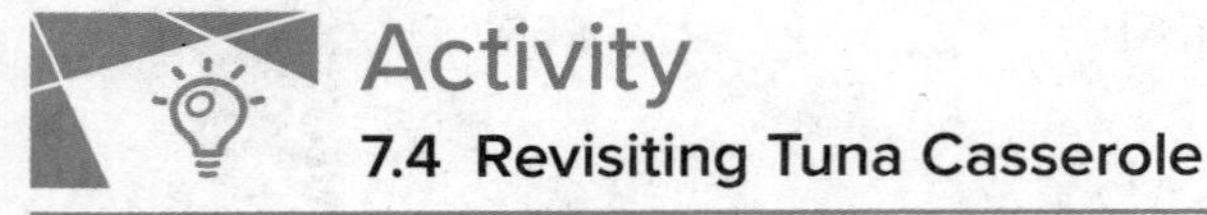

Activity

7.4 Revisiting Tuna Casserole

The other day, we looked at a recipe for tuna casserole that called for 10 ounces of cream of chicken soup for every 3 cups of elbow-shaped pasta.

1. Draw a double number line diagram that represents the amounts of soup and pasta in different-sized batches of this recipe.

2. If you made a large amount of tuna casserole by mixing 40 ounces of soup with 15 cups of pasta, would it taste the same as the original recipe? Explain or show your reasoning.

3. The original recipe called for 6 ounces of tuna for every 3 cups of pasta. Add a line to your diagram to represent the amount of tuna in different batches of casserole.

4. How many ounces of soup should you mix with 30 ounces of tuna to make a casserole that tastes the same as the original recipe?

NAME ______________________ DATE ____________ PERIOD ____________

Summary

Creating Double Number Line Diagrams

Here are some guidelines to keep in mind when drawing a double number line diagram.

- The two parallel lines should have labels that describe what the numbers represent.
- The tick marks and numbers should be spaced at equal intervals.
- Numbers that line up vertically make equivalent ratios.

For example, the ratio of the number of eggs to cups of milk in a recipe is 4 : 1. Here is a double number line that represents the situation.

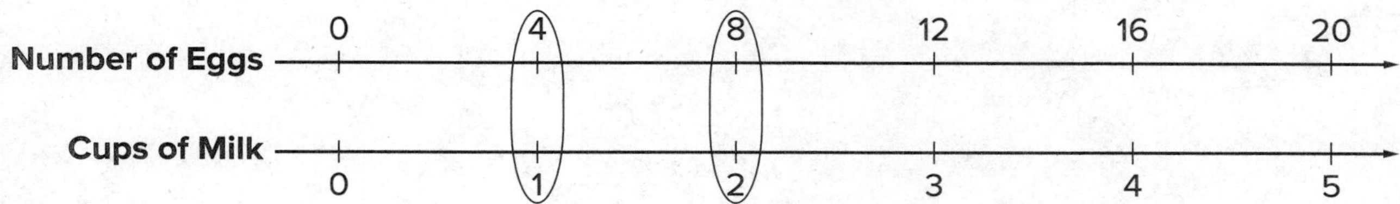

We can also say that this recipe uses "4 eggs per cup of milk" because the word **per** means "for each."

Glossary

per

Practice

Creating Double Number Line Diagrams

1. A recipe for cinnamon rolls uses 2 tablespoons of sugar per teaspoon of cinnamon for the filling. Complete the double number line diagram to show the amount of cinnamon and sugar in 3, 4, and 5 batches.

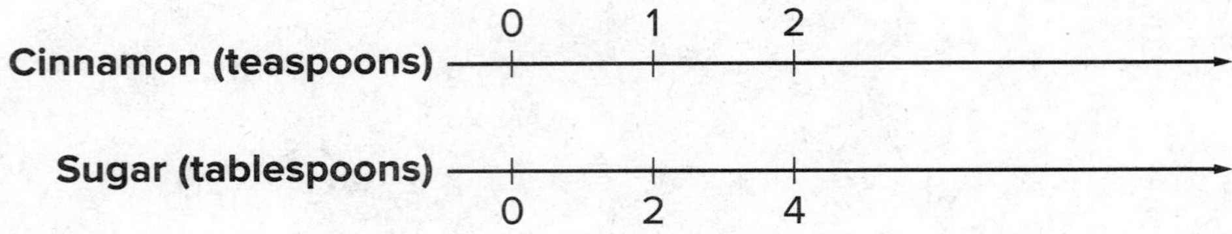

2. One batch of meatloaf contains 2 pounds of beef and $\frac{1}{2}$ cup of bread crumbs. Complete the double number line diagram to show the amounts of beef and bread crumbs needed for 1, 2, 3, and 4 batches of meatloaf.

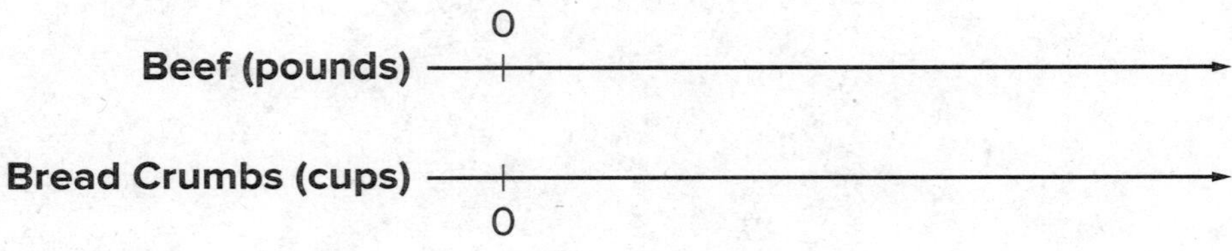

NAME ______________________ DATE ____________ PERIOD ____________

3. A recipe for tropical fruit punch says, "Combine 4 cups of pineapple juice with 5 cups of orange juice."

a. Create a double number showing the amount of each type of juice in 1, 2, 3, 4, and 5 batches of the recipe.

b. If 12 cups of pineapple juice are used with 20 cups of orange juice, will the recipe taste the same? Explain your reasoning.

c. The recipe also calls for $\frac{1}{3}$ cup of lime juice for every 5 cups of orange juice. Add a line to your diagram to represent the amount of lime juice in different batches of tropical fruit punch.

4. One batch of pink paint uses 2 cups of red paint and 7 cups of white paint. Mai made a large amount of pink paint using 14 cups of red paint. **(Lesson 2-4)**

a. How many batches of pink paint did she make?

b. How many cups of white paint did she use?

5. a. Find three different ratios that are equivalent to the ratio 3 : 11. **(Lesson 2-5)**

b. Explain why your ratios are equivalent.

6. Here is a diagram that represents the pints of red and yellow paint in a mixture. **(Lesson 2-2)**

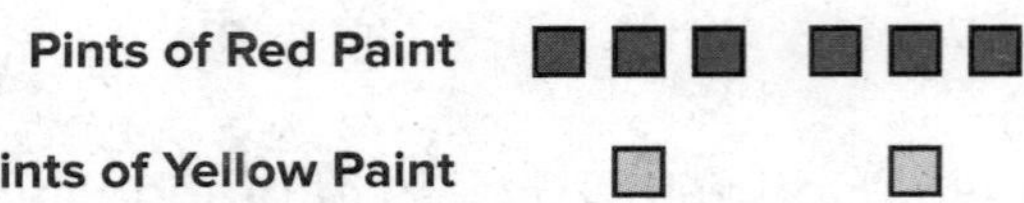

Select **all** statements that accurately describe the diagram.

(A.) The ratio of yellow paint to red paint is 2 to 6.

(B.) For every 3 pints of red paint, there is 1 pint of yellow paint.

(C.) For every pint of yellow paint, there are 3 pints of red paint.

(D.) For every pint of yellow paint there are 6 pints of red paint.

(E.) The ratio of red paint to yellow paint is 6 : 2.

Lesson 2-8

How Much for One?

NAME ______________________ DATE ______________ PERIOD ____________

Learning Goal Let's use ratios to describe how much things cost.

Warm Up

8.1 Number Talk: Remainders in Division

Find the quotient mentally.

$246 \div 12$

Activity

8.2 Grocery Shopping

Answer each question and explain or show your reasoning. If you get stuck, consider drawing a double number line diagram.

1. Eight avocados cost $4.

    a. How much do 16 avocados cost?

    b. How much do 20 avocados cost?

    c. How much do 9 avocados cost?

2. Twelve large bottles of water cost $9.

    a. How many bottles can you buy for $3?

    b. What is the cost per bottle of water?

    c. How much would 7 bottles of water cost?

3. A 10-pound sack of flour costs $8.

    a. How much does 40 pounds of flour cost?

    b. What is the cost per pound of flour?

Are you ready for more?

It is commonly thought that buying larger packages or containers, sometimes called *buying in bulk*, is a great way to save money. For example, a 6-pack of soda might cost \$3 while a 12-pack of the same brand costs \$5.

Find 3 different cases where it is not true that buying in bulk saves money. You may use the internet or go to a local grocery store and take photographs of the cases you find. Make sure the products are the same brand. For each example that you find, give the quantity or size of each, and describe how you know that the larger size is not a better deal.

Activity
8.3 More Shopping

1. Four bags of chips cost \$6.

   a. What is the cost per bag?

   b. At this rate, how much will 7 bags of chips cost?

2. At a used book sale, 5 books cost \$15.

   a. What is the cost per book?

   b. At this rate, how many books can you buy for \$21?

NAME ______________________ DATE ______________ PERIOD ____________

3. Neon bracelets cost \$1 for 4.

   a. What is the cost per bracelet?

   b. At this rate, how much will 11 neon bracelets cost?

Pause here so your teacher can review your work.

4. Your teacher will assign you one of the problems. Create a visual display that shows your solution to the problem. Be prepared to share your solution with the class.

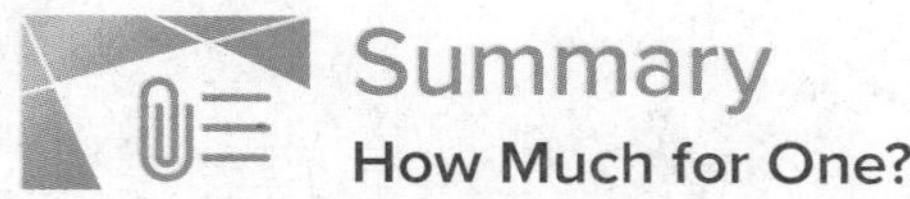

Summary

How Much for One?

The **unit price** is the price of 1 thing—for example, the price of 1 ticket, 1 slice of pizza, or 1 kilogram of peaches.

If 4 movie tickets cost \$28, then the unit price would be the cost per ticket. We can create a double number line to find the unit price.

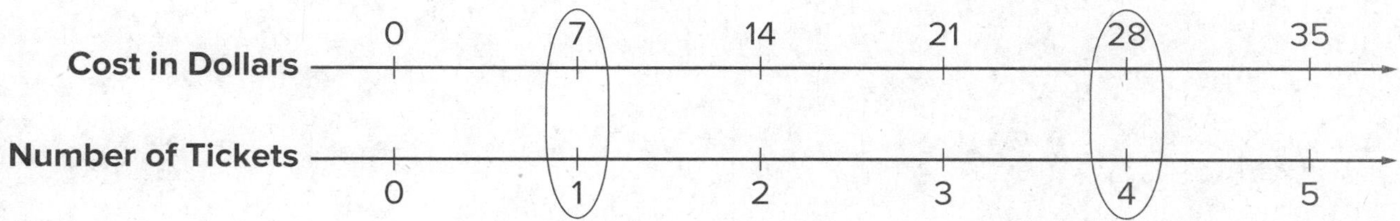

This double number line shows that the cost for 1 ticket is \$7.

We can also find the unit price by dividing $28 \div 4 = 7$, or by multiplying $28 \cdot \frac{1}{4} = 7$.

Glossary

unit price

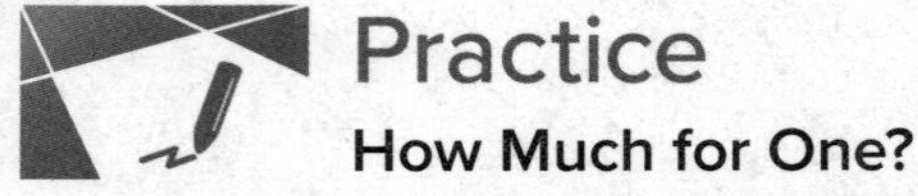

Practice
How Much for One?

1. In 2016, the cost of 2 ounces of pure gold was $2,640. Complete the double number line to show the cost for 1, 3, and 4 ounces of gold.

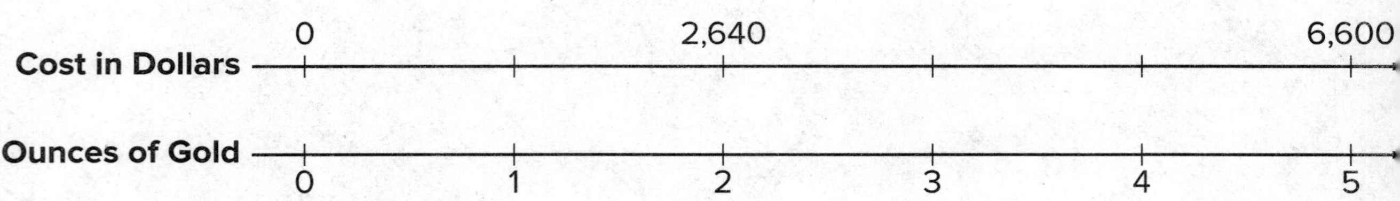

2. The double number line shows that 4 pounds of tomatoes cost $14. Draw tick marks and write labels to show the prices of 1, 2, and 3 pounds of tomatoes.

3. 4 movie tickets cost $48. At this rate, what is the cost of:

   a. 5 movie tickets?

   b. 11 movie tickets?

NAME ______________________ DATE ______________ PERIOD ____________

4. Priya bought these items at the grocery store. Find each unit price.

a. 12 eggs for \$3. How much is the cost per egg?

b. 3 pounds of peanuts for \$7.50. How much is the cost per pound?

c. 4 rolls of toilet paper for \$2. How much is the cost per roll?

d. 10 apples for \$3.50. How much is the cost per apple?

5. Clare made a smoothie with 1 cup of yogurt, 3 tablespoons of peanut butter, 2 teaspoons of chocolate syrup, and 2 cups of crushed ice. **(Lesson 2-3)**

a. Kiran tried to double this recipe. He used 2 cups of yogurt, 6 tablespoons of peanut butter, 5 teaspoons of chocolate syrup, and 4 cups of crushed ice. He didn't think it tasted right. Describe how the flavor of Kiran's recipe compares to Clare's recipe.

b. How should Kiran change the quantities that he used so that his smoothie tastes just like Clare's?

6. A drama club is building a wooden stage in the shape of a trapezoidal prism. The height of the stage is 2 feet. Some measurements of the stage are shown here. **(Lesson 1-15)**

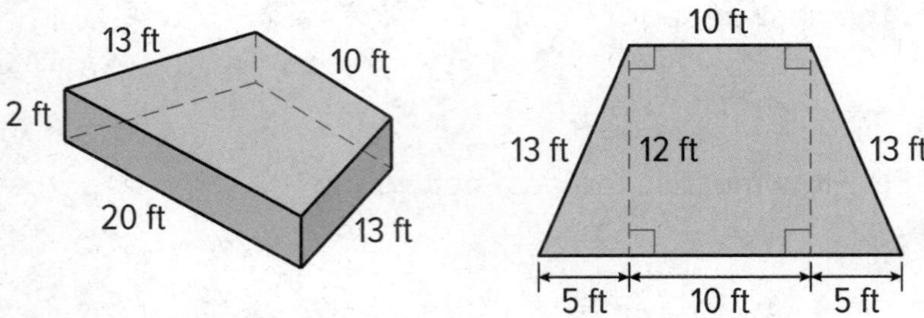

What is the area of all the faces of the stage, excluding the bottom? Show your reasoning. If you get stuck, consider drawing a net of the prism.

Lesson 2-9

Constant Speed

NAME ______________________ DATE ____________ PERIOD ____________

Learning Goal Let's use ratios to work with how fast things move.

Warm Up

9.1 Number Talk: Dividing by Powers of 10

Find the quotient mentally.

1. $30 \div 10$ **2.** $34 \div 10$ **3.** $3.4 \div 10$ **4.** $34 \div 100$

Activity

9.2 Moving 10 Meters

Your teacher will set up a straight path with a 1-meter warm-up zone and a 10-meter measuring zone. Follow the following instructions to collect the data.

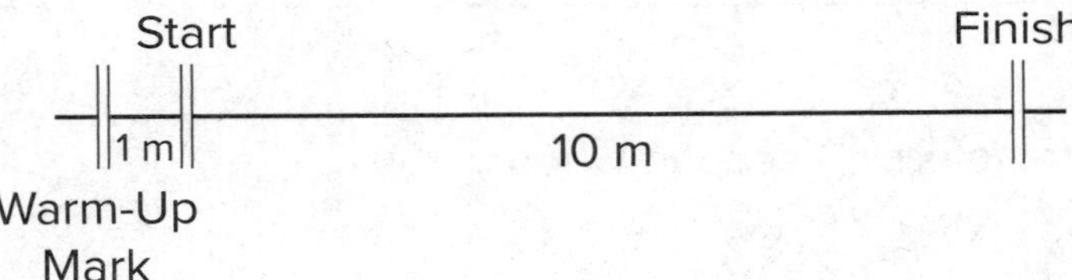

1. **a.** The person with the stopwatch (the "timer") stands at the finish line. The person being timed (the "mover") stands at the warm-up line.

   b. On the first round, the mover starts moving *at a slow, steady speed* along the path. When the mover reaches the start line, they say, "Start!" and the timer starts the stopwatch.

   c. The mover keeps moving steadily along the path. When they reach the finish line, the timer stops the stopwatch and records the time, rounded to the nearest second, in the table.

   d. On the second round, the mover follows the same instructions, but this time, moving *at a quick, steady speed.* The timer records the time the same way.

e. Repeat these steps until each person in the group has gone twice: once at a slow, steady speed, and once at a quick, steady speed.

Your Slow Moving Time (seconds)	Your Fast Moving Time (seconds)

2. After you finish collecting the data, use the double number line diagrams to answer the questions. Use the times your partner collected while you were moving.

Moving slowly:

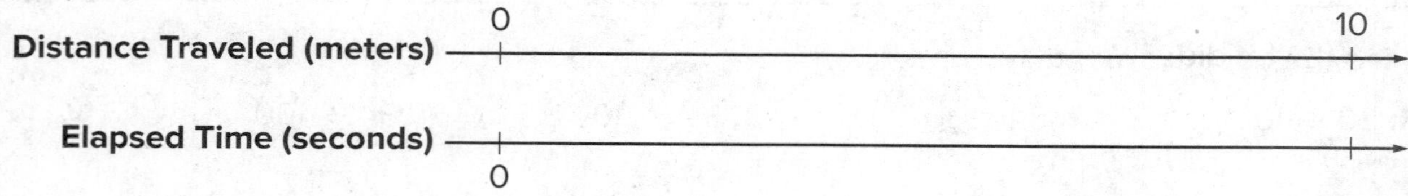

Moving quickly:

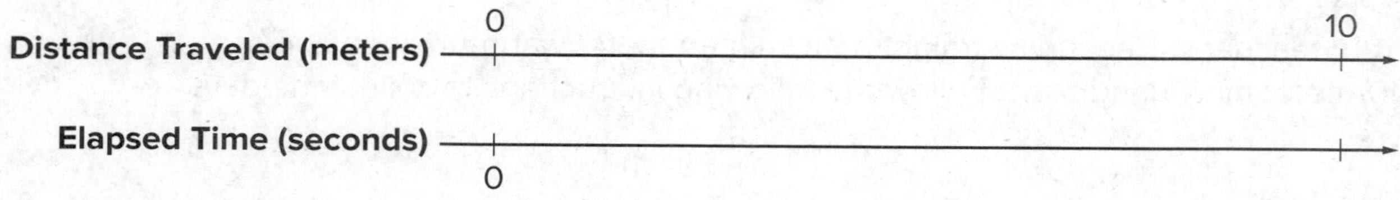

a. Estimate the distance in meters you traveled in 1 second when moving slowly.

b. Estimate the distance in meters you traveled in 1 second when moving quickly.

c. Trade diagrams with someone who is not your partner. How is the diagram representing someone moving slowly different from the diagram representing someone moving quickly?

NAME ______________________ DATE ______________ PERIOD ____________

Activity

9.3 Moving for 10 Seconds

Lin and Diego both ran for 10 seconds, each at their own constant speed. Lin ran 40 meters and Diego ran 55 meters.

1. Who was moving faster? Explain your reasoning.

2. How far did each person move in 1 second? If you get stuck, consider drawing double number line diagrams to represent the situations.

3. Use your data from the previous activity to find how far *you* could travel in 10 seconds at your quicker speed.

4. Han ran 100 meters in 20 seconds at a constant speed. Is this speed faster, slower, or the same as Lin's? Diego's? Yours?

Are you ready for more?

Lin and Diego want to run a race in which they will both finish when the timer reads exactly 30 seconds. Who should get a head start, and how long should the head start be?

Summary

Constant Speed

Suppose a train traveled 100 meters in 5 seconds at a constant speed. To find its speed in **meters per second,** we can create a double number line.

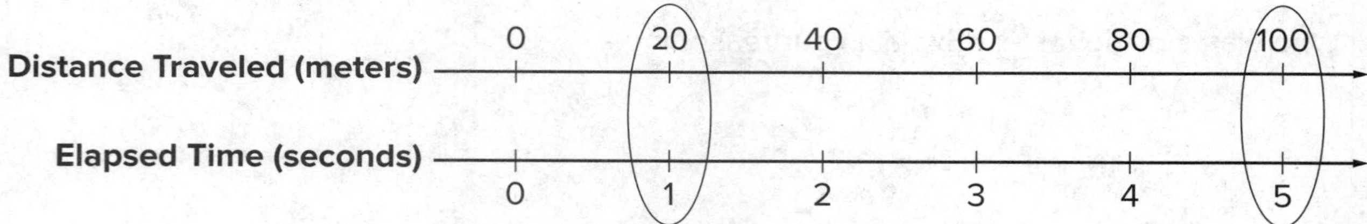

The double number line shows that the train's speed was 20 meters per second. We can also find the speed by dividing: $100 \div 5 = 20$.

Once we know the speed in meters per second, many questions about the situation become simpler to answer because we can multiply the amount of time an object travels by the speed to get the distance.

For example, at this rate, how far would the train go in 30 seconds? Since $20 \cdot 30 = 600$, the train would go 600 meters in 30 seconds.

Glossary

meters per second

NAME ______________________ DATE ______________ PERIOD ______________

Practice
Constant Speed

1. Han ran 10 meters in 2.7 seconds. Priya ran 10 meters in 2.4 seconds.

a. Who ran faster? Explain how you know.

b. At this rate, how long would it take each person to run 50 meters? Explain or show your reasoning.

2. A scooter travels 30 feet in 2 seconds at a constant speed.

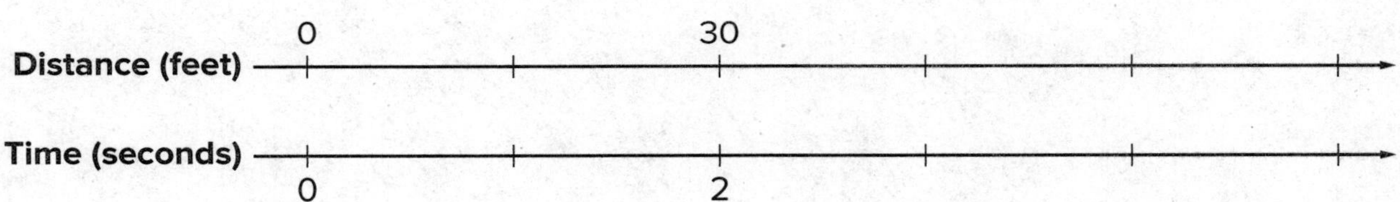

a. What is the speed of the scooter in feet per second?

b. Complete the double number line to show the distance the scooter travels after 1, 3, 4, and 5 seconds.

c. A skateboard travels 55 feet in 4 seconds. Is the skateboard going faster, slower, or the same speed as the scooter?

3. A cargo ship traveled 150 nautical miles in 6 hours at a constant speed. How far did the cargo ship travel in one hour?

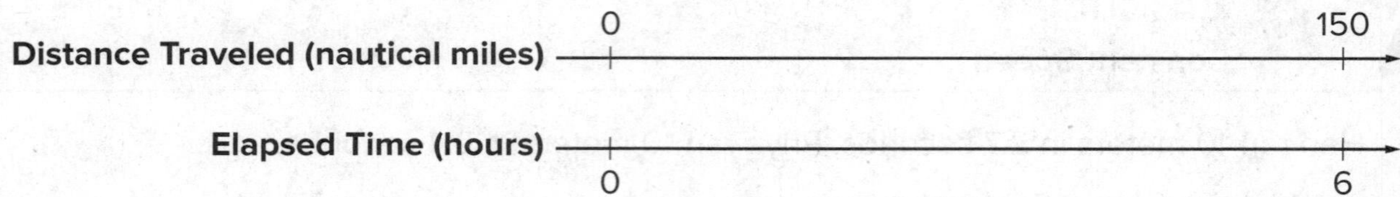

4. A recipe for pasta dough says, "Use 150 grams of flour per large egg." **(Lesson 2-3)**

   a. How much flour is needed if 6 large eggs are used?

   b. How many eggs are needed if 450 grams of flour are used?

NAME ____________________ DATE ____________ PERIOD ____________

5. The grocery store is having a sale on frozen vegetables. 4 bags are being sold for $11.96. (Lesson 2-8)

At this rate, what is the cost of:

a. 1 bag

b. 9 bags

6. A pet owner has 5 cats. Each cat has 2 ears and 4 paws. (Lesson 2-7)

a. Complete the double number line to show the numbers of ears and paws for 1, 2, 3, 4, and 5 cats.

Number of Ears 0

Number of Paws 0

b. If there are 3 cats in the room, what is the ratio of ears to paws?

c. If there are 4 cats in the room, what is the ratio of paws to ears?

d. If all 5 cats are in the room, how many more paws are there than ears?

7. Each of these is a pair of equivalent ratios. For each pair, explain why they are equivalent ratios or draw a representation that shows why they are equivalent ratios. **(Lesson 2-5)**

   a. 5 : 1 and 15 : 3

   b. 25 : 5 and 10 : 2

   c. 198 : 1,287 and 2 : 13

Lesson 2-10

Comparing Situations by Examining Ratios

NAME ______________________ DATE ______________ PERIOD ______________

Learning Goal Let's use ratios to compare situations.

Warm-Up

10.1 Treadmills

Mai and Jada each ran on a treadmill. The treadmill display shows the distance, in miles, each person ran and the amount of time it took them, in minutes and seconds.

Here is Mai's treadmill display.

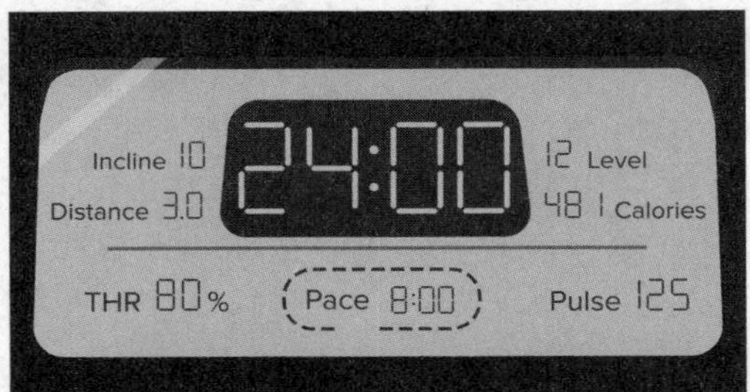

Here is Jada's treadmill display.

1. What is the same about their workouts? What is different about their workouts?

2. If each person ran at a constant speed the entire time, who was running faster? Explain your reasoning.

Activity
10.2 Concert Tickets

Diego paid $47 for 3 tickets to a concert. Andre paid $141 for 9 tickets to a concert. Did they pay at the **same rate**? Explain your reasoning.

Activity
10.3 Sparkling Orange Juice

Lin and Noah each have their own recipe for making sparkling orange juice.

- Lin mixes 3 liters of orange juice with 4 liters of soda water.
- Noah mixes 4 liters of orange juice with 5 liters of soda water.

How do the two mixtures compare in taste? Explain your reasoning.

Are you ready for more?

1. How can Lin make her sparkling orange juice taste the same as Noah's just by adding more of one ingredient? How much will she need?

2. How can Noah make his sparkling orange juice taste the same as Lin's just by adding more of one ingredient? How much will he need?

NAME _______________ DATE _______________ PERIOD _______________

Summary

Comparing Situations by Examining Ratios

Sometimes we want to know whether two situations are described by the **same rate**. To do that, we can write an equivalent ratio for one or both situations so that one part of their ratios has the same value.

Then we can compare the other part of the ratios.

For example, do these two paint mixtures make the same shade of orange?

- Kiran mixes 9 teaspoons of red paint with 15 teaspoons of yellow paint.
- Tyler mixes 7 teaspoons of red paint with 10 teaspoons of yellow paint.

Here is a double number line that represents Kiran's paint mixture.

The ratio 9 : 15 is equivalent to the ratios 3 : 5 and 6 : 10.

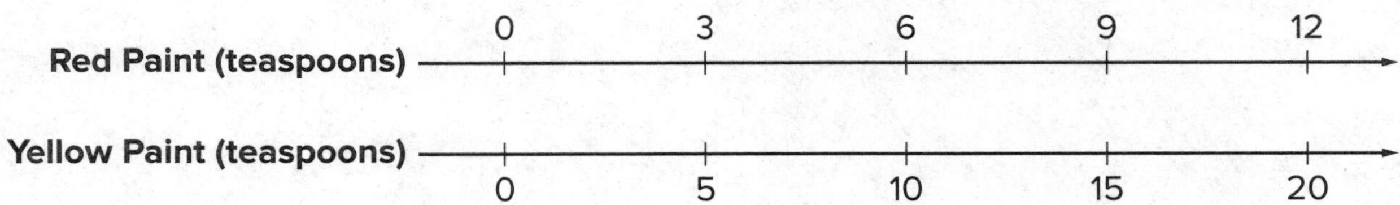

For 10 teaspoons of yellow paint, Kiran would mix in 6 teaspoons of red paint. This is less red paint than Tyler mixes with 10 teaspoons of yellow paint. The ratios 6 : 10 and 7 : 10 are not equivalent, so these two paint mixtures would not be the same shade of orange.

When we talk about two things happening at the same rate, we mean that the ratios of the quantities in the two situations are equivalent. There is also something specific about the situation that is the same.

- If two ladybugs are moving at the same rate, then they are traveling at the *same constant speed*.
- If two bags of apples are selling for the same rate, then they have the *same unit price*.
- If we mix two kinds of juice at the same rate, then the mixtures have the *same taste*.
- If we mix two colors of paint at the same rate, then the mixtures have the *same shade*.

Glossary

same rate

Practice

Comparing Situations by Examining Ratios

1. A slug travels 3 centimeters in 3 seconds. A snail travels 6 centimeters in 6 seconds. Both travel at constant speeds. Mai says, "The snail was traveling faster because it went a greater distance." Do you agree with Mai? Explain or show your reasoning.

2. If you blend 2 scoops of chocolate ice cream with 1 cup of milk, you get a milkshake with a stronger chocolate flavor than if you blended 3 scoops of chocolate ice cream with 2 cups of milk. Explain or show why.

3. There are 2 mixtures of light purple paint.

   - Mixture A is made with 5 cups of purple paint and 2 cups of white paint.
   - Mixture B is made with 15 cups of purple paint and 8 cups of white paint.

   Which mixture is a lighter shade of purple? Explain your reasoning.

NAME ______________________ DATE ____________ PERIOD ________

4. Tulip bulbs are on sale at store A, at 5 for $11.00, and the regular price at store B is 6 for $13.00. Is each store pricing tulip bulbs at the same rate? Explain how you know.

5. A plane travels at a constant speed. It takes 6 hours to travel 3,360 miles. **(Lesson 2-9)**

a. What is the plane's speed in miles per hour?

b. At this rate, how many miles can it travel in 10 hours?

6. A pound of ground beef costs $5. At this rate, what is the cost of: **(Lesson 2-8)**

a. 3 pounds?

b. $\frac{1}{2}$ pound?

c. $\frac{1}{4}$ pound?

d. $\frac{3}{4}$ pound?

e. $3\frac{3}{4}$ pounds?

7. In a triple batch of a spice mix, there are 6 teaspoons of garlic powder and 15 teaspoons of salt. Answer the following questions about the mix. If you get stuck, create a double number line. **(Lesson 2-7)**

   a. How much garlic powder is used with 5 teaspoons of salt?

   b. How much salt is used with 8 teaspoons of garlic powder?

   c. If there are 14 teaspoons of spice mix, how much salt is in it?

   d. How much more salt is there than garlic powder if 6 teaspoons of garlic powder are used?

Lesson 2-11

Representing Ratios with Tables

NAME ______________________ DATE ____________ PERIOD ________

Learning Goal Let's use tables to represent equivalent ratios.

Warm Up

11.1 How Is It Growing?

Look for a pattern in the figures.

1. How many total tiles will be in:
   a. the 4th figure?
   b. the 5th figure?
   c. the 10th figure?
2. How do you see it growing?

Activity

11.2 A Huge Amount of Sparkling Orange Juice

Noah's recipe for one batch of sparkling orange juice uses 4 liters of orange juice and 5 liters of soda water.

1. Use the double number line to show how many liters of each ingredient to use for different-sized batches of sparkling orange juice.

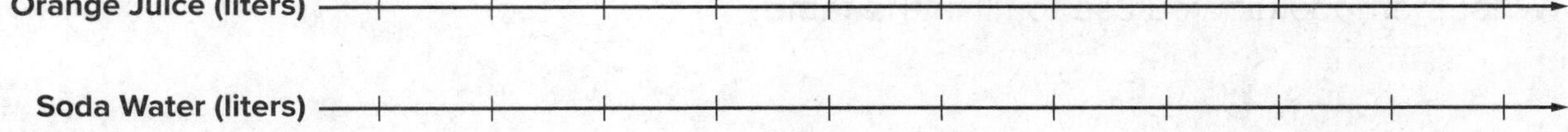

2. If someone mixes 36 liters of orange juice and 45 liters of soda water, how many batches would they make?

3. If someone uses 400 liters of orange juice, how much soda water would they need?

4. If someone uses 455 liters of soda water, how much orange juice would they need?

5. Explain the trouble with using a double number line diagram to answer the last two questions.

Activity
11.3 Batches of Trail Mix

A recipe for trail mix says: "Mix 7 ounces of almonds with 5 ounces of raisins." Here is a **table** that has been started to show how many ounces of almonds and raisins would be in different-sized batches of this trail mix.

Almonds (oz)	Raisins (oz)
7	5
28	
	10
3.5	
	250
56	

1. Complete the table so that ratios represented by each row are equivalent.
2. What methods did you use to fill in the table?

3. How do you know that each row shows a ratio that is equivalent to 7 : 5? Explain your reasoning.

NAME ______________________ DATE ______________ PERIOD ____________

Are you ready for more?

You have created a best-selling recipe for chocolate chip cookies. The ratio of sugar to flour is 2 : 5.

Create a table in which each entry represents amounts of sugar and flour that might be used at the same time in your recipe.

- One entry should have amounts where you have fewer than 25 cups of flour.
- One entry should have amounts where you have between 20–30 cups of sugar.
- One entry can have any amounts using more than 500 units of flour.

Summary

Representing Ratios with Tables

A **table** is a way to organize information. Each horizontal set of entries is called a *row*, and each vertical set of entries is called a *column*. (The table shown has 2 columns and 5 rows.) A table can be used to represent a collection of equivalent ratios.

Here is a double number line diagram and a table that both represent the situation: "The price is $2 for every 3 mangos."

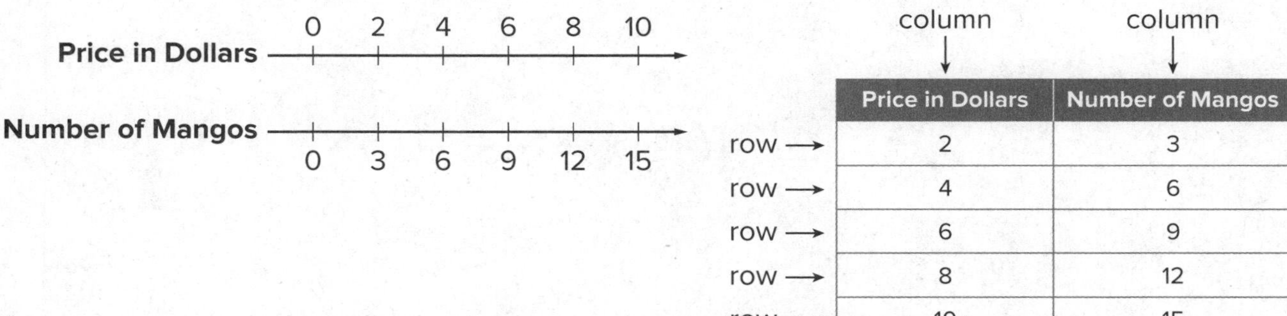

Price in Dollars	Number of Mangos
2	3
4	6
6	9
8	12
10	15

Glossary

table

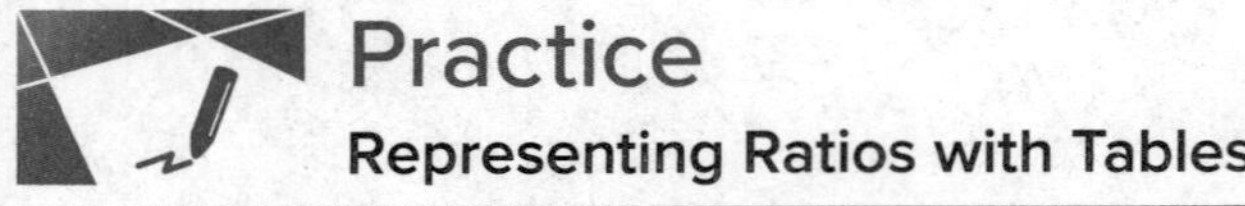

Practice

Representing Ratios with Tables

1. Complete the table to show the amounts of yellow and red paint needed for different-sized batches of the same shade of orange paint.

Yellow Paint (quarts)	Red Paint (quarts)
5	6

Explain how you know that these amounts of yellow paint and red paint will make the same shade of orange as the mixture in the first row of the table.

2. A car travels at a constant speed, as shown on the double number line.

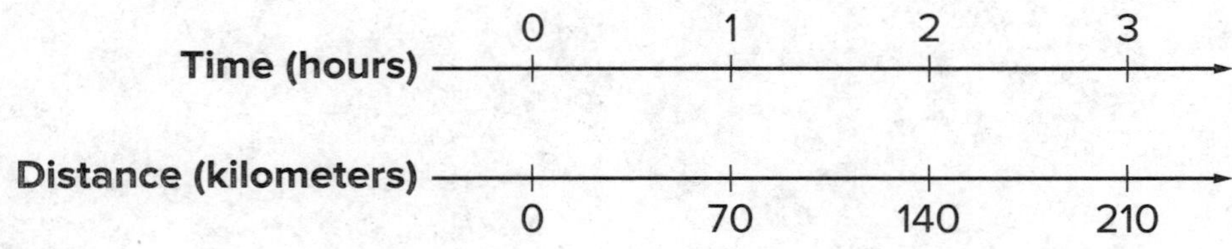

How far does the car travel in 14 hours? Explain or show your reasoning.

NAME ______________________ DATE ______________ PERIOD ________

3. The olive trees in an orchard produce 3,000 pounds of olives a year. It takes 20 pounds of olives to make 3 liters of olive oil. How many liters of olive oil can this orchard produce in a year? If you get stuck, consider using the table.

Olives (pounds)	Olive Oil (liters)
20	3
100	
3,000	

4. At a school recess, there needs to be a ratio of 2 adults for every 24 children on the playground. The double number line represents the number of adults and children on the playground at recess. **(Lesson 2-6)**

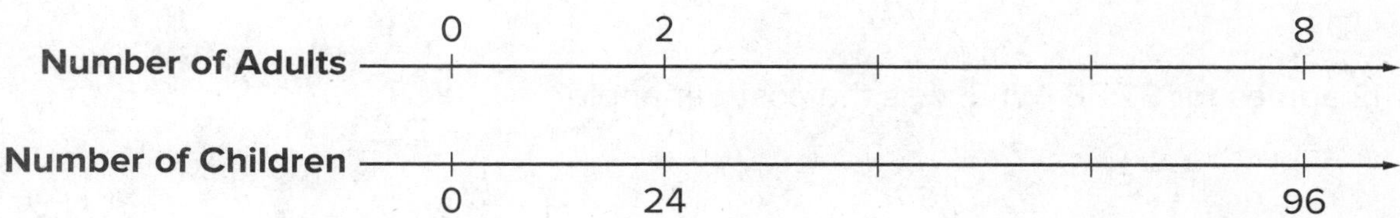

a. Label each remaining tick mark with its value.

b. How many adults are needed if there are 72 children? Circle your answer on the double number line.

5. While playing basketball, Jada's heart rate goes up to 160 beats per minute. While jogging, her heart beats 25 times in 10 seconds. Assuming her heart beats at a constant rate while jogging, which of these activities resulted in a higher heart rate? Explain your reasoning. **(Lesson 2-10)**

6. A shopper bought the following items at the farmer's market: **(Lesson 2-8)**

   a. 6 ears of corn for $1.80. What was the cost per ear?

   b. 12 apples for $2.88. What was the cost per apple?

   c. 5 tomatoes for $3.10. What was the cost per tomato?

Lesson 2-12

Navigating a Table of Equivalent Ratios

NAME ______________________ DATE ______________ PERIOD ______________

Learning Goal Let's use a table of equivalent ratios like a pro.

Warm Up

12.1 Number Talk: Multiplying by a Unit Fraction

Find each product mentally.

1. $\frac{1}{3} \cdot 21$ **2.** $\frac{1}{6} \cdot 21$ **3.** $(5.6) \cdot \frac{1}{8}$ **4.** $\frac{1}{4} \cdot (5.6)$

Activity

12.2 Comparing Taco Prices

Use the table to help you solve these problems.
Explain or show your reasoning.

Number of Tacos	Price in Dollars

1. Noah bought 4 tacos and paid \$6. At this rate, how many tacos could he buy for \$15?

2. Jada's family bought 50 tacos for a party and paid \$72. Were Jada's tacos the same price as Noah's tacos?

Activity

12.3 Hourly Wages

Lin is paid $90 for 5 hours of work. She used the table to calculate how much she would be paid at this rate for 8 hours of work.

	Amount Earned ($)	Time Worked (hours)	
	90	5	
$\cdot\frac{1}{5}$	18	1	$\cdot\frac{1}{5}$
$\cdot 8$	144	8	$\cdot 8$

1. What is the meaning of the 18 that appears in the table?

2. Why was the number $\frac{1}{5}$ used as a multiplier?

3. Explain how Lin used this table to solve the problem.

4. At this rate, how much would Lin be paid for 3 hours of work? For 2.1 hours of work?

NAME ______________________ DATE ____________ PERIOD ____________

Activity
12.4 Zeno's Memory Card

In 2016, 128 gigabytes (GB) of portable computer memory cost $32.

1. Here is a double number line that represents the situation.

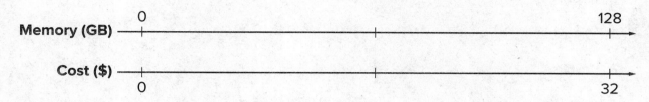

One set of tick marks has already been drawn to show the result of multiplying 128 and 32 each by $\frac{1}{2}$. Label the amount of memory and the cost for these tick marks.

Next, keep multiplying by $\frac{1}{2}$ and drawing and labeling new tick marks, until you can no longer clearly label each new tick mark with a number.

2. Here is a table that represents the situation. Find the cost of 1 gigabyte. You can use as many rows as you need.

Memory (gigabytes)	Cost (dollars)
128	32

3. Did you prefer the double number line or the table for solving this problem? Why?

Are you ready for more?

A kilometer is 1,000 meters because *kilo* is a prefix that means 1,000. The prefix *mega* means 1,000,000 and *giga* (as in gigabyte) means 1,000,000,000. One byte is the amount of memory needed to store one letter of the alphabet. About how many of each of the following would fit on a 1-gigabyte flash drive?

1. letters
2. pages
3. books
4. movies
5. songs

Summary

Navigating a Table of Equivalent Ratios

Finding a row containing a "1" is often a good way to work with tables of equivalent ratios. For example, the price for 4 lbs of granola is $5. At that rate, what would be the price for 62 lbs of granola?

Here are tables showing two different approaches to solving this problem. Both of these approaches are correct. However, one approach is more efficient.

- Less efficient

Granola (lbs)	Price ($)
4	5
8	10
16	20
32	40
64	80
62	77.50

Arrows (left): ·2, ·2, ·2, ·2, -2 lbs. Arrows (right): ·2, ·2, ·2, ·2, -$2.50

NAME ______________________ DATE ____________ PERIOD ____________

- More efficient

	Granola (lbs)	Price ($)	
	4	5	
$\cdot \frac{1}{4}$	1	1.25	$\cdot \frac{1}{4}$
$\cdot 62$	62	77.50	$\cdot 62$

Notice how the more efficient approach starts by finding the price for 1 lb of granola.

Remember that dividing by a whole number is the same as multiplying by a unit fraction. In this example, we can divide by 4 or multiply by $\frac{1}{4}$ to find the unit price.

Practice

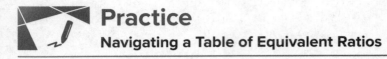

Navigating a Table of Equivalent Ratios

1. Priya collected 2,400 grams of pennies in a fundraiser. Each penny has a mass of 2.5 grams. How much money did Priya raise? If you get stuck, consider using the table.

Number of Pennies	Mass (grams)
1	2.5
	2,400

2. Kiran reads 5 pages in 20 minutes. He spends the same amount of time per page. How long will it take him to read 11 pages? If you get stuck, consider using the table.

Time (minutes)	Number of Pages
20	5
	1
	11

NAME ______________________ DATE ______________ PERIOD ____________

3. Mai is making personal pizzas. For 4 pizzas, she uses 10 ounces of cheese.

Number of Pizzas	Cheese (ounces)
4	10

a. How much cheese does Mai use per pizza?

b. At this rate, how much cheese will she need to make 15 pizzas?

4. Clare is paid $90 for 5 hours of work. At this rate, how many seconds does it take for her to earn 25 cents?

5. A car that travels 20 miles in $\frac{1}{2}$ hour at constant speed is traveling at the same speed as a car that travels 30 miles in $\frac{3}{4}$ hour at a constant speed. Explain or show why. **(Lesson 2-10)**

6. Lin makes her favorite juice blend by mixing cranberry juice with apple juice in the ratio shown on the double number line. Complete the diagram to show smaller and larger batches that would taste the same as Lin's favorite blend. **(Lesson 2-6)**

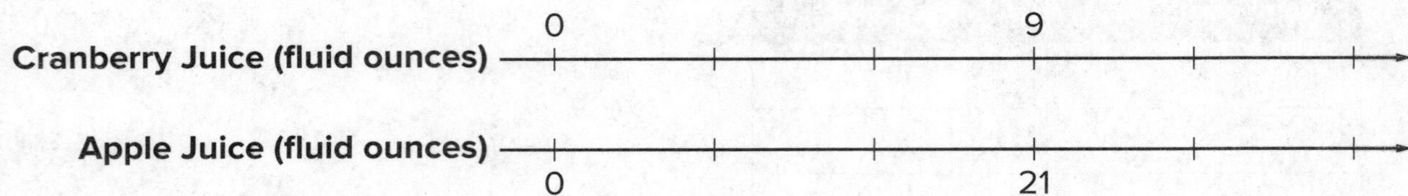

7. Each of these is a pair of equivalent ratios. For each pair, explain why they are equivalent ratios or draw a representation that shows why they are equivalent ratios. **(Lesson 2-5)**

   a. 600 : 450 and 60 : 45

   b. 60 : 45 and 4 : 3

   c. 600 : 450 and 4:3

Lesson 2-13

Tables and Double Number Line Diagrams

NAME ______________________ DATE ______________ PERIOD ____________

Learning Goal Let's contrast double number lines and tables.

Warm Up

13.1 Number Talk: Constant Dividend

1. Find the quotients mentally.

   a. $150 \div 2$ b. $150 \div 4$ c. $150 \div 8$

2. Locate and label the quotients on the number line.

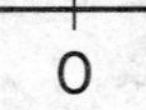

Activity

13.2 Moving 3,000 Meters

The other day, we saw that Han can run 100 meters in 20 seconds.

Han wonders how long it would take him to run 3,000 meters at this rate. He made a table of equivalent ratios.

20	100
10	50
1	5
3,000	

1. Do you agree that this table represents the situation? Explain your reasoning.

2. Complete the last row with the missing number.

3. What question about the situation does this number answer?

4. What could Han do to improve his table?

5. Priya can bike 150 meters in 20 seconds. At this rate, how long would it take her to bike 3,000 meters?

6. Priya's neighbor has a dirt bike that can go 360 meters in 15 seconds. At this rate, how long would it take them to ride 3,000 meters?

Activity
13.3 The International Space Station

The International Space Station orbits around the Earth at a constant speed. Your teacher will give you either a double number line or a table that represents this situation. Your partner will get the other representation.

Roscosmos/NASA

1. Complete the parts of your representation that you can figure out for sure.

2. Share information with your partner and use the information that your partner shares to complete your representation.

3. What is the speed of the International Space Station?

4. Place the two completed representations side by side. Discuss with your partner some ways in which they are the same and some ways in which they are different.

5. Record at least one way that they are the same and one way they are different.

NAME ______________________ DATE ____________ PERIOD ____________

Are you ready for more?

Earth's circumference is about 40,000 kilometers and the orbit of the International Space Station is just a bit more than this. About how long does it take for the International Space Station to orbit Earth?

Summary

Tables and Double Number Line Diagrams

On a double number line diagram, we put labels in front of each line to tell what the numbers represent. On a table, we put labels at the top of each column to tell what the numbers represent.

Here are two different ways we can represent the situation: "A snail is moving at a constant speed down a sidewalk, traveling 6 centimeters per minute."

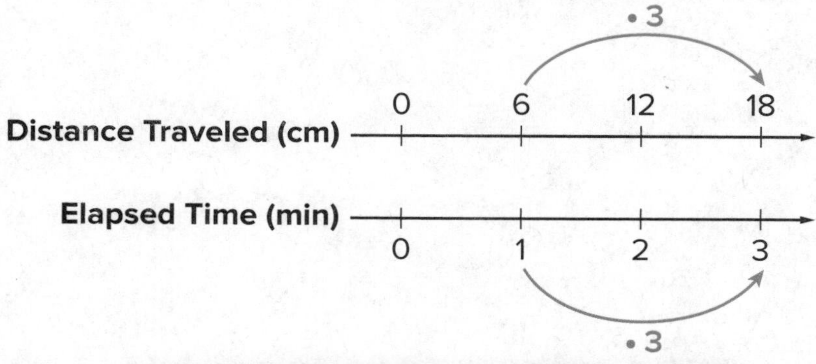

Distance Traveled (cm)	Elapsed Time (min)
12	2
6	1
60	10
18	3

• 3 • 3

Both double number lines and tables can help us use multiplication to make equivalent ratios, but there is an important difference between the two representations.

- On a double number line, the numbers on each line are listed in order.
- With a table, you can write the ratios in any order. For this reason, sometimes a table is easier to use to solve a problem.

For example, what if we wanted to know how far the snail travels in 10 minutes? Notice that 60 centimeters in 10 minutes is shown on the table, but there is not enough room for this information on the double number line.

Practice

Tables and Double Number Line Diagrams

1. The double number line shows how much water and how much lemonade powder to mix to make different amounts of lemonade.

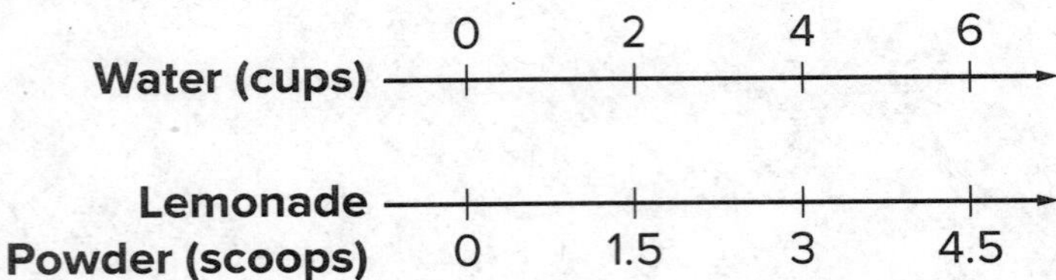

In the space at the right, make a table that represents the same situation.

2. A bread recipe uses 3 tablespoons of olive oil for every 2 cloves of crushed garlic.

   a. Complete the table to show different-sized batches of bread that taste the same as the recipe.

Olive Oil (tablespoons)	Crushed Garlic (cloves)
3	2
1	
2	
5	
10	

   b. Draw a double number line that represents the same situation.

   c. Which representation do you think works better in this situation? Explain why.

NAME ______________________ DATE ______________ PERIOD ____________

3. Clare travels at a constant speed, as shown on the double number line.

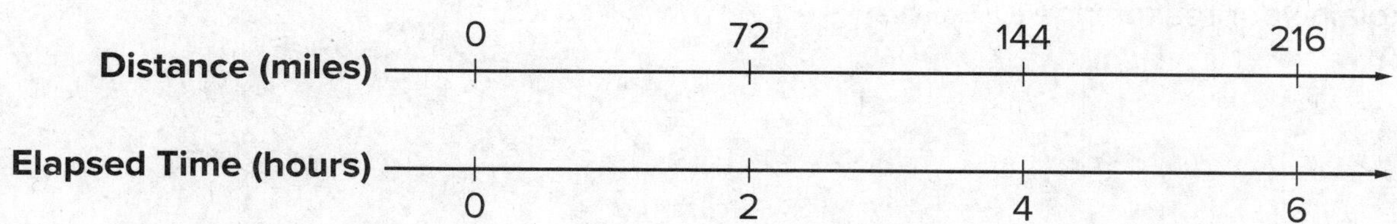

At this rate, how far does she travel in each of these intervals of time? Explain or show your reasoning. If you get stuck, consider using a table.

a. 1 hour

b. 3 hours

c. 6.5 hours

4. Lin and Diego travel in cars on the highway at constant speeds. In each case, decide who was traveling faster. Explain how you know. **(Lesson 2-9)**

a. During the first half hour, Lin travels 23 miles while Diego travels 25 miles.

b. After stopping for lunch, they travel at different speeds. To travel the next 60 miles, it takes Lin 65 minutes and it takes Diego 70 minutes.

5. A sports drink recipe calls for $\frac{5}{3}$ tablespoons of powdered drink mix for every 12 ounces of water. How many batches can you make with 5 tablespoons of drink mix and 36 ounces of water? Explain your reasoning. **(Lesson 2-3)**

6. In this cube, each small square has side length 1 unit. **(Lesson 1-18)**

   a. What is the surface area of this cube?

   b. What is the volume of this cube?

Lesson 2-14

Solving Equivalent Ratio Problems

NAME ______________________ DATE ______________ PERIOD ____________

Learning Goal Let's practice getting information from our partner.

Warm Up

14.1 What Do You Want to Know?

Consider the problem. A red car and a blue car enter the highway at the same time and travel at a constant speed. How far apart are they after 4 hours?

What information would you need to solve the problem?

Activity

14.2 Info Gap: Hot Chocolate and Potatoes

Your teacher will give you either a *problem card* or a *data card*. Do not show or read your card to your partner.

If your teacher gives you the *problem card*:	If your teacher gives you the *data card:*
1. Silently read your card and think about what information you need to be able to answer the question. **2.** Ask your partner for the specific information that you need. **3.** Explain how you are using the information to solve the problem. Continue to ask questions until you have enough information to solve the problem. **4.** Share the *problem card* and solve the problem independently. **5.** Read the *data card* and discuss your reasoning.	**1.** Silently read your card. **2.** Ask your partner "*What specific information do you need*?" and wait for them to ask for information. If your partner asks for information that is not on the card, do not do the calculations for them. Tell them you don't have that information. **3.** Before sharing the information, ask, "*Why do you need that information*?" Listen to your partner's reasoning and ask clarifying questions. **4.** Read the *problem card* and solve the problem independently. **5.** Share the *data card* and discuss your reasoning.

Pause here so your teacher can review your work. Ask your teacher for a new set of cards and repeat the activity, trading roles with your partner.

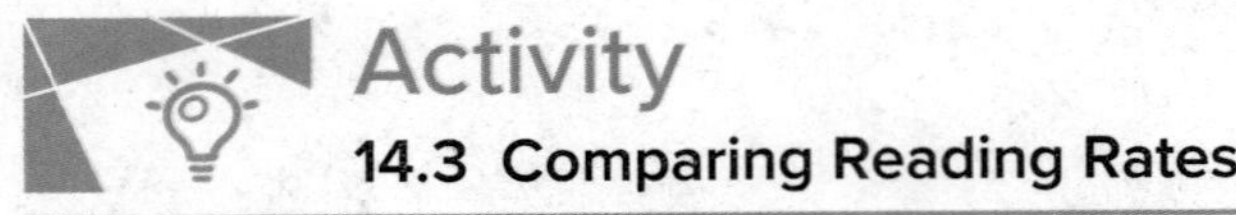

Activity

14.3 Comparing Reading Rates

- Lin read the first 54 pages from a 270-page book in the last 3 days.
- Diego read the first 100 pages from a 320-page book in the last 4 days.
- Elena read the first 160 pages from a 480-page book in the last 5 days.

If they continue to read every day at these rates, who will finish first, second, and third? Explain or show your reasoning.

Are you ready for more?

The ratio of cats to dogs in a room is 2 : 3. Five more cats enter the room, and then the ratio of cats to dogs is 9 : 11. How many cats and dogs were in the room to begin with?

NAME ______________________ DATE ____________ PERIOD ____________

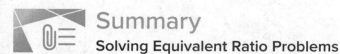

Summary
Solving Equivalent Ratio Problems

To solve problems about something happening at the same rate, we often need:

- Two pieces of information that allow us to write a ratio that describes the situation.
- A third piece of information that gives us one number of an equivalent ratio. Solving the problem often involves finding the other number in the equivalent ratio.

Suppose we are making a large batch of fizzy juice and the recipe says, “Mix 5 cups of cranberry juice with 2 cups of soda water.” We know that the ratio of cranberry juice to soda water is 5 : 2, and that we need 2.5 cups of cranberry juice per cup of soda water.

We still need to know something about the size of the large batch. If we use 16 cups of soda water, what number goes with 16 to make a ratio that is equivalent to 5 : 2?

To make this large batch taste the same as the original recipe, we would need to use 40 cups of cranberry juice.

Cranberry Juice (cups)	Soda Water (cups)
5	2
2.5	1
40	16

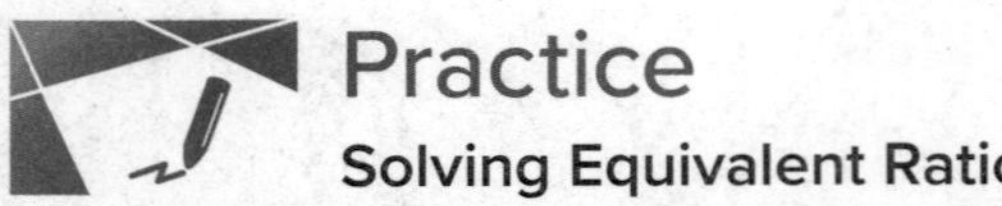

Practice

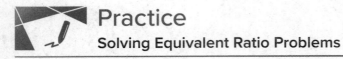

Solving Equivalent Ratio Problems

1. A chef is making pickles. He needs 15 gallons of vinegar. The store sells 2 gallons of vinegar for $3.00 and allows customers to buy any amount of vinegar. Decide whether each of the following ratios correctly represents the price of vinegar.

   a. 4 gallons to $3.00

   b. 1 gallon to $1.50

   c. 30 gallons to $45.00

   d. $2.00 to 30 gallons

   e. $1.00 to $\frac{2}{3}$ gallon

2. A caterer needs to buy 21 pounds of pasta to cater a wedding. At a local store, 8 pounds of pasta cost $12. How much will the caterer pay for the pasta there?

   a. Write a ratio for the given information about the cost of pasta.

   b. Would it be more helpful to write an equivalent ratio with 1 pound of pasta as one of the numbers, or with $1 as one of the numbers? Explain your reasoning, and then write that equivalent ratio.

   c. Find the answer and explain or show your reasoning.

NAME ______________________ DATE ______________ PERIOD ____________

3. Lin is reading a 47-page book. She read the first 20 pages in 35 minutes.

   a. If she continues to read at the same rate, will she be able to complete this book in under 1 hour?

   b. If so, how much time will she have left? If not, how much more time is needed? Explain or show your reasoning.

4. Diego can type 140 words in 4 minutes.

   a. At this rate, how long will it take him to type 385 words?

   b. How many words can he type in 15 minutes?

   If you get stuck, consider creating a table.

5. A train that travels 30 miles in $\frac{1}{3}$ hour at a constant speed is going faster than a train that travels 20 miles in $\frac{1}{2}$ hour at a constant speed. Explain or show why. **(Lesson 2-10)**

6. Find the surface area of the polyhedron that can be assembled from this net. Show your reasoning. **(Lesson 1-14)**

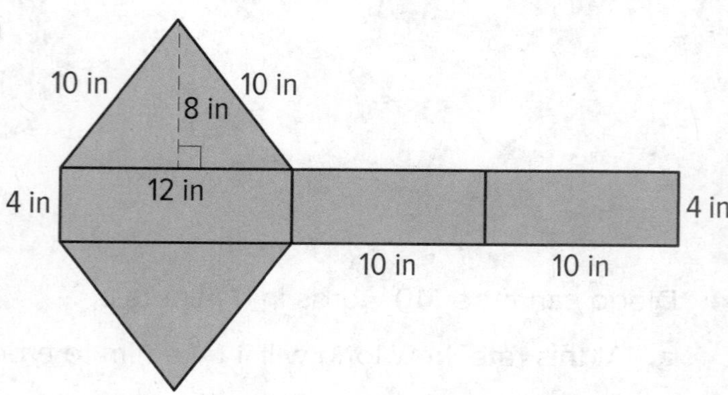

Lesson 2-15

Part-Part-Whole Ratios

NAME ______________________ DATE ____________ PERIOD ____________

Learning Goal Let's look at situations where you can add the quantities in a ratio together.

Warm Up

15.1 True or False: Multiplying by a Unit Fraction

True or false?

1. $\frac{1}{5} \cdot 45 = \frac{45}{5}$
2. $\frac{1}{5} \cdot 20 = \frac{1}{4} \cdot 24$
3. $42 \cdot \frac{1}{6} = \frac{1}{6} \cdot 42$
4. $486 \cdot \frac{1}{12} = \frac{480}{12} + \frac{6}{12}$

Activity

15.2 Cubes of Paint

A recipe for maroon paint says, "Mix 5 ml of red paint with 3 ml of blue paint."

1. Use snap cubes to represent the amounts of red and blue paint in the recipe. Then, draw a sketch of your snap-cube representation of the maroon paint.
   a. What amount does each cube represent?
   b. How many milliliters of maroon paint will there be?
2. a. Suppose each cube represents 2 ml. How much of each color paint is there?

   Red: __________ ml Blue: __________ ml Maroon: __________ ml

   b. Suppose each cube represents 5 ml. How much of each color paint is there?

   Red: __________ ml Blue: __________ ml Maroon: __________ ml

3. a. Suppose you need 80 ml of maroon paint. How much red and blue paint would you mix? Be prepared to explain your reasoning.

   Red: __________ ml Blue: __________ ml Maroon: 80 ml

   b. If the original recipe is for one batch of maroon paint, how many batches are in 80 ml of maroon paint?

Activity
15.3 Sneakers, Chicken, and Fruit Juice

Solve each of the following problems and show your thinking. If you get stuck, consider drawing a **tape diagram** to represent the situation.

1. The ratio of students wearing sneakers to those wearing boots is 5 to 6. If there are 33 students in the class, and all of them are wearing either sneakers or boots, how many of them are wearing sneakers?

2. A recipe for chicken marinade says, "Mix 3 parts oil with 2 parts soy sauce and 1 part orange juice." If you need 42 cups of marinade in all, how much of each ingredient should you use?

3. Elena makes fruit punch by mixing 4 parts cranberry juice to 3 parts apple juice to 2 parts grape juice. If one batch of fruit punch includes 30 cups of apple juice, how large is this batch of fruit punch?

NAME ______________________ DATE ____________ PERIOD ____________

Are you ready for more?

Using the recipe from earlier, how much fruit punch can you make if you have 50 cups of cranberry juice, 40 cups of apple juice, and 30 cups of grape juice?

Activity

15.4 Invent Your Own Ratio Problem

1. Invent another ratio problem that can be solved with a tape diagram and solve it. If you get stuck, consider looking back at the problems you solved in the earlier activity.

2. Create a visual display that includes:
   - The new problem that you wrote, without the solution.
   - Enough work space for someone to show a solution.

3. Trade your display with another group and solve each other's problem. Include a tape diagram as part of your solution. Be prepared to share the solution with the class.

4. When the solution to the problem you invented is being shared by another group, check their answer for accuracy.

Summary

Part-Part-Whole Ratios

A **tape diagram** is another way to represent a ratio.

All the parts of the diagram that are the same size have the same value.

For example, this tape diagram represents the ratio of ducks to swans in a pond, which is 4 : 5.

The first tape represents the number of ducks. It has 4 parts.

The second tape represents the number of swans. It has 5 parts.

There are 9 parts in all, because $4 + 5 = 9$.

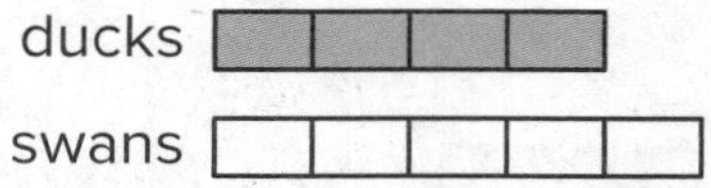

ducks
swans

Suppose we know there are 18 of these birds in the pond, and we want to know how many are ducks.

The 9 equal parts on the diagram need to represent 18 birds in all. This means that each part of the tape diagram represents 2 birds, because $18 \div 9 = 2$.

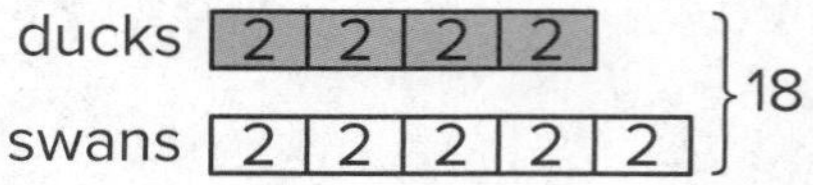

ducks 2 2 2 2
swans 2 2 2 2 2
18

There are 4 parts of the tape representing ducks, and $4 \cdot 2 = 8$, so there are 8 ducks in the pond.

Glossary

tape diagram

NAME ________________ DATE ________ PERIOD ________

Practice
Part-Part-Whole Ratios

1. Here is a tape diagram representing the ratio of red paint to yellow paint in a mixture of orange paint.

cups of red paint	3	3	3
cups of yellow paint	3	3	

   a. What is the ratio of yellow paint to red paint?

   b. How many total cups of orange paint will this mixture yield?

2. At the kennel, the ratio of cats to dogs is 4 : 5. There are 27 animals in all. Here is a tape diagram representing this ratio.

number of cats [ ][ ][ ][ ]
number of dogs [ ][ ][ ][ ][ ]

   a. What is the value of each small rectangle?

   b. How many dogs are at the kennel?

   c. How many cats are at the kennel?

3. Last month, there were 4 sunny days for every rainy day. If there were 30 days in the month, how many days were rainy? Explain your reasoning. If you get stuck, consider using a tape diagram.

4. Noah entered a 100-mile bike race. He knows he can ride 32 miles in 160 minutes. At this rate, how long will it take him to finish the race? Use each table to find the answer. **(Lesson 2-12)**

Table A:

Distance (miles)	Elapsed Time (minutes)
32	160
1	
100	

Table B:

Distance (miles)	Elapsed Time (minutes)
32	160
96	
4	
100	

Next, explain which table you think works better in finding the answer.

5. A cashier worked an 8-hour day and earned $58.00. The double number line shows the amount she earned for working different numbers of hours. For each question, explain your reasoning. **(Lesson 2-13)**

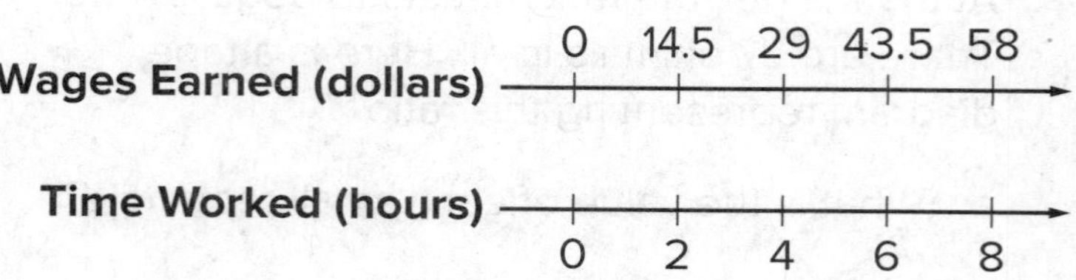

a. How much does the cashier earn per hour?

b. How much does the cashier earn if she works 3 hours?

6. A grocery store sells bags of oranges in two different sizes. **(Lesson 2-10)**

- The 3-pound bags of oranges cost $4.
- The 8-pound bags of oranges for $9.

Which oranges cost less per pound? Explain or show your reasoning.

Lesson 2-16

Solving More Ratio Problems

NAME ______________________ DATE ____________ PERIOD ____________

Learning Goal Let's compare all our strategies for solving ratio problems.

Warm Up

16.1 You Tell the Story

Describe a situation with two quantities that this tape diagram could represent.

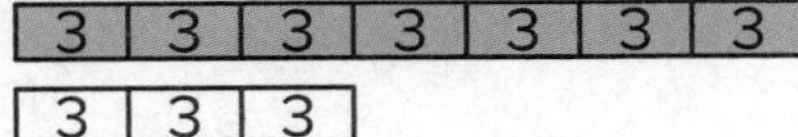

Activity

16.2 A Trip to the Aquarium

Consider the problem: A teacher is planning a class trip to the aquarium.
The aquarium requires 2 chaperones for every 15 students.
The teacher plans accordingly and orders a total of 85 tickets.
How many tickets are for chaperones, and how many are for students?

1. Solve this problem in *one* of three ways:
   - Use a triple number line.

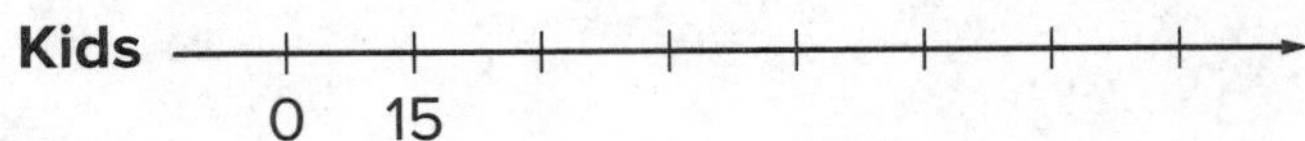

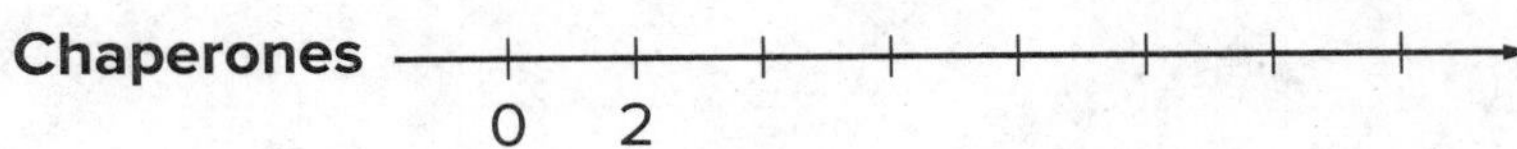

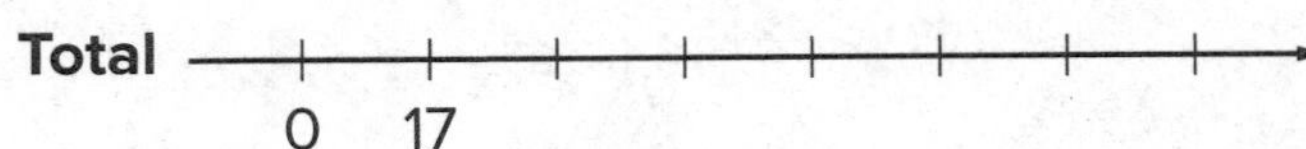

- Use a table. (Fill rows as needed.)

Kids	Chaperones	Total
15	2	17

- Use a tape diagram.

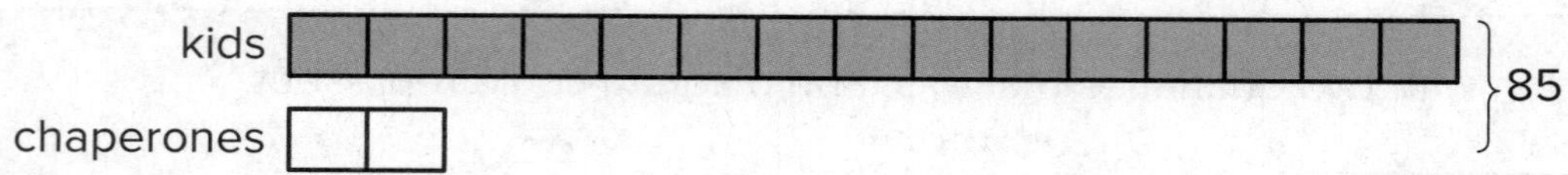

2. After your class discusses all three strategies, which do you prefer for this problem and why?

Are you ready for more?

Use the digits 1 through 9 to create three equivalent ratios. Use each digit only one time.

☐ : ☐ is equivalent to ☐☐ : ☐ and ☐☐ : ☐☐

NAME ______________________ DATE ____________ PERIOD ____________

Activity

16.3 Salad Dressing and Moving Boxes

Solve each problem and show your thinking. Organize it so it can be followed by others. If you get stuck, consider drawing a double number line, table, or tape diagram on a separate sheet of paper.

1. A recipe for salad dressing calls for 4 parts oil for every 3 parts vinegar. How much oil should you use to make a total of 28 teaspoons of dressing?

2. Andre and Han are moving boxes. Andre can move 4 boxes every half hour. Han can move 5 boxes every half hour. How long will it take Andre and Han to move all 72 boxes?

Summary

Solving More Ratio Problems

When solving a problem involving equivalent ratios, it is often helpful to use a diagram. Any diagram is fine as long as it correctly shows the mathematics and you can explain it.

Let's compare three different ways to solve the same problem: The ratio of adults to kids in a school is 2 : 7. If there is a total of 180 people, how many of them are adults?

- *Tape diagrams* are especially useful for this type of problem because both parts of the ratio have the same units ("number of people") and we can see the total number of parts.

number of adults
number of kids

This tape diagram has 9 equal parts, and they need to represent 180 people total. That means each part represents $180 \div 9$, or 20 people.

number of adults | 20 | 20 |
number of kids | 20 | 20 | 20 | 20 | 20 | 20 | 20 |
180

Two parts of the tape diagram represent adults.
There are 40 adults in the school because $2 \cdot 20 = 40$.

- *Double or triple number lines* are useful when we want to see how far apart the numbers are from one another. They are harder to use with very big or very small numbers, but they could support our reasoning.

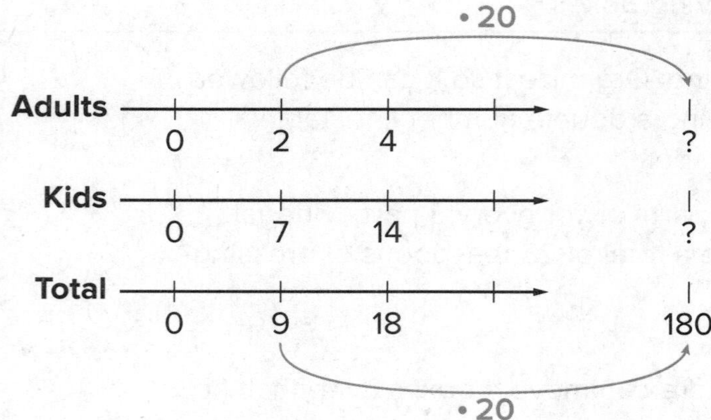

- *Tables* are especially useful when the problem has very large or very small numbers.

Adults	Kids	Total
2	?	9
?		180

• 20 (Adults column) • 20 (Total column)

We ask ourselves, "9 times what is 180?" The answer is 20.
Next, we multiply 2 by 20 to get the total number of adults in the school.

Another reason to make diagrams is to communicate our thinking to others. Here are some good habits when making diagrams.

- Label each part of the diagram with what it represents.
- Label important amounts.
- Make sure you read what the question is asking and answer it.
- Make sure you make the answer easy to find.
- Include units in your answer. For example, write "4 cups" instead of just "4."
- Double check that your ratio language is correct and matches your diagram.

NAME ______________________ DATE ____________ PERIOD ____________

Practice

Solving More Ratio Problems

1. Describe a situation that could be represented with this tape diagram.

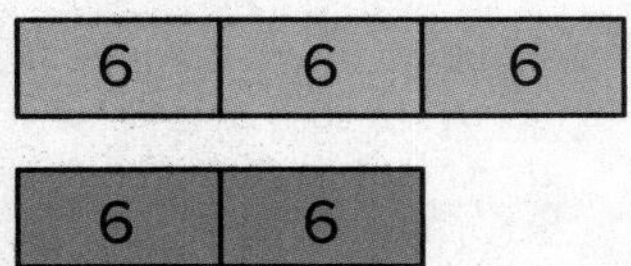

2. There are some nickels, dimes, and quarters in a large piggy bank. For every 2 nickels there are 3 dimes. For every 2 dimes there are 5 quarters. There are 500 coins total.

   a. How many nickels, dimes, and quarters are in the piggy bank? Explain your reasoning.

   b. How much are the coins in the piggy bank worth?

3. Two horses start a race at the same time. Horse A gallops at a steady rate of 32 feet per second and Horse B gallops at a steady rate of 28 feet per second. After 5 seconds, how much farther will Horse A have traveled? Explain or show your reasoning.

4. Andre paid \$13 for 3 books. Diego bought 12 books priced at the same rate. How much did Diego pay for the 12 books? Explain your reasoning. **(Lesson 2-10)**

5. Which polyhedron can be assembled from this net? **(Lesson 1-15)**

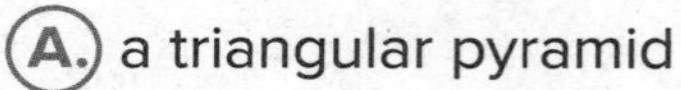
A. a triangular pyramid

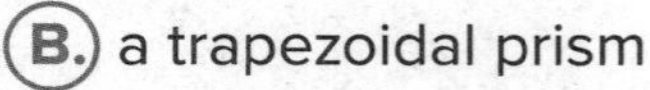
B. a trapezoidal prism

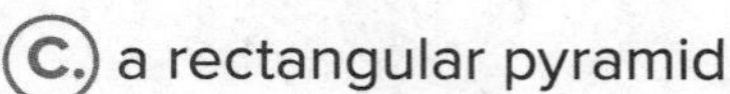
C. a rectangular pyramid

D. a triangular prism

6. Find the area of the triangle. Show your reasoning. If you get stuck, consider drawing a rectangle around the triangle. **(Lesson 1-10)**

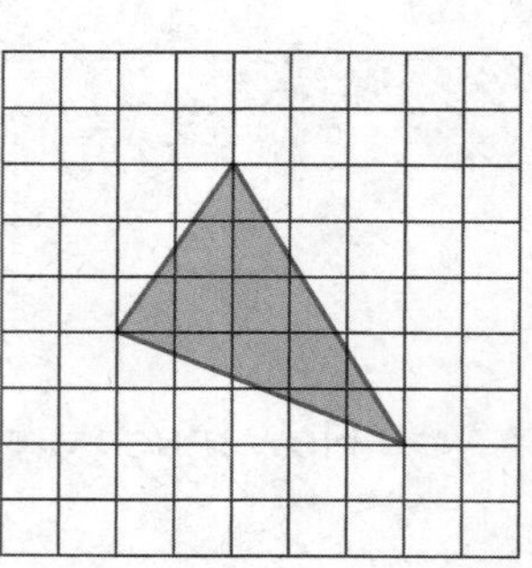

Lesson 2-17

A Fermi Problem

NAME _______________ DATE _______________ PERIOD _______________

Learning Goal Let's solve a Fermi problem.

Warm Up

17.1 Fix It!

Andre likes a hot cocoa recipe with 1 cup of milk and 3 tablespoons of cocoa. He poured 1 cup of milk but accidentally added 5 tablespoons of cocoa.

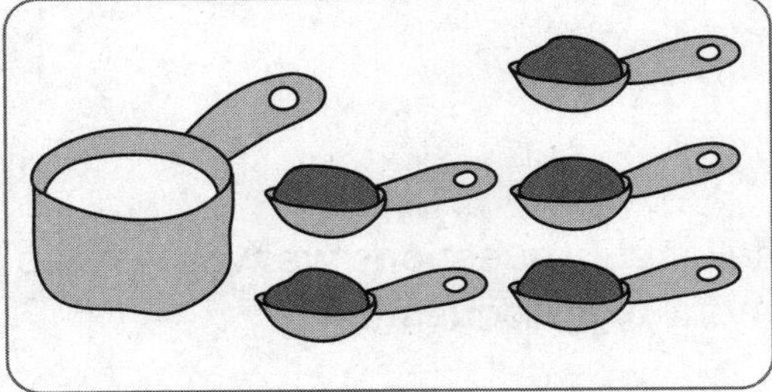

1. How can you fix Andre's mistake and make his hot cocoa taste like the recipe?

2. Explain how you know your adjustment will make Andre's hot cocoa taste the same as the one in the recipe.

Activity

17.2 Who Was Fermi?

1. Record the Fermi question that your class will explore together.

2. Make an estimate of the answer. If making an estimate is too hard, consider writing down a number that would definitely be too low and another number that would definitely be too high.

3. What are some smaller sub-questions we would need to figure out to reasonably answer our bigger question?

4. Think about how the smaller questions above should be organized to answer the big question.
   - Label each smaller question with a number to show the order in which they should be answered.
   - If you notice a gap in the set of sub-questions (i.e., there is an unlisted question that would need to be answered before the next one could be tackled), write another question to fill the gap.

NAME ______________________ DATE ____________ PERIOD ________

Activity

17.3 Researching Your Own Fermi Problem

1. Brainstorm at least five Fermi problems that you want to research and solve. If you get stuck, consider starting with...
   - "How much would it cost to . . .?" or
   - "How long would it take to . . .?"

2. Pause here so your teacher can review your questions and approve one of them.

3. Use the graphic organizer to break your problem down into sub-questions.

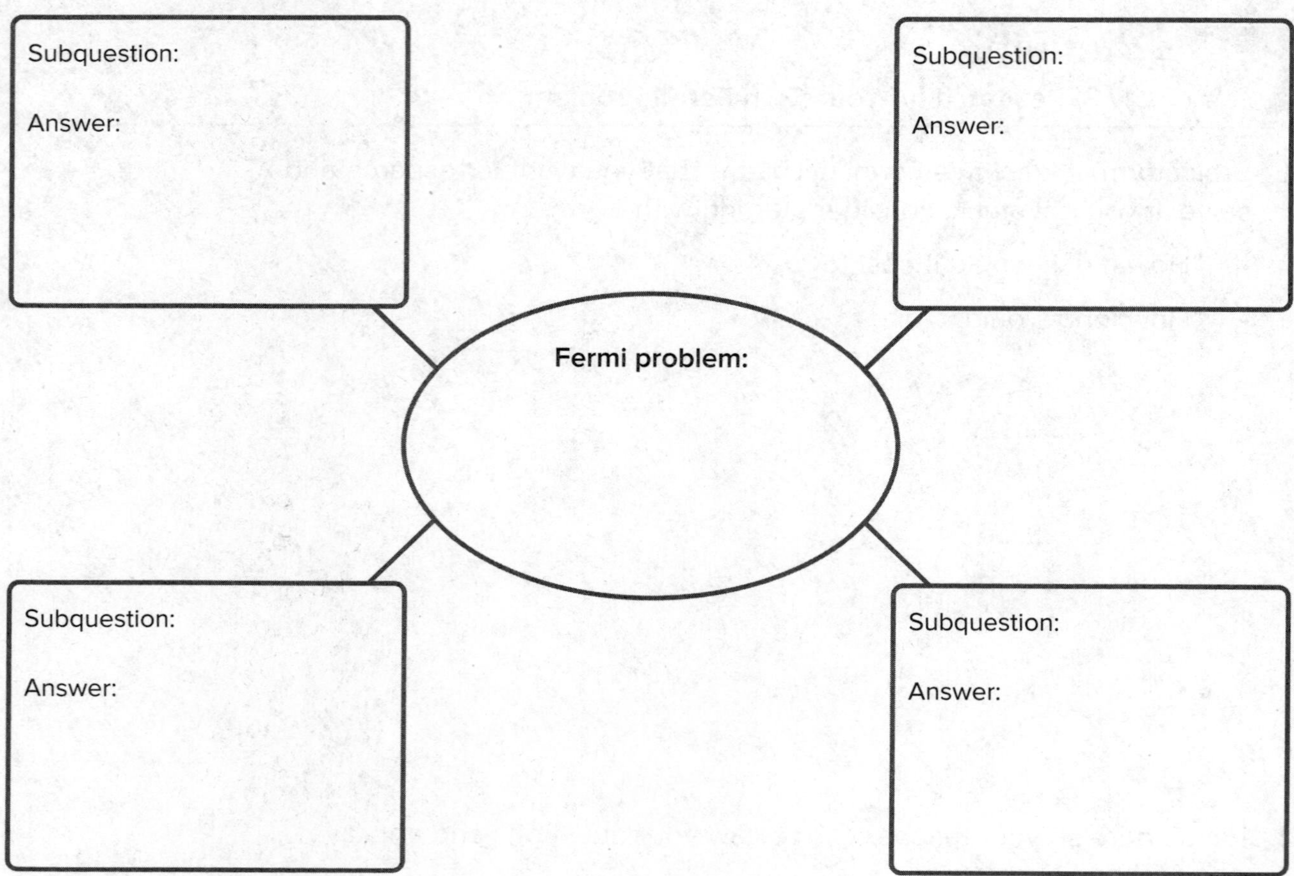

4. Find the information you need to get closer to answering your question. Measure, make estimates, and perform any necessary calculations. If you get stuck, consider using tables or double number line diagrams.

5. Create a visual display that includes your Fermi problem and your solution. Organize your thinking so it can be followed by others.

Learning Targets

Lesson	Learning Target(s)
2-1 Introducing Ratios and Ratio Language	• I can write or say a sentence that describes a ratio. • I know how to say words and numbers in the correct order to accurately describe the ratio.
2-2 Representing Ratios with Diagrams	• I can draw a diagram that represents a ratio and explain what the diagram means. • I include labels when I draw a diagram representing a ratio, so that the meaning of the diagram is clear.
2-3 Recipes	• I can explain the meaning of equivalent ratios using a recipe as an example. • I can use a diagram to represent a recipe, a double batch, and a triple batch of a recipe. • I know what it means to double or triple a recipe.

(continued on the next page)

(continued from the previous page)

Lesson	Learning Target(s)
2-4 Color Mixtures	• I can explain the meaning of equivalent ratios using a color mixture as an example. • I can use a diagram to represent a recipe, a double batch, and a triple batch of a color mixture. • I know what it means to double or triple a color mixture.
2-5 Defining Equivalent Ratios	• If I have a ratio, I can create a new ratio that is equivalent to it. • If I have two ratios, I can decide whether they are equivalent to each other.
2-6 Introducing Double Number Line Diagrams	• I can label a double number line diagram to represent batches of a recipe or color mixture. • When I have a double number line that represents a situation, I can explain what it means.
2-7 Creating Double Number Line Diagrams	• I can create a double number line diagram and correctly place and label tick marks to represent equivalent ratios. • I can explain what the word *per* means.

Lesson	Learning Target(s)
2-8 How Much for One?	• I can choose and create diagrams to help me reason about prices. • I can explain what the phrase "at this rate" means, using prices as an example. • If I know the price of multiple things, I can find the price per thing.
2-9 Constant Speed	• I can choose and create diagrams to help me reason about constant speed. • If I know an object is moving at a constant speed, and I know two of these things: the distance it travels, the amount of time it takes, and its speed, I can find the other thing.
2-10 Comparing Situations by Examining Ratios	• I can decide whether or not two situations are happening at the same rate. • I can explain what it means when two situations happen at the same rate. • I know some examples of situations where things can happen at the same rate.

(*continued on the next page*)

(continued from the previous page)

Lesson	Learning Target(s)
2-11 Representing Ratios with Tables	• If I am looking at a table of values, I know where the rows are and where the columns are. • When I see a table representing a set of equivalent ratios, I can come up with numbers to make a new row. • When I see a table representing a set of equivalent ratios, I can explain what the numbers mean.
2-12 Navigating a Table of Equivalent Ratios	• I can solve problems about situations happening at the same rate by using a table and finding a "1" row. • I can use a table of equivalent ratios to solve problems about unit price.
2-13 Tables and Double Number Line Diagrams	• I can create a table that represents a set of equivalent ratios. • I can explain why sometimes a table is easier to use than a double number line to solve problems involving equivalent ratios. • I include column labels when I create a table, so that the meaning of the numbers is clear.
2-14 Solving Equivalent Ratio Problems	• I can decide what information I need to know to be able to solve problems about situations happening at the same rate. • I can explain my reasoning using diagrams that I choose.

Lesson	Learning Target(s)
2-15 Part-Part-Whole Ratios	• I can create tape diagrams to help me reason about problems involving a ratio and a total amount. • I can solve problems when I know a ratio and a total amount.
2-16 Solving More Ratio Problems	• I can choose and create diagrams to help think through my solution. • I can solve all kinds of problems about equivalent ratios. • I can use diagrams to help someone else understand why my solution makes sense.
2-17 A Fermi Problem	• I can apply what I have learned about ratios and rates to solve a more complicated problem. • I can decide what information I need to know to be able to solve a real-world problem about ratios and rates.

(continued on the next page)

(continued from the previous page)

Notes:

Unit 3

Unit Rates and Percentages

Easy Production/Image Source

At the end of this unit, you'll apply what you learned about unit rates to find how much it will cost to paint all the walls in a room.

Topics

- Units of Measurement
- Unit Conversion
- Rates
- Percentages
- Let's Put It to Work

Unit 3

Unit Rates and Percentages

Lesson 3-1

The Burj Khalifa

NAME ______________________ DATE ____________ PERIOD ____________

Learning Goal Let's investigate the Burj Khalifa building.

Warm Up

1.1 Estimating Height

Use the picture to estimate the height of Hyperion, the tallest known tree.

Activity

1.2 Window Washing

A window-washing crew can finish 15 windows in 18 minutes.

If this crew was assigned to wash all the windows on the outside of the Burj Khalifa, how long will the crew be washing at this rate?

Activity

1.3 Climbing the Burj Khalifa

In 2011, a professional climber scaled the outside of the Burj Khalifa, making it all the way to 828 meters (the highest point on which a person can stand) in 6 hours.

Assuming they climbed at the same rate the whole way:

1. How far did they climb in the first 2 hours?

2. How far did they climb in 5 hours?

3. How far did they climb in the final 15 minutes?

Are you ready for more?

Have you ever seen videos of astronauts on the Moon jumping really high?

An object on the Moon weighs less than it does on Earth because the Moon has much less mass than Earth.

1. A person who weighs 100 pounds on Earth weighs 16.5 pounds on the Moon. If a boy weighs 60 pounds on Earth, how much does he weigh on the Moon?

2. Every 100 pounds on Earth are the equivalent to 38 pounds on Mars. If the same boy travels to Mars, how much would he weigh there?

NAME ______________________ DATE ______________ PERIOD ____________

Summary

The Burj Khalifa

There are many real-world situations in which something keeps happening at the same rate.

For example:

- a bus stop that is serviced by 4 buses per hour
- a washing machine that takes 45 minutes per load of laundry
- a school cafeteria that serves 15 students per minute

In situations like these, we can use equivalent ratios to predict how long it will take for something to happen some number of times, or how many times it will happen in a particular length of time.

For example, how long will it take the school cafeteria to serve 600 students?

The table shows that it will take the cafeteria 40 minutes to serve 600 students.

Number of Students	Time in Minutes
15	1
60	4
600	40

How many students can the cafeteria serve in 1 hour?

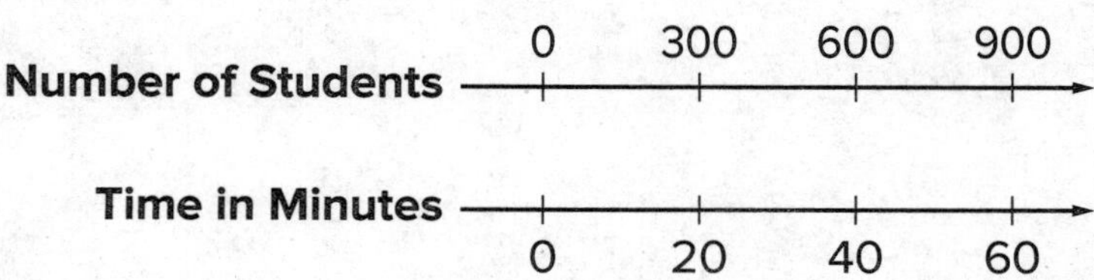

The double number line shows that the cafeteria can serve 900 students in 1 hour.

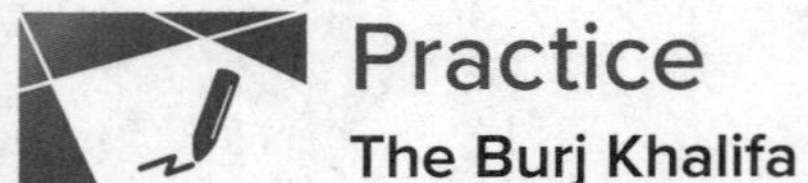

Practice

The Burj Khalifa

1. An elevator travels 310 feet in 10 seconds. At that speed, how far can this elevator travel in 12 seconds? Explain your reasoning.

2. Han earns $33.00 for babysitting 4 hours. At this rate, how much will he earn if he babysits for 7 hours? Explain your reasoning.

3. The cost of 5 cans of dog food is $4.35. At this price, how much do 11 cans of dog food cost? Explain your reasoning.

NAME ______________________ DATE ____________ PERIOD ________

4. A restaurant has 26 tables in its dining room. It takes the waitstaff 10 minutes to clear and set 4 tables. At this rate, how long will it take the waitstaff to clear and set all the tables in the dining room? Explain or show your reasoning.

5. A sandwich shop serves 4 ounces of meat and 3 ounces of cheese on each sandwich. After making sandwiches for an hour, the shop owner has used 91 combined ounces of meat and cheese. **(Lesson 2-16)**

a. How many combined ounces of meat and cheese are used on each sandwich?

b. How many sandwiches were made in the hour?

c. How many ounces of meat were used?

d. How many ounces of cheese were used?

6. Here is a flower made up of yellow hexagons, red trapezoids, and green triangles. **(Lesson 2-14)**

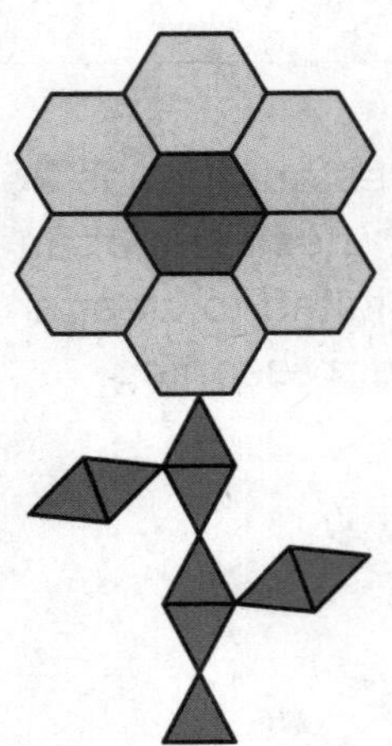

   a. How many copies of this flower pattern could you build if you had 30 yellow hexagons, 50 red trapezoids, and 60 green triangles?

   b. Of which shape would you have the most left over?

7. Match each quantity in the first list with an appropriate unit of measurement from the second list. **(Lesson 1-16)**

Quantity	Unit of Measurement
a. the perimeter of a baseball field	centimeters (cm)
b. the area of a bed sheet	cubic feet (cu ft)
c. the volume of a refrigerator	cubic kilometers (cu km)
d. the surface area of a tissue box	meters (m)
e. the length of a spaghetti noodle	square feet (sq ft)
f. the volume of a large lake	square inches (sq in)
g. the surface area of the moon	square kilometers (sq km)

Lesson 3-2

Anchoring Units of Measurement

NAME ______________________ DATE ______________ PERIOD ______________

Learning Goal Let's see how big different things are.

Warm Up

2.1 Estimating Volume

Estimate the volume of the tiny salt shaker.

Activity

2.2 Cutting String

Your teacher will assign you one of the following lengths: 1 centimeter, 1 foot, 1 inch, 1 meter, or 1 yard.

Estimate and cut a piece of string as close to your assigned length as you can without using a measurement tool.

Activity

2.3 Card Sort: Measurements

Your teacher will give you some cards with the names of different units of measurement and other cards with pictures of objects.

1. Sort the units of measurement into groups based on the attribute they measure. Pause here so your teacher can review your groups.
2. Match each picture card that has "L" in the top right corner with the closest unit to the length of the object.
3. Match each picture card that has "V" in the top right corner with the closest unit to the volume of the object.

3. Match each picture card that has "V" in the top right corner with the closest unit to the volume of the object.
4. Match each picture card that has "WM" in the top right corner with the closest unit to the weight or mass of the object.

Your teacher will assign you a new group to discuss how you matched the objects. If you disagree, work to reach an agreement.

Summary
Anchoring Units of Measurement

We can use everyday objects to estimate standard units of measurement.

For units of length:
- 1 millimeter is about the thickness of a dime.
- 1 centimeter is about the width of a pinky finger.
- 1 inch is about the length from the tip of your thumb to the first knuckle.
- 1 foot is the length of a football.
- 1 yard is about the length of a baseball bat.
- 1 meter is about the length of a baseball bat and ball.
- 1 kilometer is about the distance someone walks in ten minutes.
- 1 mile is about the distance someone runs in ten minutes.

For units of volume:
- 1 milliliter is about the volume of a raindrop.
- 1 cup is about the volume of a school milk carton.
- 1 quart is about the volume of a large sports drink bottle.
- 1 liter is about the volume of a reusable water bottle.
- 1 gallon is about the volume of a large milk jug.

For units of weight and mass:
- 1 gram is about the mass of a raisin.
- 1 ounce is about the weight of a slice of bread.
- 1 pound is about the weight of a loaf of bread.
- 1 kilogram is about the mass of a textbook.
- 1 ton is about the weight of a small car.

NAME ______________________ DATE ____________ PERIOD ____________

Practice

Anchoring Units of Measurement

1. Select the unit from the list that you would use to measure each object.

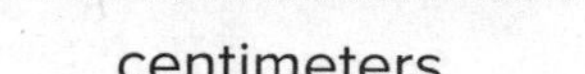

Quantity	Unit of Measurement
a. the length of a pencil	centimeters
b. the weight or mass of a pencil	cups
c. the volume of a pencil	feet
d. the weight or mass of a hippopotamus	gallons
e. the length of a hippopotamus	grams
f. the length of a fingernail clipping	inches
g. the weight or mass of a fingernail clipping	kilograms
h. the volume of a sink	kilometers
i. the volume of a bowl	liters
j. the length of a chalkboard or whiteboard	meters
k. the weight or mass of a chalkboard or whiteboard	miles
l. the length of the border between the United States and Canada	milliliters
	millimeters
	ounces
	pounds
	quarts
	tons
	yards

2. When this pet hamster is placed on a digital scale, the scale reads 1.5. What could be the units?

4. Circle the larger unit of measure. Then, determine if the unit measures distance, volume, or weight (or mass).

a. meter or kilometer

b. yard or foot

c. cup or quart

d. pound or ounce

e. liter or milliliter

f. gram or kilogram

4. Elena mixes 5 cups of apple juice with 2 cups of sparkling water to make sparkling apple juice. For a party, she wants to make 35 cups of sparkling apple juice. How much of each ingredient should Elena use? Explain or show your reasoning. **(Lesson 2-15)**

5. Lin bought 3 hats for $22.50. At this rate, how many hats could she buy with $60.00? If you get stuck, consider using the table. **(Lesson 2-12)**

Number of Hats	Price in Dollars

6. Light travels about 180 million kilometers in 10 minutes. How far does it travel in 1 minute? How far does it travel in 1 second? Show your reasoning. **(Lesson 2-9)**

Lesson 3-3

Measuring with Different-Sized Units

NAME ______________________ DATE ______________ PERIOD ______________

Learning Goal Let's measure things.

Warm Up

3.1 Width of a Paper

Your teacher will show you two rods. Does it take more green rods or blue rods lined up end to end to measure the width of a piece of printer paper?

Activity

3.2 Measurement Stations

Station 1

- Each large cube is 1 cubic inch. Count how many cubic inches completely pack the box without gaps.
- Each small cube is 1 cubic centimeter. Each rod is composed of 10 cubic centimeters. Count how many cubic centimeters completely fill the box.

	Cubic Inches	Cubic Centimeters
Volume of the Box		

Station 2

Your teacher showed you a length.

- Use the meter stick to measure the length to the nearest meter.
- Use a ruler to measure the length to the nearest foot.

	Meters	Feet
Length of ______		

Station 3

If not using real water, watch the video online for this activity.

- Count how many times you can fill the quart bottle from the gallon jug.
- Count how many times you can fill the liter bottle from the gallon jug.

	Quarts	Liters
1 Gallon of Water		

Station 4

If not using a real scale, use the digital tool online for this activity.

- Select 2 to 3 different objects to measure on the scale.
- Record the weights in ounces, pounds, grams, and kilograms.

	Object	Ounces	Pounds	Grams	Kilograms
Object 1					
Object 2					
Object 3					

Station 5

- Count how many level teaspoons of salt fill the graduated cylinder to 20 milliliters, 40 milliliters, and 50 milliliters.
- Pour the salt back into the original container.

	Milliliters	Teaspoons
Small Amount of Salt	20	
Medium Amount of Salt	40	
Large Amount of Salt	50	

After you finish all five stations, answer these questions with your group.

1. a. Which is larger, a cubic inch or a cubic centimeter?

b. Did more cubic inches or cubic centimeters fit in the cardboard box? Why?

2. Did it take more feet or meters to measure the indicated length? Why?

NAME ______________________ DATE ____________ PERIOD ____________

3. Which is larger, a quart or a liter? Explain your reasoning.

4. Use the data from Station 4 to put the units of weight and mass in order from smallest to largest. Explain your reasoning.

5. a. About how many teaspoons of salt would it take to fill the graduated cylinder to 100 milliliters?

   b. If you poured 15 teaspoons of salt into an empty graduated cylinder, about how many milliliters would it fill?

   c. How many milliliters per teaspoon are there?

   d. How many teaspoons per milliliter are there?

Are you ready for more?

People in the medical field use metric measurements when working with medicine. For example, a doctor might prescribe medication in 10 mg tablets.

Brainstorm a list of reasons why healthcare workers would do this. Organize your thinking so it can be followed by others.

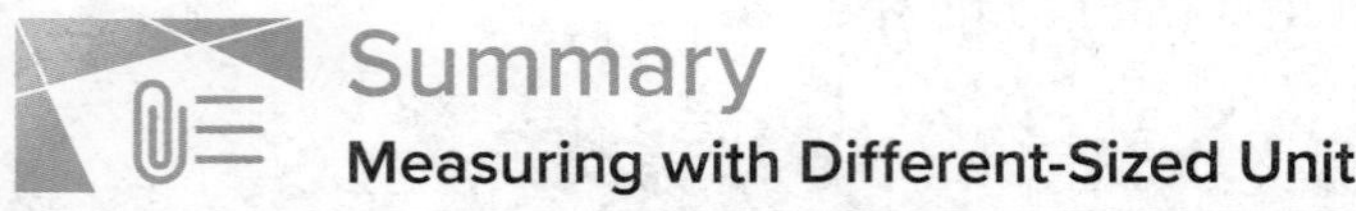

Summary

Measuring with Different-Sized Units

The size of the unit we use to measure something affects the measurement.

If we measure the same quantity with different units, it will take more of the smaller unit and fewer of the larger unit to express the measurement. For example, a room that measures 4 yards in length will measure 12 feet.

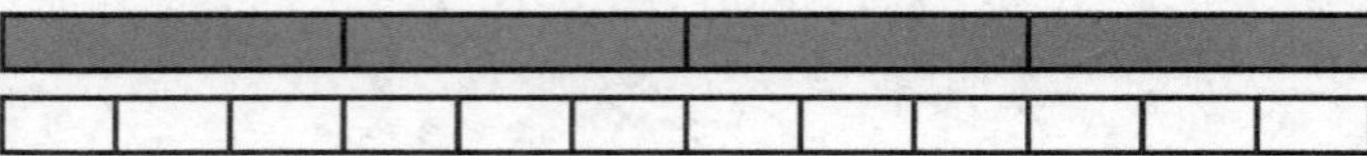

There are 3 feet in a yard, so one foot is $\frac{1}{3}$ of a yard.

- It takes 3 times as many feet to measure the same length as it does with yards.
- It takes $\frac{1}{3}$ as many yards to measure the same length as it does with feet.

NAME ______________________ DATE ____________ PERIOD ____________

Practice

Measuring with Different-Sized Units

1. Decide if each is a measurement of length, area, volume, or weight (or mass). **(Lesson 3-2)**
   a. How many centimeters across a handprint
   b. How many square inches of paper needed to wrap a box
   c. How many gallons of water in a fish tank
   d. How many pounds in a bag of potatoes
   e. How many feet across a swimming pool
   f. How many ounces in a bag of grapes
   g. How many liters in a punch bowl
   h. How many square feet of grass in a lawn

2. Clare says, "This classroom is 11 meters long. A meter is longer than a yard, so if I measure the length of this classroom in yards, I will get less than 11 yards." Do you agree with Clare? Explain your reasoning.

3. Tyler's height is 57 inches. What could be his height in centimeters?

   (A.) 22.4 (C.) 144.8

   (B.) 57 (D.) 3,551

4. A large soup pot holds 20 quarts. What could be its volume in liters?

   (A.) 7.57 (C.) 21

   (B.) 19 (D.) 75.7

5. Clare wants to mail a package that weighs $4\frac{1}{2}$ pounds. What could this weight be in kilograms?

(A.) 2.04

(B.) 4.5

(C.) 9.92

(D.) 4,500

6. Noah bought 15 baseball cards for $9.00. Assuming each baseball card costs the same amount, answer the following questions. **(Lesson 2-13)**

   a. At this rate, how much will 30 baseball cards cost? Explain your reasoning.

   b. At this rate, how much will 12 baseball cards cost? Explain your reasoning.

   c. Do you think this information would be better represented using a table or a double number line? Explain your reasoning.

7. Jada traveled 135 miles in 3 hours. Andre traveled 228 miles in 6 hours. Both Jada and Andre traveled at a constant speed. **(Lesson 2-9)**

   a. How far did Jada travel in 1 hour?

   b. How far did Andre travel in 1 hour?

   c. Who traveled faster? Explain or show your reasoning.

Lesson 3-4

Converting Units

NAME ______________________ DATE ____________ PERIOD ____________

Learning Goal Let's convert measurements to different units.

Warm Up

4.1 Number Talk: Fractions of a Number

Find the values mentally.

1. $\frac{1}{4}$ of 32 **2.** $\frac{3}{4}$ of 32 **3.** $\frac{3}{8}$ of 32 **4.** $\frac{3}{8}$ of 64

Activity

4.2 Road Trip

Elena and her mom are on a road trip outside the United States. Elena sees this road sign.

Elena's mom is driving 75 miles per hour when she gets pulled over for speeding.

MAXIMUM

80

1. The police officer explains that 8 kilometers is approximately 5 miles.

a. How many kilometers are in 1 mile?

b. How many miles are in 1 kilometer?

2. If the speed limit is 80 kilometers per hour, and Elena's mom was driving 75 miles per hour, was she speeding? By how much?

Activity

4.3 Veterinary Weights

A veterinarian uses weights in kilograms to figure out what dosages of medicines to prescribe for animals. For every 10 kilograms, there are 22 pounds.

1. Calculate each animal's weight in kilograms. Explain or show your reasoning. If you get stuck, consider drawing a double number line or table.

   a. Fido the Labrador weighs 88 pounds.

   b. Spot the Beagle weighs 33 pounds.

   c. Bella the Chihuahua weighs $5\frac{1}{2}$ pounds.

2. A certain medication says it can only be given to animals over 25 kilograms. How much is this in pounds?

NAME _______________ DATE _______________ PERIOD _______________

Activity
4.4 Cooking with a Tablespoon

Diego is trying to follow a recipe, but he cannot find any measuring cups! He only has a tablespoon. In the cookbook, it says that 1 cup equals 16 tablespoons.

1. How could Diego use the tablespoon to measure out these ingredients?

   a. $\frac{1}{2}$ cup almonds

   b. $1\frac{1}{4}$ cup of oatmeal

   c. $2\frac{3}{4}$ cup of flour

2. Diego also adds the following ingredients. How many cups of each did he use?

   a. 28 tablespoons of sugar

   b. 6 tablespoons of cocoa powder

When we measure something in two different units, the measurements form an equivalent ratio. We can reason with these equivalent ratios to convert measurements from one unit to another.

Suppose you cut off 20 inches of hair. Your Canadian friend asks how many centimeters of hair that was. Since 100 inches equal 254 centimeters, we can use equivalent ratios to find out how many centimeters equal 20 inches.

Using a double number line:

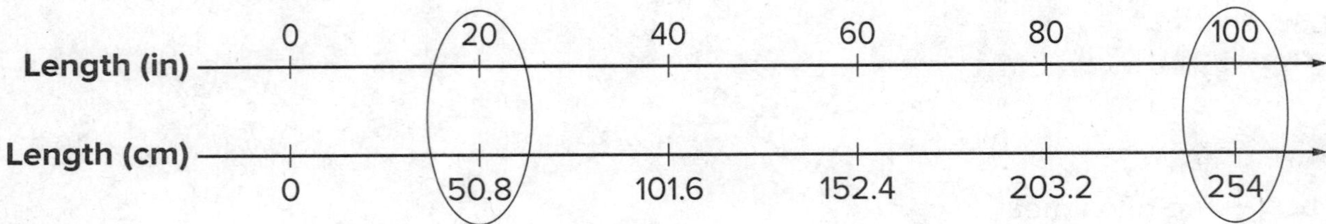

Using a table:

Length (in)	Length (cm)
100	254
1	2.54
20	50.8

One quick way to solve the problem is to start by finding out how many centimeters are in 1 inch. We can then multiply 2.54 and 20 to find that 20 inches equal 50.8 centimeters.

NAME _______________ DATE _______________ PERIOD _______________

Practice

Converting Units

1. Priya's family exchanged 250 dollars for 4,250 pesos. Priya bought a sweater for 510 pesos. How many dollars did the sweater cost?

Pesos	Dollars
4,250	250
	25
	1
	3
510	

2. There are 3,785 milliliters in 1 gallon, and there are 4 quarts in 1 gallon. For each question, explain or show your reasoning.

a. How many milliliters are in 3 gallons?

b. How many milliliters are in 1 quart?

3. Lin knows that there are 4 quarts in a gallon. She wants to convert 6 quarts to gallons, but cannot decide if she should multiply 6 by 4 or divide 6 by 4 to find her answer. What should she do? Explain or show your reasoning. If you get stuck, consider drawing a double number line or using a table.

4. Tyler has a baseball bat that weighs 28 ounces. Find this weight in kilograms and in grams. (Note: 1 kilogram ≈ 35 ounces)

5. Identify whether each unit measures length, volume, or weight (or mass). **(Lesson 3-1)**

a. mile

b. cup

c. pound

d. centimeter

e. liter

f. gram

g. pint

h. yard

i. kilogram

j. teaspoon

k. milliliter

6. A recipe for trail mix uses 7 ounces of almonds with 5 ounces of raisins. (Almonds and raisins are the only ingredients.) How many ounces of almonds would be in a one-pound bag of this trail mix? Explain or show your reasoning. **(Lesson 2-11)**

7. An ant can travel at a constant speed of 980 inches every 5 minutes. **(Lesson 2-9)**

a. How far does the ant travel in 1 minute?

b. At this rate, how far can the ant travel in 7 minutes?

Lesson 3-5

Comparing Speeds and Prices

NAME ______________________ DATE ____________ PERIOD ____________

Learning Goal Let's compare some speeds and some prices.

Warm Up

5.1 Closest Quotient

Is the value of each expression closer to $\frac{1}{2}$, 1, or $1\frac{1}{2}$?

1. $20 \div 18$

2. $9 \div 20$

3. $7 \div 5$

Activity

5.2 More Treadmills

Some students did treadmill workouts, each one running at a constant speed. Answer the questions about their workouts. Explain or show your reasoning.

- Tyler ran 4,200 meters in 30 minutes.
- Kiran ran 6,300 meters in $\frac{1}{2}$ hour.
- Mai ran 6.3 kilometers in 45 minutes.

1. What is the same about the workouts done by:

a. Tyler and Kiran?

b. Kiran and Mai?

c. Mai and Tyler?

2. At what rate did each of them run?

3. How far did Mai run in her first 30 minutes on the treadmill?

Are you ready for more?

Tyler and Kiran each started running at a constant speed at the same time.

Tyler ran 4,200 meters in 30 minutes and Kiran ran 6,300 meters in $\frac{1}{2}$ hour.

Eventually, Kiran ran 1 kilometer more than Tyler. How much time did it take for this to happen?

NAME ______________________ DATE ____________ PERIOD ____________

Activity
5.3 The Best Deal on Beans

Four different stores posted ads about special sales on 15-oz cans of baked beans.

BAKED BEANS
8 FOR $6
BAKED BEANS
10 FOR $10
BAKED BEANS
2 FOR $3
BAKED BEANS
80¢ EACH

1. Which store is offering the best deal? Explain your reasoning.

2. The last store listed is also selling 28-oz cans of baked beans for $1.40 each. How does that price compare to the other prices?

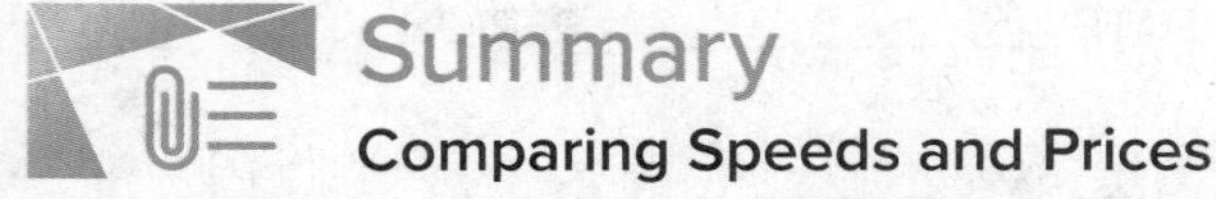

Summary

Comparing Speeds and Prices

Diego ran 3 kilometers in 20 minutes. Andre ran 2,550 meters in 17 minutes. Who ran faster? Since neither their distances nor their times are the same, we have two possible strategies:

- Find the time each person took to travel the *same distance*. The person who traveled that distance in less time is faster.
- Find the distance each person traveled in the *same time*. The person who traveled a longer distance in the same amount of time is faster.

It is often helpful to compare distances traveled in *1 unit* of time (1 minute, for example), which means finding the speed such as meters per minute.

Let's compare Diego and Andre's speeds in meters per minute.

Distance (meters)	Time (minutes)
3,000	20
1,500	10
150	1

Distance (meters)	Time (minutes)
2,550	17
150	1

Both Diego and Andre ran 150 meters per minute, so they ran at the same speed.

Finding ratios that tell us how much of quantity A per 1 unit of quantity B is an efficient way to compare rates in different situations.

Here are some familiar examples:

- Car speeds in *miles per hour*.
- Fruit and vegetable prices in *dollars per pound*.

NAME ______________________ DATE ____________ PERIOD ____________

Practice

Comparing Speeds and Prices

1. Mai and Priya were on scooters. Mai traveled 15 meters in 6 seconds. Priya traveled 22 meters in 10 seconds. Who was moving faster? Explain your reasoning.

2. Here are the prices for cans of juice that are the same brand and the same size at different stores. Which store offers the best deal? Explain your reasoning.

   Store X: 4 cans for $2.48 **Store Y:** 5 cans for $3.00 **Store Z:** 59 cents per can

3. Costs of homes can be very different in different parts of the United States.

   a. A 450-square-foot apartment in New York City costs $540,000. What is the price per square foot? Explain or show your reasoning.

   b. A 2,100-square-foot home in Cheyenne, Wyoming, costs $110 per square foot. How much does this home cost? Explain or show your reasoning.

4. Respond to the following questions. **(Lesson 3-4)**

   a. There are 33.8 fluid ounces in a liter. There are 128 fluid ounces in a gallon. About how many liters are in a gallon?

      (A.) 2

      (B.) 3

      (C.) 4

      (D.) 5

   b. Is your estimate larger or smaller than the actual number of liters in a gallon? Explain how you know.

5. Diego is 165 cm tall. Andre is 1.7 m tall. Who is taller, Diego or Andre? Explain your reasoning. **(Lesson 3-3)**

6. Name an object that could be about the same length as each measurement. **(Lesson 3-2)**

   a. 4 inches

   b. 6 feet

   c. 1 meter

   d. 5 yards

   e. 6 centimeters

   f. 2 millimeters

   g. 3 kilometers

Lesson 3-6

Interpreting Rates

NAME ______________________ DATE ______________ PERIOD ______________

Learning Goal Let's explore unit rates.

Warm Up

6.1 Something per Something

1. Think of two things you have heard described in terms of "something per something."

2. Share your ideas with your group, and listen to everyone else's idea. Make a group list of all unique ideas. Be prepared to share these with the class.

Activity

6.2 Cooking Oatmeal

Priya, Han, Lin, and Diego are all on a camping trip with their families. The first morning, Priya and Han make oatmeal for the group. The instructions for a large batch say, "Bring 15 cups of water to a boil, and then add 6 cups of oats."

- Priya says, "The ratio of the cups of oats to the cups of water is 6 : 15. That's 0.4 cups of oats per cup of water."
- Han says, "The ratio of the cups of water to the cups of oats is 15 : 6. That's 2.5 cups of water per cup of oats."

1. Who is correct? Explain your reasoning. If you get stuck, consider using the table.

Water (cups)	Oats (cups)
15	6
1	
	1

2. The next weekend after the camping trip, Lin and Diego each decide to cook a large batch of oatmeal to have breakfasts ready for the whole week.

   a. Lin decides to cook 5 cups of oats. How many cups of water should she boil?

   b. Diego boils 10 cups of water. How many cups of oats should he add into the water?

3. Did you use Priya's rate (0.4 cups of oats per cup of water) or Han's rate (2.5 cups of water per cup of oats) to help you answer each of the previous two questions? Why?

Activity

6.3 Cheesecake, Milk, and Raffle Tickets

For each situation, find the **unit rates**.

1. A cheesecake recipe says, "Mix 12 oz of cream cheese with 15 oz of sugar."
   a. How many ounces of cream cheese are there for every ounce of sugar?
   b. How many ounces of sugar is that for every ounce of cream cheese?

2. Mai's family drinks a total of 10 gallons of milk every 6 weeks.
   a. How many gallons of milk does the family drink per week?
   b. How many weeks does it take the family to consume 1 gallon of milk?

3. Tyler paid $16 for 4 raffle tickets.
   a. What is the price per ticket?
   b. How many tickets is that per dollar?

4. For each problem, decide which unit rate from the previous situations you prefer to use. Next, solve the problem, and show your thinking.
   a. If Lin wants to make extra cheesecake filling, how much cream cheese will she need to mix with 35 ounces of sugar?
   b. How many weeks will it take Mai's family to finish 3 gallons of milk?
   c. How much would all 1,000 raffle tickets cost?

Are you ready for more?

Write a "deal" on tickets for Tyler's raffle that sounds good, but is actually a little worse than just buying tickets at the normal price.

NAME ______________________ DATE ______________ PERIOD ______________

Summary

Interpreting Rates

Suppose a farm lets us pick 2 pounds of blueberries for 5 dollars. We can say:

- We get $\frac{2}{5}$ pound of blueberries per dollar.
- The blueberries cost $\frac{5}{2}$ dollars per pound.

The "cost per pound" and the "number of pounds per dollar" are the two *unit rates* for this situation.

Blueberries (pounds)	Price (dollars)
2	5
1	$\frac{5}{2}$
$\frac{2}{5}$	1

A **unit rate** tells us how much of one quantity for 1 of the other quantity. Each of these numbers is useful in the right situation.

If we want to find out how much 8 pounds of blueberries will cost, it helps to know how much 1 pound of blueberries will cost.

Blueberries (pounds)	Price (dollars)
1	$\frac{5}{2}$
8	$8 \cdot \frac{5}{2}$

If we want to find out how many pounds we can buy for 10 dollars, it helps to know how many pounds we can buy for 1 dollar.

Blueberries (pounds)	Price (dollars)
$\frac{2}{5}$	1
$10 \cdot \frac{2}{5}$	10

Which unit rate is most useful depends on what question we want to answer, so be ready to find either one!

Glossary

unit rate

Practice

Interpreting Rates

1. A pink paint mixture uses 4 cups of white paint for every 3 cups of red paint.

   The table shows different quantities of red and white paint for the same shade of pink. Complete the table.

White Paint (cups)	Red Paint (cups)
4	3
	1
1	
	4
5	

2. A farm lets you pick 3 pints of raspberries for $12.00.

   a. What is the cost per pint?

   b. How many pints do you get per dollar?

   c. At this rate, how many pints can you afford for $20.00?

   d. At this rate, how much will 8 pints of raspberries cost?

NAME ______________________ DATE ______________ PERIOD ____________

3. Han and Tyler are following a polenta recipe that uses 5 cups of water for every 2 cups of cornmeal.

- Han says, "I am using 3 cups of water. I will need $1\frac{1}{5}$ cups of cornmeal."
- Tyler says, "I am using 3 cups of cornmeal. I will need $7\frac{1}{2}$ cups of water."

Do you agree with either of them? Explain your reasoning.

4. A large art project requires enough paint to cover 1,750 square feet. Each gallon of paint can cover 350 square feet. Each square foot requires $\frac{1}{350}$ of a gallon of paint.

Andre thinks he should use the rate $\frac{1}{350}$ gallons of paint per square foot to find how much paint they need. Do you agree with Andre? Explain or show your reasoning.

5. Andre types 208 words in 4 minutes. Noah types 342 words in 6 minutes. Who types faster? Explain your reasoning. **(Lesson 3-5)**

6. A corn vendor at a farmer's market was selling a bag of 8 ears of corn for $2.56. Another vendor was selling a bag of 12 for $4.32. Which bag is the better deal? Explain or show your reasoning. **(Lesson 3-5)**

7. A soccer field is 100 meters long. What could be its length in yards? **(Lesson 3-3)**

   A. 33.3

   B. 91

   C. 100

   D. 109

Lesson 3-7

Equivalent Ratios Have the Same Unit Rates

NAME ______________________ DATE ______________ PERIOD ____________

Learning Goal Let's revisit equivalent ratios.

Warm Up

7.1 Which One Doesn't Belong: Comparing Speeds

Which one doesn't belong? Be prepared to explain your reasoning.

5 miles in 15 minutes

20 miles per hour

3 minutes per mile

32 kilometers per hour

Activity

7.2 Price of Burritos

1. Two burritos cost $14. Complete the table to show the cost for 4, 5, and 10 burritos at that rate. Next, find the cost for a single burrito in each case.

Number of Burritos	Cost in Dollars	Unit Price (dollars per burrito)
2	14	
4		
5		
10		
b		

2. What do you notice about the values in this table?

3. Noah bought b burritos and paid c dollars. Lin bought twice as many burritos as Noah and paid twice the cost he did. How much did Lin pay per burrito?

	Number of Burritos	Cost in Dollars	Unit Price (dollars per burrito)
Noah	b	c	$\frac{c}{b}$
Lin	$2 \cdot b$	$2 \cdot c$	

4. Explain why, if you can buy b burritos for c dollars, or buy $2 \cdot b$ burritos for $2 \cdot c$ dollars, the cost per item is the same in either case.

Activity

7.3 Making Bracelets

1. Complete the table. Then, explain the strategy you used to do so.

Time in Hours	Number of Bracelets	Speed (bracelets per hour)
2		6
5		6
7		6
	66	6
	100	6

2. Here is a partially filled table from an earlier activity. Use the same strategy you used for the bracelet problem to complete this table.

Number of Burritos	Cost in Dollars	Unit Price (dollars per burrito)
	14	7
	28	7
5		7
10		7

3. Next, compare your results with those in the first table in the previous activity. Do they match? Explain why or why not.

NAME ______________________ DATE ____________ PERIOD ____________

Activity

7.4 How Much Applesauce?

It takes 4 pounds of apples to make 6 cups of applesauce.

1. At this rate, how much applesauce can you make with:
   a. 7 pounds of apples?
   b. 10 pounds of apples?

2. How many pounds of apples would you need to make:
   a. 9 cups of applesauce?
   b. 20 cups of applesauce?

Pounds of Apples	Cups of Applesauce
4	6
7	
10	
	9
	20

Are you ready for more?

1. Jada eats 2 scoops of ice cream in 5 minutes. Noah eats 3 scoops of ice cream in 5 minutes. How long does it take them to eat 1 scoop of ice cream working together (if they continue eating ice cream at the same rate they do individually)?

2. The garden hose at Andre's house can fill a 5-gallon bucket in 2 minutes. The hose at his next-door neighbor's house can fill a 10-gallon bucket in 8 minutes. If they use both their garden hoses at the same time, and the hoses continue working at the same rate they did when filling a bucket, how long will it take to fill a 750-gallon pool?

Summary

Equivalent Ratios Have the Same Unit Rates

The table shows different amounts of apples selling at the same rate, which means all of the ratios in the table are equivalent. In each case, we can find the *unit price* in dollars per pound by dividing the price by the number of pounds.

The unit price is always the same. Whether we buy 4 pounds of apples for 10 dollars or 8 pounds of apples for 20 dollars, the apples cost 2.50 dollars per pound.

Apples (pounds)	Price (dollars)	Unit Price (dollars per pound)
4	10	$10 \div 4 = 2.50$
8	20	$20 \div 8 = 2.50$
20	50	$50 \div 20 = 2.50$

We can also find the number of pounds of apples we can buy per dollar by dividing the number of pounds by the price.

The number of pounds we can buy for a dollar is the same as well! Whether we buy 4 pounds of apples for 10 dollars or 8 pounds of apples for 20 dollars, we are getting 0.4 pounds per dollar.

Apples (pounds)	Price (dollars)	Pounds per Dollar
4	10	$4 \div 10 = 0.4$
8	20	$8 \div 20 = 0.4$
20	50	$20 \div 50 = 0.4$

This is true in all contexts: when two ratios are equivalent, their unit rates will be equal.

Quantity x	Quantity y	Unit Rate 1	Unit Rate 2
a	b	$\frac{a}{b}$	$\frac{b}{a}$
$s \cdot a$	$s \cdot b$	$\frac{s \cdot a}{s \cdot b} = \frac{a}{b}$	$\frac{s \cdot b}{s \cdot a} = \frac{b}{a}$

NAME ______________________ DATE ______________ PERIOD ________

Practice

Equivalent Ratios Have the Same Unit Rates

1. A car travels 55 miles per hour for 2 hours. Complete the table.

Time (hours)	Distance (miles)	Miles per Hour
1	55	55
$\frac{1}{2}$		
$1\frac{1}{2}$		
	110	

2. The table shows the amounts of onions and tomatoes in different-sized batches of a salsa recipe.

Onions (ounces)	Tomatoes (ounces)
2	16
4	32
6	48

Elena notices that if she takes the number in the Tomatoes column and divides it by the corresponding number in the Onions column, she always gets the same result.

What is the meaning of the number that Elena has calculated?

3. A restaurant is offering 2 specials: 10 burritos for $12, or 6 burritos for $7.50. Noah needs 60 burritos for his party. Should he buy 6 orders of the 10-burrito special or 10 orders of the 6-burrito special? Explain your reasoning.

4. Complete the table so that the cost per banana remains the same.

Number of Bananas	Cost in Dollars	Unit Price (Dollars per Banana)
4		0.50
6		0.50
7		0.50
10		0.50
	10.00	0.50
	16.50	0.50

5. Two planes travel at a constant speed. Plane A travels 2,800 miles in 5 hours. Plane B travels 3,885 miles in 7 hours. Which plane is faster? Explain your reasoning. **(Lesson 3-5)**

6. A car has 15 gallons of gas in its tank. The car travels 35 miles per gallon of gas. It uses $\frac{1}{35}$ of a gallon of gas to go 1 mile. **(Lesson 3-6)**

   a. How far can the car travel with 15 gallons? Show your reasoning.

   b. How much gas does the car use to go 100 miles? Show your reasoning.

7. A box of cereal weighs 600 grams. How much is this weight in pounds? Explain or show your reasoning. (Note: 1 kilogram = 2.2 pounds) **(Lesson 3-4)**

Lesson 3-8

More about Constant Speed

NAME ______________________ DATE ______________ PERIOD ____________

Learning Goal Let's investigate constant speed some more.

Warm Up

8.1 Back on the Treadmill Again

While training for a race, Andre's dad ran 12 miles in 75 minutes on a treadmill. If he runs at that rate:

1. How long would it take him to run 8 miles?

2. How far could he run in 30 minutes?

Activity

8.2 Picnics on the Rail Trail

Kiran and Clare live 24 miles away from each other along a rail trail. One Saturday, the two friends started walking toward each other along the trail at 8:00 a.m. with a plan to have a picnic when they meet.

Kiran walks at a **speed** of 3 miles per hour while Clare walks 3.4 miles per hour.

1. After one hour, how far apart will they be?

2. Make a table showing how far apart the two friends are after 0 hours, 1 hour, 2 hours, and 3 hours.

3. At what time will the two friends meet and have their picnic?

4. Kiran says "If I walk 3 miles per hour toward you, and you walk 3.4 miles per hour toward me, it's the same as if you stay put and I jog 6.4 miles per hour." What do you think Kiran means by this? Is he correct?

5. Several months later, they both set out at 8:00 a.m. again, this time with Kiran jogging and Clare still walking at 3.4 miles per hour. This time, they meet at 10:30 a.m. How fast was Kiran jogging?

Are you ready for more?

1. On his trip to meet Clare, Kiran brought his dog with him. At the same time Kiran and Clare started walking, the dog started running 6 miles per hour. When it got to Clare it turned around and ran back to Kiran. When it got to Kiran, it turned around and ran back to Clare, and continued running in this fashion until Kiran and Clare met. How far did the dog run?

2. The next Saturday, the two friends leave at the same time again, and Kiran jogs twice as fast as Clare walks. Where on the rail trail do Kiran and Clare meet?

NAME _______________ DATE _______________ PERIOD _______________

Activity
8.3 Swimming and Biking

Jada bikes 2 miles in 12 minutes. Jada's cousin swims 1 mile in 24 minutes.

1. Who is moving faster? How much faster?

2. One day Jada and her cousin line up on the end of a swimming pier on the edge of a lake. At the same time, they start swimming and biking in opposite directions.
   a. How far apart will they be after 15 minutes?

   b. How long will it take them to be 5 miles apart?

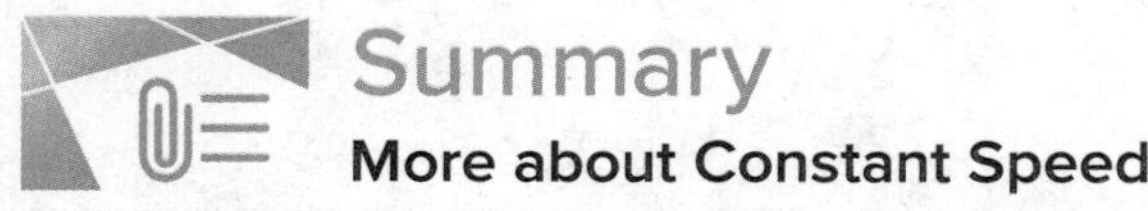

Summary
More about Constant Speed

When two objects are each moving at a constant speed and their distance-to-time ratios are equivalent, we say that they are moving at the *same speed.* If their time-distance ratios are not equivalent, they are not moving at the same speed.

We describe **speed** in units of distance per unit of time, like *miles per hour* or *meters per second.*

We can also use **pace** to describe distance and time. We measure pace in units such as *hours per mile* or *seconds per meter.*

	Speed	Pace
A snail that crawls 5 centimeters in 2 minutes	is traveling at a rate of 2.5 centimeters per minute.	has a pace of 0.4 minutes per centimeter.
A toddler that walks 9 feet in 6 seconds	is traveling at a rate of 1.5 feet per second.	has a pace of $\frac{2}{3}$ seconds per foot.
A cyclist who bikes 20 kilometers in 2 hours	is traveling at a rate of 10 kilometers per hour.	has a pace of 0.1 hours per kilometer.

Speed and pace are reciprocals. Both can be used to compare whether one object is moving faster or slower than another object.

- An object with the higher speed is *faster* than one with a lower speed because the former travels a greater distance in the same amount of time.
- An object with the greater pace is *slower* than one with a smaller pace because the former takes more time to travel the same distance.

Because speed is a *rate per 1 unit of time* for ratios that relate distance and time, we can multiply the amount of time traveled by the speed to find the distance traveled.

Time (minutes)	Distance (centimeters)
2	5
1	2.5
4	4 • (2.5)

Glossary

pace
speed

NAME ______________________ DATE ____________ PERIOD ____________

Practice

More about Constant Speed

1. A kangaroo hops 2 kilometers in 3 minutes. At this rate:
   a. How long does it take the kangaroo to travel 5 kilometers?
   b. How far does the kangaroo travel in 2 minutes?

2. Mai runs around a 400-meter track at a constant speed of 250 meters per minute. How many minutes does it take Mai to complete 4 laps of the track? Explain or show your reasoning.

3. At 10:00 a.m., Han and Tyler both started running toward each other from opposite ends of a 10-mile path along a river. Han runs at a pace of 12 minutes per mile. Tyler runs at a pace of 15 minutes per mile.
   a. How far does Han run after a half hour? After an hour?
   b. Do Han and Tyler meet on the path within 1 hour? Explain or show your reasoning.

4. Two skateboarders start a race at the same time. Skateboarder A travels at a steady rate of 15 feet per second. Skateboarder B travels at a steady rate of 22 feet per second. After 4 minutes, how much farther will Skateboarder B have traveled? Explain your reasoning. **(Lesson 2-16)**

5. There are 4 tablespoons in $\frac{1}{4}$ cup. There are 2 cups in 1 pint. How many tablespoons are there in 1 pint? If you get stuck, consider drawing a double number line or making a table. **(Lesson 3-4)**

6. Two larger cubes are made out of unit cubes. Cube A is 2 by 2 by 2. Cube B is 4 by 4 by 4. The side length of Cube B is twice that of Cube A. **(Lesson 1-12)**

Cube A

Cube B

a. Is the surface area of Cube B also twice that of Cube A? Explain or show your reasoning.

b. Is the volume of Cube B also twice that of Cube A? Explain or show your reasoning.

Lesson 3-9

Solving Rate Problems

NAME ______________________ DATE ____________ PERIOD ____________

Learning Goal Let's use unit rates like a pro.

Warm Up

9.1 Grid of 100

How much is shaded in each one?

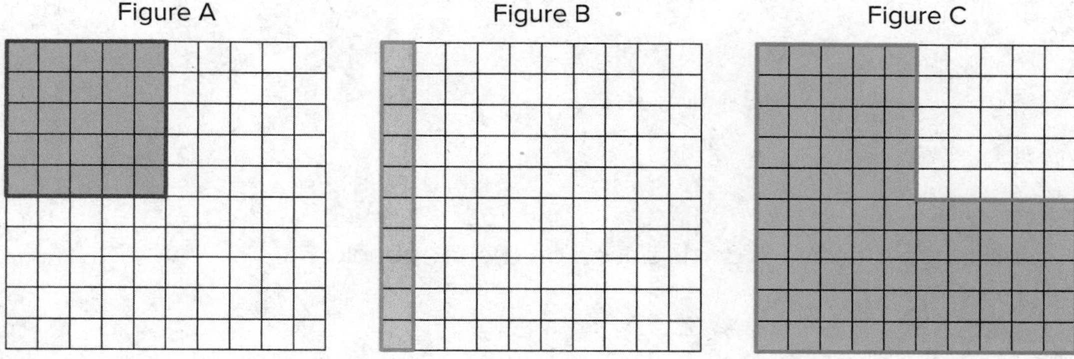

Activity

9.2 Card Sort: Is it a Deal?

Your teacher will give you a set of cards showing different offers.

1. Find card A and work with your partner to decide whether the offer on card A is a good deal. Explain or show your reasoning.

2. Next, split cards B–E so you and your partner each have two.

   a. Decide individually if your two cards are good deals. Explain your reasoning.

   b. For each of your cards, explain to your partner if you think it is a good deal and why. Listen to your partner's explanations for their cards. If you disagree, explain your thinking.

   c. Revise any decisions about your cards based on the feedback from your partner.

3. When you and your partner are in agreement about cards B–E, place all the cards you think are a good deal in one stack and all the cards you think are a bad deal in another stack. Be prepared to explain your reasoning.

NAME ______________________ DATE ____________ PERIOD ________

Are you ready for more?

Time to make your own deal! Read the information on card F and then decide what you would charge if you were the clerk. When your teacher signals, trade cards with another group and decide whether or not you would take the other group's offer.

Keep in mind that you may offer a fair deal or an unfair deal, but the goal is to set a price close enough to the value it should be that the other group cannot immediately tell if the deal you offer is a good one.

Activity
9.3 The Fastest of All

Wild animals from around the world wanted to hold an athletic competition, but no one would let them on an airplane. They decided to just measure how far each animal could sprint in one minute and send the results to you to decide the winner.

You look up the following information about converting units of length:

1 inch = 2.54 centimeters

1. Which animal sprinted the farthest?

Animal	Sprint Distance
Cougar	1,408 yards
Antelope	1 mile
Hare	49,632 inches
Kangaroo	1,073 meters
Ostrich	1.15 kilometers
Coyote	3,773 feet

2. What are the place rankings for all of the animals?

Summary
Solving Rate Problems

Sometimes we can find and use more than one unit rate to solve a problem.

Suppose a grocery store is having a sale on shredded cheese. A small bag that holds 8 ounces is sold for \$2. A large bag that holds 2 kilograms is sold for \$16. How do you know which is a better deal?

Here are two different ways to solve this problem.

1. Compare dollars per kilogram.

- The large bag costs \$8 per kilogram, because $16 \div 2 = 8$.
- The small bag holds $\frac{1}{2}$ pound of cheese, because there are 16 ounces in 1 pound, and $8 \div 16 = \frac{1}{2}$.
- The small bag costs \$4 per pound, because $2 \div \frac{1}{2} = 4$. This is about \$8.80 per kilogram, because there are about 2.2 pounds in 1 kilogram, and $4.00 \cdot 2.2 = 8.80$.

The large bag is a better deal, because it costs less money for the same amount of cheese.

2. Compare ounces per dollar.

- With the small bag, we get 4 ounces per dollar, because $8 \div 2 = 4$.
- The large bag holds 2,000 grams of cheese. There are 1,000 grams in 1 kilogram, and $2 \cdot 1{,}000 = 2{,}000$. This means 125 grams per dollar, because $2{,}000 \div 16 = 125$.
- There are about 28.35 grams in 1 ounce, and $125 \div 28.35 \approx 4.4$, so this is about 4.4 ounces per dollar.

The large bag is a better deal, because you get more cheese for the same amount of money.

Another way to solve the problem would be to compare the unit prices of each bag in dollars per ounce. Try it!

NAME _______________ DATE _______________ PERIOD _______________

Practice

Solving Rate Problems

1. This package of sliced cheese costs \$2.97.

How much would a package with 18 slices cost at the same price per slice? Explain or show your reasoning.

2. A copy machine can print 480 copies every 4 minutes. For each question, explain or show your reasoning.

a. How many copies can it print in 10 minutes?

b. A teacher printed 720 copies. How long did it take to print?

3. Order these objects from heaviest to lightest. (Note: 1 pound = 16 ounces, 1 kilogram $\approx$ 2.2 pounds, and 1 ton = 2,000 pounds)

Item	Weight
School Bus	9 tons
Horse	1,100 pounds
Elephant	5,500 kilograms
Grand Piano	15,840 ounces

4. Andre sometimes mows lawns on the weekend to make extra money. Two weeks ago, he mowed a neighbor's lawn for $\frac{1}{2}$ hour and earned $10. Last week, he mowed his uncle's lawn for $\frac{3}{2}$ hours and earned $30. This week, he mowed the lawn of a community center for 2 hours and earned $30. **(Lesson 3-5)**

Which jobs paid better than others? Explain your reasoning.

5. Calculate and express your answer in decimal form. **(Lesson 3-1)**

a. $\frac{1}{2} \cdot 17$

b. $\frac{3}{4} \cdot 200$

c. $(0.2) \cdot 40$

d. $(0.25) \cdot 60$

6. Respond to the following statements. **(Lesson 1-11)**

a. Decompose this polygon so that its area can be calculated. All measurements are in centimeters.

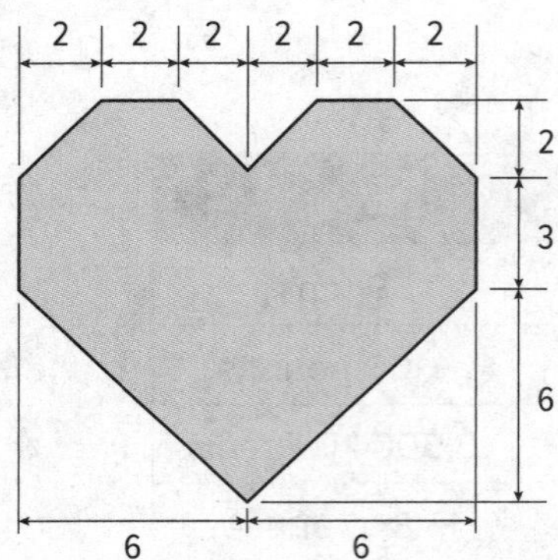

b. Calculate its area. Organize your work so that it can be followed by others.

Lesson 3-10

What Are Percentages?

NAME ______________________________ DATE ______________ PERIOD ____________

Learning Goal Let's learn about percentages.

Warm Up

10.1 Dollars and Cents

Find each answer mentally.

1. A sticker costs 25 cents. How many dollars is that?
2. A pen costs 1.50 dollars. How many cents is that?
3. How many cents are in one dollar?
4. How many dollars are in one cent?

Activity

10.2 Coins

1. Complete the table to show the values of these U.S. coins.

Coin	Penny	Nickel	Dime	Quarter	Half Dollar	Dollar
Value (cents)						

The value of a quarter is 25% of the value of a dollar because there are 25 cents for every 100 cents.

1 Quarter: 25¢

1 Dollar: 100¢

2. Write the name of the coin that matches each expression.

a. 25% of a dollar

b. 5% of a dollar

c. 1% of a dollar

d. 100% of a dollar

e. 10% of a dollar

f. 50% of a dollar

3. The value of 6 dimes is what percent of the value of a dollar?

4. The value of 6 quarters is what percent of the value of a dollar?

Are you ready for more?

Find two different sets of coins that each make 120% of a dollar, where no type of coin is in both sets.

NAME ______________________ DATE ______________ PERIOD ______________

Activity

10.3 Coins on a Number Line

A $1 coin is worth 100% of the value of a dollar.

Here is a double number line that shows this.

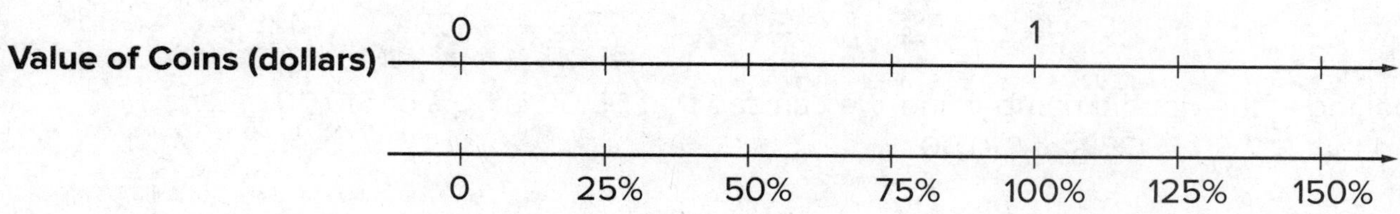

1. The coins in Jada's pocket are worth 75% of a dollar. How much are they worth (in dollars)?

2. The coins in Diego's pocket are worth 150% of a dollar. How much are they worth (in dollars)?

3. Elena has 3 quarters and 5 dimes. What percentage of a dollar does she have?

Summary

What Are Percentages?

A **percentage** is a *rate per 100*.

We can find percentages of $10 using a double number line where 10 and 100% are aligned, as shown here:

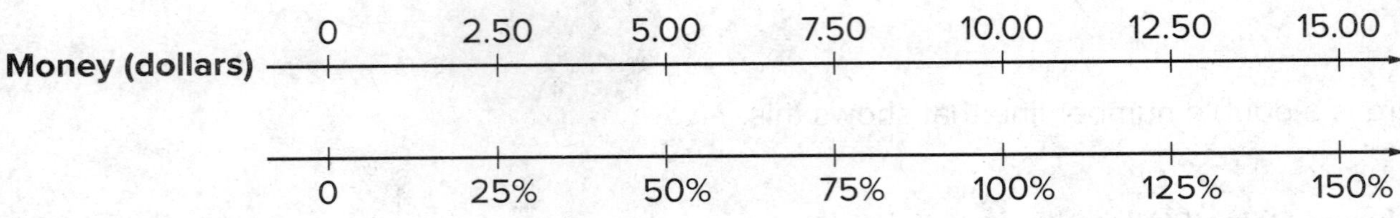

Looking at the double number line, we can see that $5.00 is 50% of $10.00 and that $12.50 is 125% of $10.00.

Glossary

percent
percentage

NAME ______________________ DATE ____________ PERIOD ____________

Practice

What Are Percentages?

1. What percentage of a dollar is the value of each coin combination?
   a. 4 dimes
   b. 1 nickel and 3 pennies
   c. 5 quarters and 1 dime

2. Respond to each of the following.
   a. List three different combinations of coins, each with a value of 30% of a dollar.
   b. List two different combinations of coins, each with a value of 140% of a dollar.

3. The United States government used to make coins of many different values. For each coin, state its worth as a percentage of $1.

   a. $\frac{1}{2}$ cent
   b. 3 cents
   c. 20 cents
   d. $\$2\frac{1}{2}$
   e. $5

4. Complete the double number to line show percentages of $50.

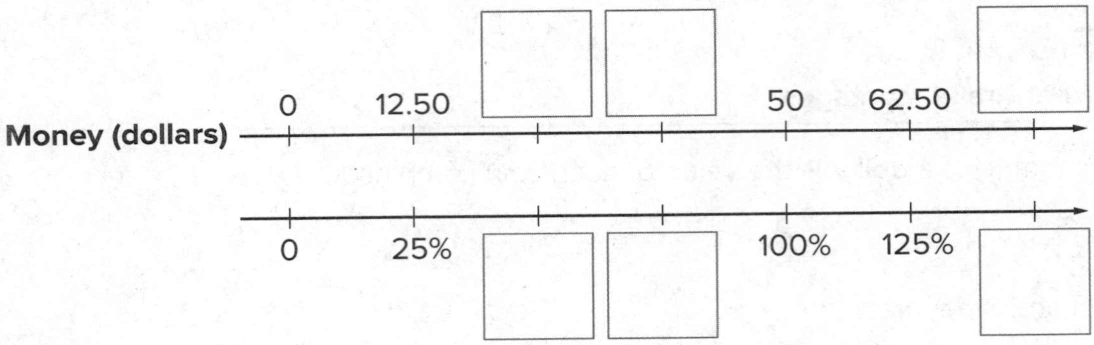

5. Elena bought 8 tokens for $4.40. At this rate: **(Lesson 3-9)**

   a. How many tokens could she buy with $6.05?

   b. How much do 19 tokens cost?

6. A snail travels 10 cm in 4 minutes. At this rate: **(Lesson 3-8)**

   a. How long will it take the snail to travel 24 cm?

   b. How far does the snail travel in 6 minutes?

7. Respond to the following. **(Lesson 3-7)**

   a. 3 tacos cost $18. Complete the table to show the cost of 4, 5, and 6 tacos at the same rate.

Number of Tacos	Cost in Dollars	Rate in Dollars per Taco
3	18	
4		
5		
6		

   b. If you buy *t* tacos for *c* dollars, what is the unit rate?

Lesson 3-11

Percentages and Double Number Lines

NAME ______________________ DATE ____________ PERIOD ____________

Learning Goal Let's use double number lines to represent percentages.

Warm Up

11.1 Fundraising Goal

Each of three friends—Lin, Jada, and Andre—had the goal of raising $40. How much money did each person raise? Be prepared to explain your reasoning.

1. Lin raised 100% of her goal.

2. Jada raised 50% of her goal.

3. Andre raised 150% of his goal.

Activity

11.2 Three-Day Biking Trip

Elena biked 8 miles on Saturday. Use the double number line to answer the questions. Be prepared to explain your reasoning.

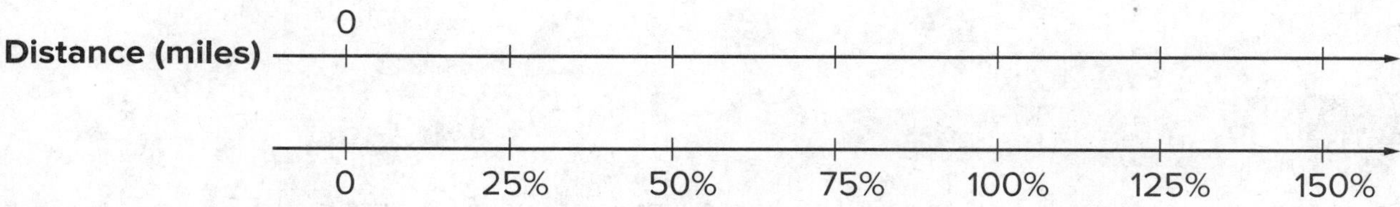

1. What is 100% of her Saturday distance?

2. On Sunday, she biked 75% of her Saturday distance. How far was that?

3. On Monday, she biked 125% of her Saturday distance. How far was that?

Activity

11.3 Puppies Grow Up

1. Jada has a new puppy that weighs 9 pounds. The vet says that the puppy is now at about 20% of its adult weight. What will be the adult weight of the puppy?

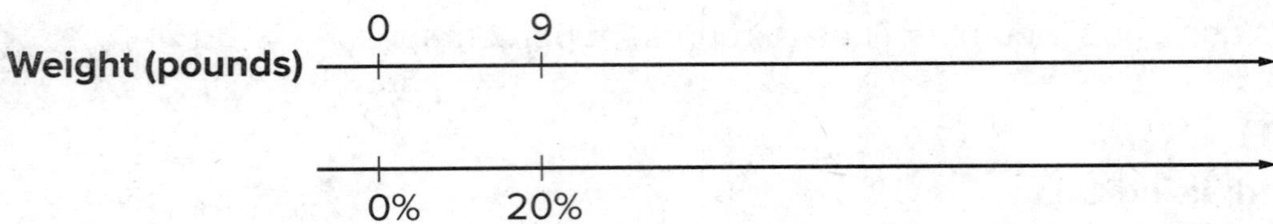

2. Andre also has a puppy that weighs 9 pounds. The vet says that this puppy is now at about 30% of its adult weight. What will be the adult weight of Andre's puppy?

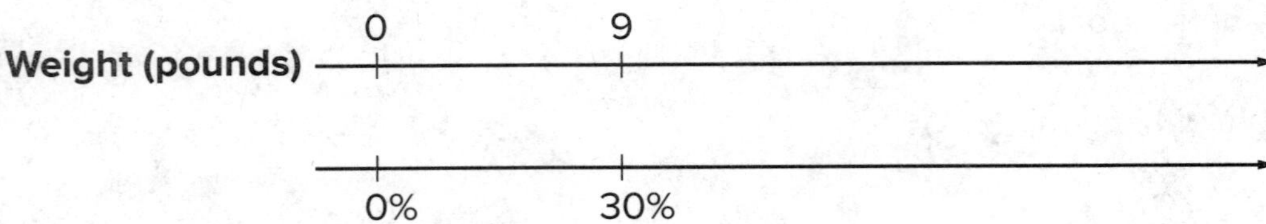

3. What is the same about Jada and Andre's puppies? What is different?

Are you ready for more?

A loaf of bread costs $2.50 today. The same size loaf cost 20 cents in 1955.

1. What percentage of today's price did someone in 1955 pay for bread?

2. A job pays $10.00 an hour today. If the same percentage applies to income as well, how much would that job have paid in 1955?

NAME ______________________ DATE ______________ PERIOD ____________

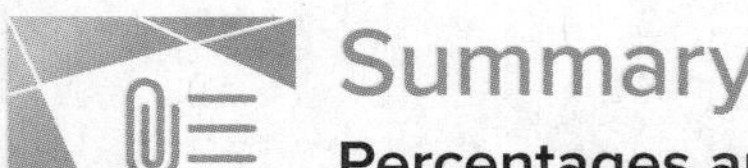

Summary

Percentages and Double Number Lines

We can use a double number line to solve problems about percentages. For example, what is 30% of 50 pounds?

We can draw a double number line like this:

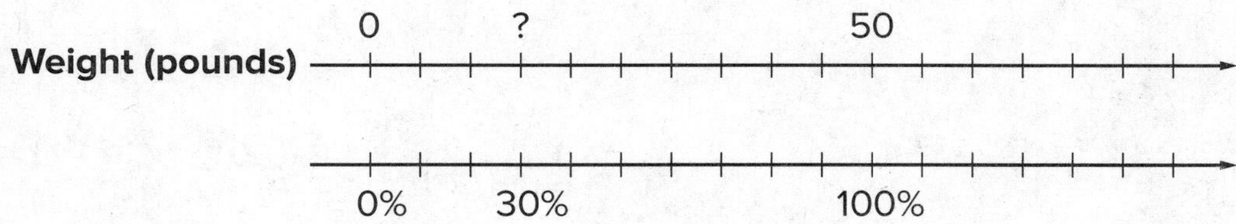

- We divide the distance between 0% and 100% and that between 0 and 50 pounds into ten equal parts.
- We label the tick marks on the top line by counting by 5s ($50 \div 10 = 5$) and on the bottom line counting by 10% ($100 \div 10 = 10$).
- We can then see that 30% of 50 pounds is 15 pounds.

We can also use a table to solve this problem.

	Weight (pounds)	Percentage	
	50	100	
$\cdot \frac{1}{10}$	5	10	$\cdot \frac{1}{10}$
$\cdot 3$	15	30	$\cdot 3$

Suppose we know that 140% of an amount is \$28. What is 100% of that amount? Let's use a double number line to find out.

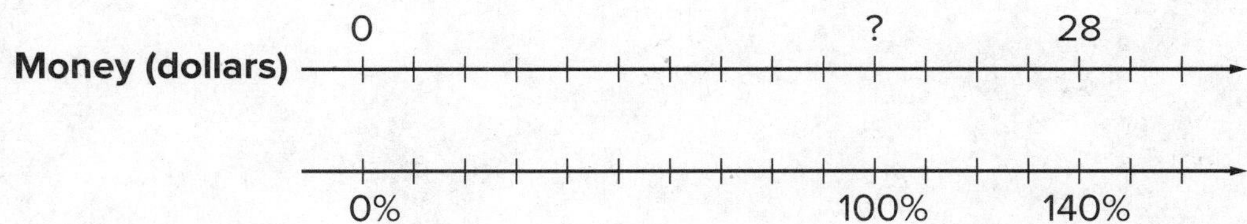

- We divide the distance between 0% and 140% and that between 0 and 28 into fourteen equal intervals.
- We label the tick marks on the top line by counting by 2s and on the bottom line counting by 10%.
- We would then see that 100% is \$20.

Or we can use a table as shown.

	Money (dollars)	Percentage	
	28	140	
$\cdot \frac{1}{14}$	2	10	$\cdot \frac{1}{14}$
$\cdot 10$	20	100	$\cdot 10$

Practice

Percentages and Double Number Lines

1. Solve each problem. If you get stuck, consider using the double number lines.

   a. During a basketball practice, Mai attempted 40 free throws and was successful on 25% of them. How many successful free throws did she make?

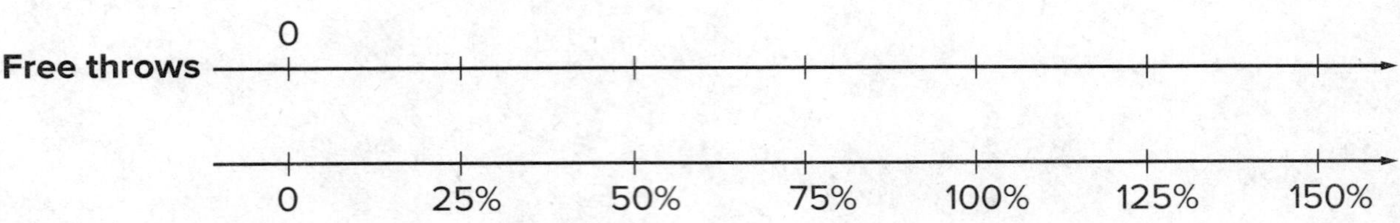

   b. Yesterday, Priya successfully made 12 free throws. Today, she made 150% as many. How many successful free throws did Priya make today?

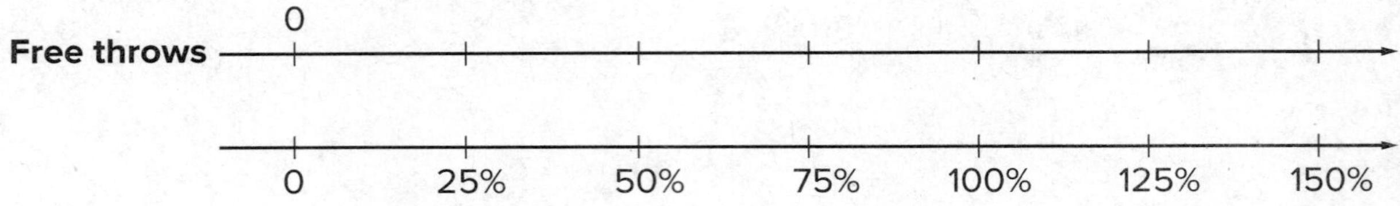

2. A 16-ounce bottle of orange juice says it contains 200 milligrams of vitamin C, which is 250% of the daily recommended allowance of vitamin C for adults. What is 100% of the daily recommended allowance of vitamin C for adults?

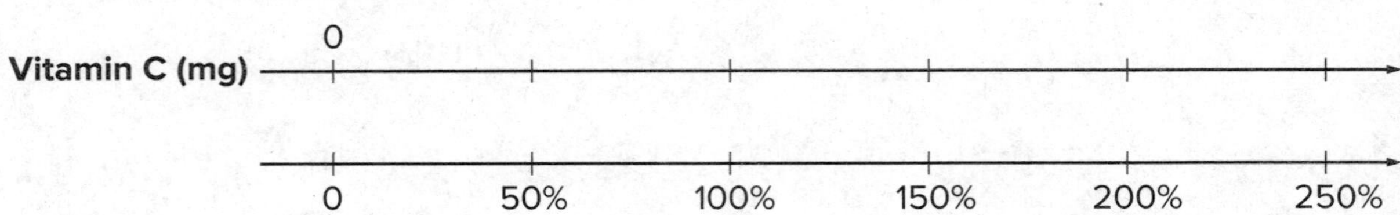

NAME ______________________ DATE ______________ PERIOD ____________

3. At a school, 40% of the sixth-grade students said that hip-hop is their favorite kind of music. If 100 sixth-grade students prefer hip hop music, how many sixth-grade students are at the school? Explain or show your reasoning.

4. Diego has a skateboard, scooter, bike, and go-cart. He wants to know which vehicle is the fastest. A friend records how far Diego travels on each vehicle in 5 seconds. For each vehicle, Diego travels as fast as he can along a straight, level path. **(Lesson 3-9)**

Vehicle	Distance Traveled
Skateboard	90 feet
Scooter	1,020 inches
Bike	4,800 centimeters
Go-Cart	0.03 kilometers

a. What is the distance each vehicle traveled in centimeters?

b. Rank the vehicles in order from fastest to slowest.

5. It takes 10 pounds of potatoes to make 15 pounds of mashed potatoes. At this rate: **(Lesson 3-7)**

   a. How many pounds of mashed potatoes can they make with 15 pounds of potatoes?

   b. How many pounds of potatoes are needed to make 50 pounds of mashed potatoes?

Lesson 3-12

Percentages and Tape Diagrams

NAME ______________________ DATE ______________ PERIOD ______________

Learning Goal Let's use tape diagrams to understand percentages.

Warm Up

12.1 Notice and Wonder: Tape Diagrams

What do you notice? What do you wonder?

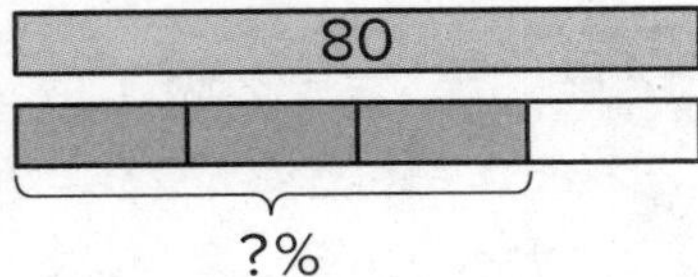

Activity

12.2 Revisiting Jada's Puppy

Jada has a new puppy that weighs 9 pounds.
It is now at about 20% of its adult weight.

1. Here is a diagram that Jada drew about the weight of her puppy.

9	9	9	9	9

20%

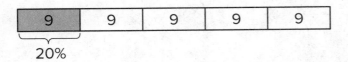

   a. The adult weight of the puppy will be 45 pounds. How can you see that in the diagram?

   b. What fraction of its adult weight is the puppy now? How can you see that in the diagram?

2. Jada's friend has a dog that weighs 90 pounds. Here is a diagram Jada drew that represents the weight of her friend's dog and the weight of her puppy.

9	9	9	9	9	9	9	9	9	9

9

a. How many times greater is the dog's weight than the puppy's?

b. Compare the weight of the puppy and the dog using fractions.

c. Compare the weight of the puppy and the dog using percentages.

Activity

12.3 5 Dollars

Noah has $5.

1. a. Elena has 40% as much as Noah. How much does Elena have?

   b. Compare Elena's and Noah's money using fractions. Draw a diagram to illustrate.

2. a. Diego has 150% as much as Noah. How much does Diego have?

   b. Compare Diego's and Noah's money using fractions. Draw a diagram to illustrate.

NAME ______________________ DATE ______________ PERIOD ______________

Activity

12.4 Staying Hydrated

During the first part of a hike, Andre drank 1.5 liters of the water he brought.

1. If this is 50% of the water he brought, how much water did he bring?

2. If he drank 80% of his water on his entire hike, how much did he drink?

Are you ready for more?

Decide if each scenario is possible.

1. Andre plans to bring his dog on his next hike, along with 150% as much water as he brought on this hike.

2. Andre plans to drink 150% of the water he brought on his hike.

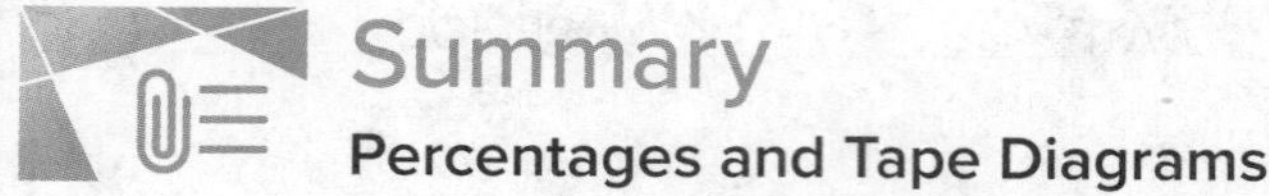

Summary

Percentages and Tape Diagrams

Tape diagrams can help us make sense of percentages.

Consider two problems that we solved earlier using double number lines and tables:

- "What is 30% of 50 pounds?" and
- "What is 100% of a number if 140% of it is 28?"

Here is a tape diagram that shows that 30% of 50 pounds is 15 pounds.

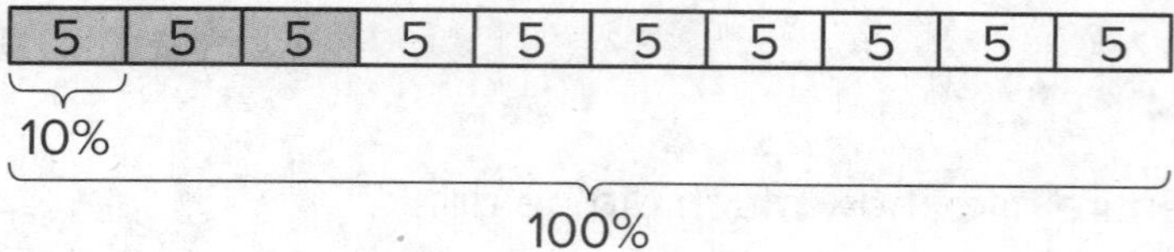

This diagram shows that if 140% of some number is 28, then that number must be 20.

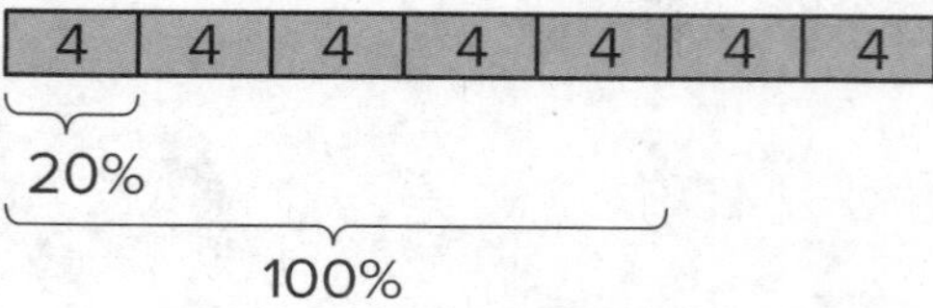

NAME ________________ DATE ________ PERIOD ________

Practice

Percentages and Tape Diagrams

1. Here is a tape diagram that shows how far two students walked.

Priya's distance (km)	2	2	2	2	2
Tyler's distance (km)	2	2	2	2	

   a. What percentage of Priya's distance did Tyler walk?

   b. What percentage of Tyler's distance did Priya walk?

2. A bakery makes 40 different flavors of muffins. 25% of the flavors have chocolate as one of the ingredients. Draw a tape diagram to show how many flavors have chocolate and how many don't.

3. There are 70 students in the school band. 40% of them are sixth graders, 20% are seventh graders, and the rest are eighth graders.

   a. How many band members are sixth graders?

   b. How many band members are seventh graders?

   c. What percentage of the band members are eighth graders? Explain your reasoning.

4. Jada has a monthly budget for her cell phone bill. Last month she spent 120% of her budget, and the bill was $60. What is Jada's monthly budget? Explain or show your reasoning. **(Lesson 3-11)**

5. Which is a better deal, 5 tickets for $12.50 or 8 tickets for $20.16? Explain your reasoning. **(Lesson 3-9)**

6. An athlete runs 8 miles in 50 minutes on a treadmill. **(Lesson 3-8)** At this rate:

   a. How long will it take the athlete to run 9 miles?

   b. How far can the athlete run in 1 hour?

Lesson 3-13

Benchmark Percentages

NAME _______________ DATE _______________ PERIOD _______________

Learning Goal Let's contrast percentages and fractions.

Warm Up

13.1 What Percentage Is Shaded?

What percentage of each diagram is shaded?

Diagram A

Diagram B

Diagram C

Activity

13.2 Liters, Meters, and Hours

1. a. How much is 50% of 10 liters of milk?

   b. How far is 50% of a 2,000-kilometer trip?

   c. How long is 50% of a 24-hour day?

   d. How can you find 50% of any number?

10

2,000

?

50%

2. a. How far is 10% of a 2,000-kilometer trip?

   b. How much is 10% of 10 liters of milk?

   c. How long is 10% of a 24-hour day?

   d. How can you find 10% of any number?

3. a. How long is 75% of a 24-hour day?

b. How far is 75% of a 2,000-kilometer trip?

c. How much is 75% of 10 liters of milk?

d. How can you find 75% of any number?

Activity

13.3 Nine is . . .

Explain how you can calculate each value mentally.

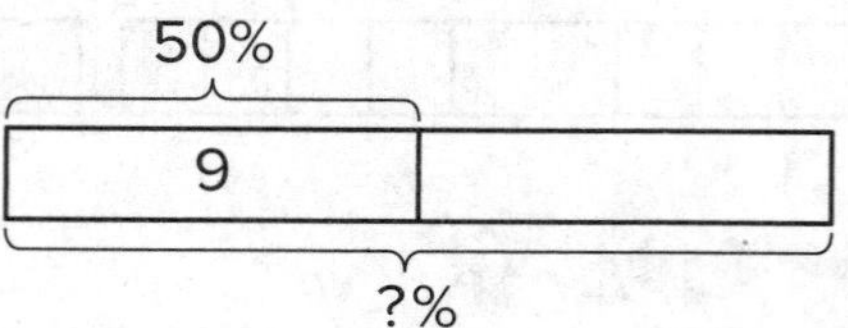

1. 9 is 50% of what number?

2. 9 is 25% of what number?

3. 9 is 10% of what number?

4. 9 is 75% of what number?

5. 9 is 150% of what number?

NAME ________________________ DATE ____________ PERIOD ____________

Activity

13.4 Matching the Percentage

Match the percentage that describes the relationship between each pair of numbers. One percentage will be left over. Be prepared to explain your reasoning.

Pairs of Numbers	Percentage
1. 7 is what percentage of 14?	4%
	10%
2. 5 is what percentage of 20?	25%
	50%
3. 3 is what percentage of 30?	75%
	400%
4. 6 is what percentage of 8?	
5. 20 is what percentage of 5?	

Are you ready for more?

1. What percentage of the world's current population is under the age of 14?
2. How many people is that?
3. How many people are 14 or older?

Summary

Benchmark Percentages

Certain percentages are easy to think about in terms of fractions.

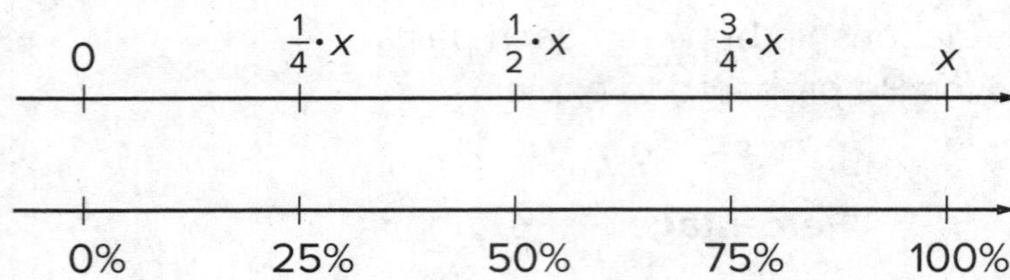

- 25% of a number is always $\frac{1}{4}$ of that number.
  For example, 25% of 40 liters is $\frac{1}{4} \cdot 40$ or 10 liters.
- 50% of a number is always $\frac{1}{2}$ of that number.
  For example, 50% of 82 kilometers $\frac{1}{2} \cdot 82$ or 41 kilometers.
- 75% of a number is always $\frac{3}{4}$ of that number.
  For example, 75% of 1 pound is $\frac{3}{4}$ pound.
- 10% of a number is always $\frac{1}{10}$ of that number.
  For example, 10% of 95 meters is 9.5 meters.

We can also find multiples of 10% using tenths.

For example, 70% of a number is always $\frac{7}{10}$ of that number, so 70% of 30 days is $\frac{7}{10} \cdot 30$ or 21 days.

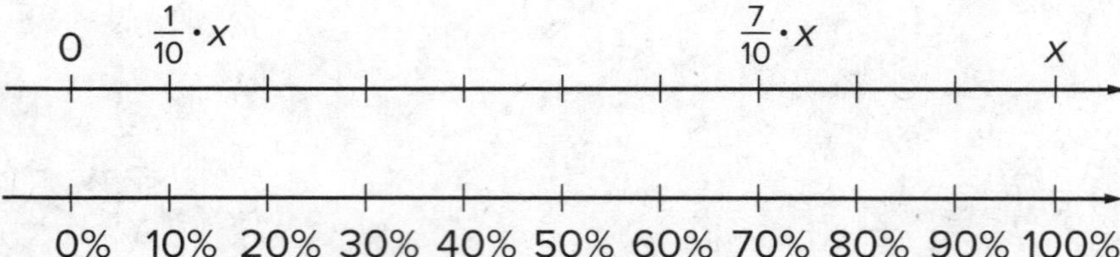

NAME ______________________ DATE ____________ PERIOD ____________

Practice

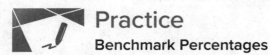

Benchmark Percentages

1. Respond to each question.

   a. How can you find 50% of a number quickly in your head?

   b. Andre lives 1.6 km from school. What is 50% of 1.6 km?

   c. Diego lives $\frac{1}{2}$ mile from school. What is 50% of $\frac{1}{2}$ mile?

2. There is a 10% off sale on laptop computers. If someone saves \$35 on a laptop, what was its original cost? If you get stuck, consider using the table.

Savings (dollars)	Percentage
35	10
?	100

3. Explain how to calculate these mentally.

   a. 15 is what percentage of 30?

   b. 3 is what percentage of 12?

   c. 6 is what percentage of 10?

4. Noah says that to find 20% of a number he divides the number by 5. For example, 20% of 60 is 12, because 60 ÷ 5 = 12. Does Noah's method always work? Explain why or why not.

5. Diego has 75% of $10. Noah has 25% of $30. Diego thinks he has more money than Noah, but Noah thinks they have an equal amount of money. Who is right? Explain your reasoning. **(Lesson 3-10)**

6. Lin and Andre start walking toward each other at the same time from opposite ends of 22-mile walking trail. Lin walks at a speed of 2.5 miles per hour. Andre walks at a speed of 3 miles per hour.

Here is a table showing the distances traveled and how far apart Lin and Andre were over time. Use the table to find how much time passes before they meet. **(Lesson 3-8)**

Elapsed Time (hour)	Lin's Distance (miles)	Andre's Distance (miles)	Distance Apart (miles)
0	0	0	22
1	2.5	3	16.5
			0

Lesson 3-14

Solving Percentage Problems

NAME ______________________ DATE ______________ PERIOD ____________

Learning Goal Let's solve more percentage problems.

Warm Up

14.1 Number Talk: Multiplication with Decimals

Find the products mentally.

1. $6 \cdot (0.8) \cdot 2$

2. $(4.5) \cdot (0.6) \cdot 4$

Activity

14.2 Coupons

klotz/123RF

Han and Clare go shopping, and they each have a coupon. Answer each question and show your reasoning.

1. Han buys an item with a normal price of \$15, and uses a 10% off coupon. How much does he save by using the coupon?

2. Clare buys an item with a normal price of \$24, but saves \$6 by using a coupon. For what percentage off is this coupon?

Are you ready for more?

Clare paid full price for an item. Han bought the same item for 80% of the full price. Clare said, "I can't believe I paid 125% of what you paid, Han!" Is what she said true? Explain.

Activity

14.3 Info Gap: Music Devices

Your teacher will give you either a *problem card* or a *data card*. Do not show or read your card to your partner.

If your teacher gives you the *problem card:*	If your teacher gives you the *data card:*
1. Silently read your card and think about what information you need to be able to answer the question. **2.** Ask your partner for the specific information that you need. **3.** Explain how you are using the information to solve the problem. Continue to ask questions until you have enough information to solve the problem. **4.** Share the *problem card* and solve the problem independently. **5.** Read the *data card* and discuss your reasoning.	**1.** Silently read your card. **2.** Ask your partner *"What specific information do you need?"* and wait for them to ask for information. If your partner asks for information that is not on the card, do not do the calculations for them. Tell them you don't have that information. **3.** Before sharing the information, ask *"Why do you need that information?"* Listen to your partner's reasoning and ask clarifying questions. **4.** Read the *problem card* and solve the problem independently. **5.** Share the *data card* and discuss your reasoning.

Summary

Solving Percentage Problems

A pot can hold 36 liters of water. What percentage of the pot is filled when it contains 9 liters of water?

Here are two different ways to solve this problem:

- Using a double number line:

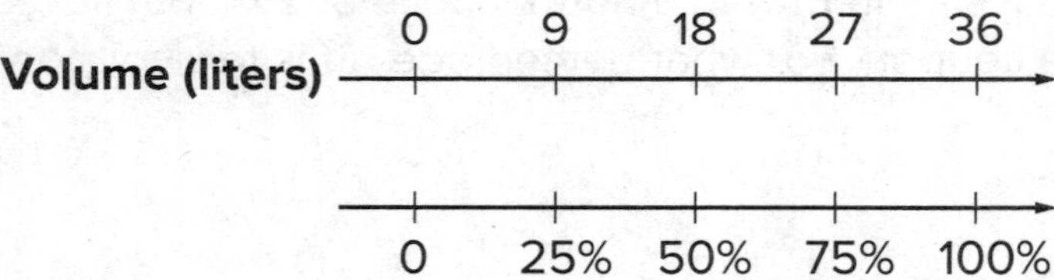

  We can divide the distance between 0 and 36 into four equal intervals, so 9 is $\frac{1}{4}$ of 36, or 9 is 25% of 36.

- Using a table:

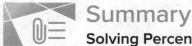

Volume (liters)	Percentage
36	100
9	25

$\cdot \frac{1}{4}$ (from 36 to 9) $\quad$ $\cdot \frac{1}{4}$ (from 100 to 25)

NAME ______________________ DATE ______________ PERIOD ____________

Practice

Solving Percentage Problems

1. For each problem, explain or show your reasoning.

   a. 160 is what percentage of 40?

   b. 40 is 160% of what number?

   c. What number is 40% of 160?

2. A store is having a 20%-off sale on all merchandise. If Mai buys one item and saves $13, what was the original price of her purchase? Explain or show your reasoning.

3. The original price of a scarf was $16. During a store-closing sale, a shopper saved $12 on the scarf. What percentage discount did she receive? Explain or show your reasoning.

4. Select **all** the expressions whose value is larger than 100.

   (A.) 120% of 100

   (B.) 50% of 150

   (C.) 150% of 50

   (D.) 20% of 800

   (E.) 200% of 30

   (F.) 500% of 400

   (G.) 1% of 1,000

5. An ant travels at a constant rate of 30 cm every 2 minutes. **(Lesson 3-8)**

   a. At what pace does the ant travel per centimeter?

   b. At what speed does the ant travel per minute?

6. Is $3\frac{1}{2}$ cups more or less than 1 liter? Explain or show your reasoning. (Note: 1 cup $\approx$ 236.6 milliliters) **(Lesson 3-4)**

7. Name a unit of measurement that is about the same size as each object. **(Lesson 3-2)**

   a. The distance of a doorknob from the floor is about 1 ________________.

   b. The thickness of a fingernail is about 1 ________________.

   c. The volume of a drop of honey is about 1 ________________.

   d. The weight or mass of a pineapple is about 1 ________________.

   e. The thickness of a picture book is about 1 ________________.

   f. The weight or mass of a buffalo is about 1 ________________.

   g. The volume of a flower vase is about 1 ________________.

   h. The weight or mass of 20 staples is about 1 ________________.

   i. The volume of a melon is about 1 ________________.

   j. The length of a piece of printer paper is about 1 ________________.

Lesson 3-15

Finding This Percent of That

NAME ______________________ DATE ______________ PERIOD ____________

Learning Goal Let's solve percentage problems like a pro.

Warm Up

15.1 Number Talk: Decimals

Find the value of each expression mentally.

1. $(0.23) \cdot 100$
2. $50 \div 100$
3. $145 \cdot \frac{1}{100}$
4. $7 \div 100$

Activity

15.2 Audience Size

A school held several evening activities last month—a music concert, a basketball game, a drama play, and literacy night. The music concert was attended by 250 people. How many people came to each of the other activities?

1. Attendance at a basketball game was 30% of attendance at the concert.

2. Attendance at the drama play was 140% of attendance at the concert.

3. Attendance at literacy night was 44% of attendance at the concert.

Are you ready for more?

50% of the people who attended the drama play also attended the music concert. What percentage of the people who attended the music concert also attended the drama play?

Activity

15.3 Everything Is On Sale

During a sale, every item in a store is 80% of its regular price.

1. If the regular price of a T-shirt is $10, what is its sale price?
2. The regular prices of five items are shown here. Find the sale price of each item.

	Item 1	Item 2	Item 3	Item 4	Item 5
Regular Price	$1	$4	$10	$55	$120
Sale Price					

3. You found 80% of many values. Was there a process you repeated over and over to find the sale prices? If so, describe it.

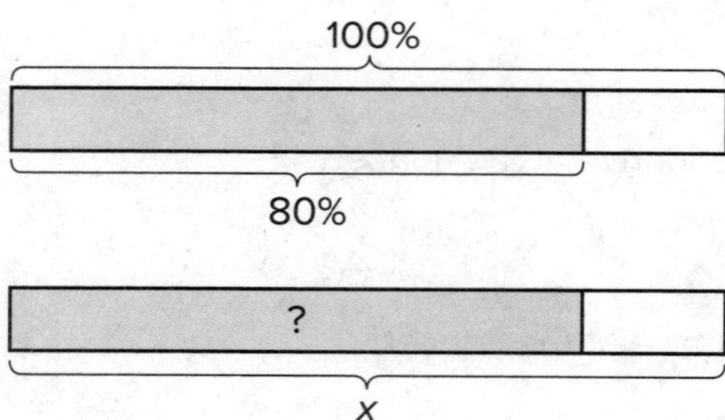

4. Select **all** of the expressions that could be used to find 80% of x. Be prepared to explain your reasoning.

A. $\frac{8}{100} \cdot x$
B. $\frac{80}{100} \cdot x$
C. $\frac{8}{10} \cdot x$
D. $\frac{4}{10} \cdot x$
E. $\frac{8}{5} \cdot x$
F. $\frac{4}{5} \cdot x$
G. $80 \cdot x$
H. $8 \cdot x$
I. $(0.8) \cdot x$
J. $(0.08) \cdot x$

NAME ______________________ DATE ______________ PERIOD ______________

Summary

Finding This Percent of That

To find 49% of a number, we can multiply the number by $\frac{49}{100}$ or 0.49.

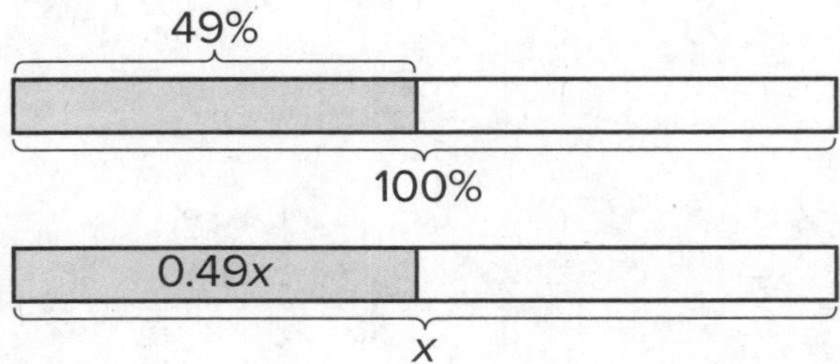

To find 135% of a number, we can multiply the number by $\frac{135}{100}$ or 1.35.

To find 6% of a number, we can multiply the number by $\frac{6}{100}$ or 0.06.

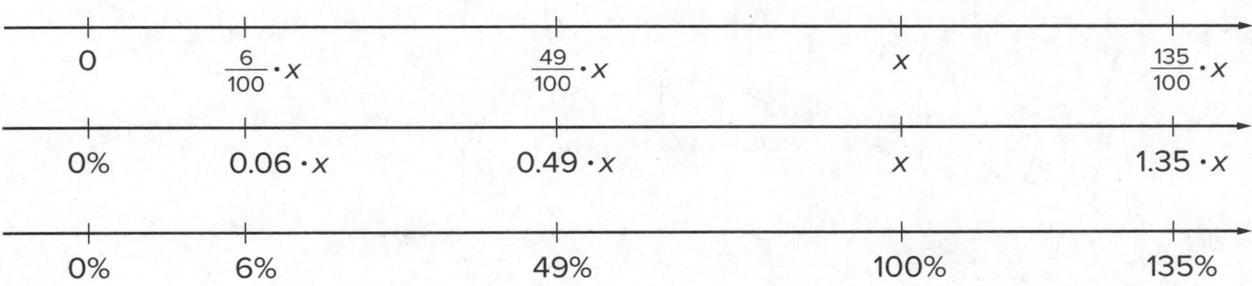

In general, to find P% of x, we can multiply:

$\frac{P}{100} \cdot x$

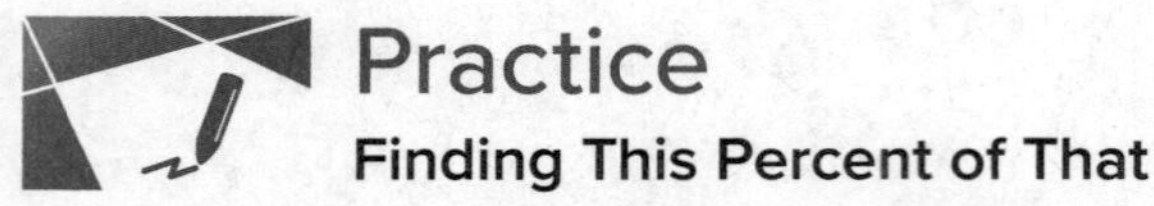

Practice
Finding This Percent of That

1. Respond to each question.

   a. To find 40% of 75, Priya calculates $\frac{2}{5} \cdot 75$. Does her calculation give the correct value for 40% of 75? Explain or show how you know.

   b. If x represents a number, does $\frac{2}{5} \cdot x$ always represent 40% of that number? Explain your reasoning.

2. Han spent 75 minutes practicing the piano over the weekend. For each question, explain or show your reasoning.

   a. Priya practiced the violin for 152% as much time as Han practiced the piano. How long did she practice?

   b. Tyler practiced the clarinet for 64% as much time as Han practiced the piano. How long did he practice?

NAME ________________ DATE ________ PERIOD ________

3. Last Sunday 1,575 people visited the amusement park. Fifty-six percent of the visitors were adults, 16% were teenagers, and 28% were children ages 12 and under. Find the number of adults, teenagers, and children that visited the park.

4. Order from greatest to least:

- 55% of 180
- 300% of 26
- 12% of 700

5. Complete each statement. **(Lesson 3-14)**

a. 20% of 60 is ______

b. 25% of ______ is 6

c. ______ % of 100 is 14

d. 50% of 90 is ______

e. 10% of ______ is 7

f. 30% of 70 is ____

6. A shopper needs 24 sandwich rolls. The store sells identical rolls in 2 differently sized packages. They sell a six-pack for $5.28 and a four-pack for $3.40. Should the shopper buy 4 six-packs or 6 four-packs? Explain your reasoning. **(Lesson 3-9)**

7. On a field trip, there are 3 chaperones for every 20 students. There are 92 people on the trip. Answer these questions. If you get stuck, consider using a tape diagram. **(Lesson 2-15)**

   a. How many chaperones are there?

   b. How many students are there?

Lesson 3-16

Finding the Percentage

NAME ______________________ DATE ______________ PERIOD ______________

Learning Goal Let's find percentages in general.

Warm Up

16.1 True or False: Percentages

Is each statement true or false? Be prepared to explain your reasoning.

1. 25% of 512 is equal to $\frac{1}{4} \cdot 500$.

2. 90% of 133 is equal to $(0.9) \cdot 133$.

3. 30% of 44 is equal to 3% of 440.

4. The percentage 21 is of 28 is equal to the percentage 30 is of 40.

Activity
16.2 Jumping Rope

A school held a jump-roping contest. Diego jumped rope for 20 minutes.

1. Jada jumped rope for 15 minutes. What percentage of Diego's time is that?

2. Lin jumped rope for 24 minutes. What percentage of Diego's time is that?

3. Noah jumped rope for 9 minutes. What percentage of Diego's time is that?

4. Record your answers in this table. Write the quotients in the last column as decimals.

	Time (minutes)	Percentage	Time ÷ 20
Diego	20	100	$\frac{20}{20} = 1$
Jada	15		$\frac{15}{20} =$
Lin	24		$\frac{24}{20} =$
Noah	9		$\frac{9}{20} =$

5. What do you notice about the numbers in the last two columns of the table?

NAME ______________________ DATE ____________ PERIOD ____________

Activity

16.3 Restaurant Capacity

A restaurant has a sign by the front door that says, "Maximum occupancy: 75 people." Answer each question. Explain or show your reasoning.

1. What percentage of its capacity is 9 people?

2. What percentage of its capacity is 51 people?

3. What percentage of its capacity is 84 people?

Are you ready for more?

Water makes up about 71% of Earth's surface, while the other 29% consists of continents and islands. 96% of all Earth's water is contained within the oceans as salt water, while the remaining 4% is fresh water located in lakes, rivers, glaciers, and the polar ice caps.

If the total volume of water on Earth is 1,386 million cubic kilometers, what is the volume of salt water? What is the volume of fresh water?

Summary

Finding the Percentage

What percentage of 90 kg is 36 kg? One way to solve this problem is to first find what percentage 1 kg is of 90, and then multiply by 36.

Mass (kg)	Percentage
90	100
1	$\frac{1}{90} \cdot 100$
36	$\frac{36}{90} \cdot 100$

(Both columns: $\cdot \frac{1}{90}$ from 90 to 1, then $\cdot 36$ from 1 to 36.)

From the table we can see that 1 kg is $\left(\frac{1}{90} \cdot 100\right)$%, so 36 kg is $\left(\frac{36}{90} \cdot 100\right)$% or 40% of 90. We can confirm this on a double number line:

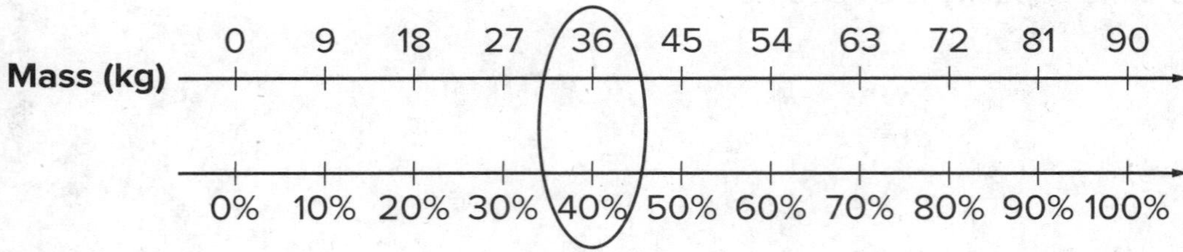

In general, to find what percentage a number C is of another number B is to calculate $\frac{C}{B}$ of 100%. We can do that by multiplying:

$\frac{C}{B} \cdot 100$

Suppose a school club has raised $88 for a project but needs a total of $160. What percentage of its goal has the club raised?

To find what percentage $88 is of $160, we find $\frac{88}{160}$ of 100% or $\frac{88}{160} \cdot 100$, which equals $\frac{11}{20} \cdot 100$ or 55.

The club has raised 55% of its goal.

NAME ______________________ DATE ______________ PERIOD ____________

Practice

Finding the Percentage

1. A sign in front of a roller coaster says "You must be 40 inches tall to ride." What percentage of this height is:

   a. 34 inches?

   b. 54 inches?

2. At a hardware store, a tool set normally costs $80. During a sale this week, the tool set costs $12 less than usual. What percentage of the usual price is the savings? Explain or show your reasoning.

3. A bathtub can hold 80 gallons of water. The faucet flows at a rate of 4 gallons per minute. What percentage of the tub will be filled after 6 minutes?

4. The sale price of every item in a store is 85% of its usual price. **(Lesson 3-15)**

   a. The usual price of a backpack is \$30, what is its sale price?

   b. The usual price of a sweatshirt is \$18, what is its sale price?

   c. The usual price of a soccer ball is \$24.80, what is its sale price?

5. A shopper needs 48 hot dogs. The store sells identical hot dogs in 2 differently sized packages. They sell a six-pack of hot dogs for \$2.10, and an eight-pack of hot dogs for \$3.12. Should the shopper buy 8 six-packs, or 6 eight-packs? Explain your reasoning. **(Lesson 3-9)**

6. Elena is 56 inches tall. **(Lesson 3-4)**

   a. What is her height in centimeters? (Note: 100 inches = 254 centimeters)

   b. What is her height in meters?

Lesson 3-17

Painting a Room

NAME ______________________ DATE ____________ PERIOD ____________

Learning Goal Let's see what it takes to paint a room.

Warm Up

17.1 Getting Ready to Paint

What are some tools that are helpful when painting a room?

Activity

17.2 How Much It Costs to Paint

Here is the floor plan for a bedroom:

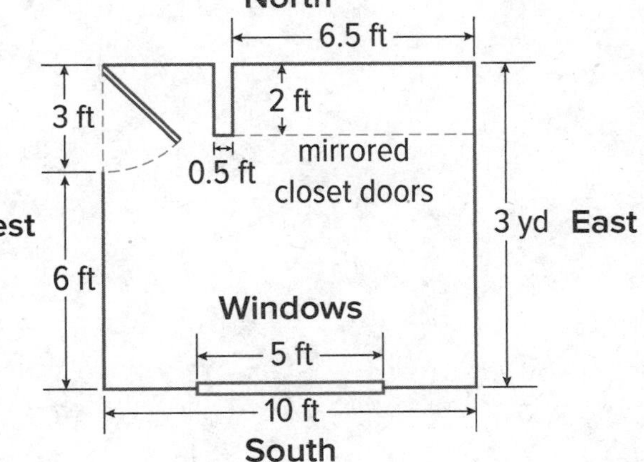

Imagine you are planning to repaint all the walls in this room, including inside the closet.

- The east wall is 3 yards long.
- The south wall is 10 feet long but has a window, 5 feet by 3 feet, that does not need to be painted.
- The west wall is 3 yards long but has a door, 7 feet tall and 3 feet wide, that does not need to be painted.
- The north wall includes a closet, 6.5 feet wide, with floor-to-ceiling mirrored doors that do not need to be painted. There is, however, a smaller wall between the west wall and the closet that does need to be painted on all sides. The wall is 0.5 feet wide and extends 2 feet into the room.
- The ceiling in this room is 8 feet high.
- All of the corners are right angles.

1. If you paint all the walls in the room, how many square feet do you need to cover?

2. An advertisement about the paint that you want to use reads: "Just 2 quarts covers 175 square feet!" If you need to apply two coats of paint on all the walls, how much paint do you need to buy?

NAME ______________________ DATE ____________ PERIOD ________

3. Paint can only be purchased in 1-quart, 1-gallon, and 5-gallon containers. How much will all supplies for the project cost if the cans of paint cost $10.90 for a quart, $34.90 for a gallon, and $165.00 for 5 gallons?

4. You have a coupon for 20% off all quart-sized paint cans. How does that affect the cost of the project?

Activity

17.3 How Long It Takes to Paint

After buying the supplies, you start painting the east wall. It takes you 96 minutes to put two coats of paint on that wall (not including a lunch break between the two coats).

1. Your friend stops by to see how you are doing and comments that you are 25% finished with the painting. Are they correct?

2. Your friend offers to help you with the rest of the painting. It takes the two of you 150 more minutes of painting time to finish the entire room. How much time did your friend save you?

Learning Targets

Lesson	Learning Target(s)
3-1 The Burj Khalifa	• I can see that thinking about "how much for 1" is useful for solving different types of problems.
3-2 Anchoring Units of Measurement	• I can name common objects that are about as long as 1 inch, foot, yard, mile, millimeter, centimeter, meter, or kilometer. • I can name common objects that weigh about 1 ounce, pound, ton, gram, or kilogram, or that hold about 1 cup, quart, gallon, milliliter, or liter. • When I read or hear a unit of measurement, I know whether it is used to measure length, weight, or volume.
3-3 Measuring with Different-Sized Units	• When I know a measurement in one unit, I can decide whether it takes more or less of a different unit to measure the same quantity.

(continued on the next page)

(continued from the previous page)

Lesson	Learning Target(s)
3-4 Converting Units	• I can convert measurements from one unit to another, using double number lines, tables, or by thinking about "how much for 1." • I know that when we measure things in two different units, the pairs of measurements are equivalent ratios.
3-5 Comparing Speeds and Prices	• I understand that if two ratios have the same rate per 1, they are equivalent ratios. • When measurements are expressed in different units, I can decide who is traveling faster or which item is the better deal by comparing "how much for 1" of the same unit.
3-6 Interpreting Rates	• I can choose which unit rate to use based on how I plan to solve the problem. • When I have a ratio, I can calculate its two unit rates and explain what each of them means in the situation.
3-7 Equivalent Ratios Have the Same Unit Rates	• I can give an example of two equivalent ratios and show that they have the same unit rates. • I can multiply or divide by the unit rate to calculate missing values in a table of equivalent ratios.

Lesson	Learning Target(s)
3-8 More about Constant Speed	• I can solve more complicated problems about constant speed situations.
3-9 Solving Rate Problems	• I can choose how to use unit rates to solve problems.
3-10 What Are Percentages?	• I can create a double number line with percentages on one line and dollar amounts on the other line. • I can explain the meaning of percentages using dollars and cents as an example.
3-11 Percentages and Double Number Lines	• I can use double number line diagrams to solve different problems like "What is 40% of 60?" or "60 is 40% of what number?"

(continued on the next page)

(continued from the previous page)

Lesson	Learning Target(s)
3-12 Percentages and Tape Diagrams	• I can use tape diagrams to solve different problems like "What is 40% of 60?" or "60 is 40% of what number?"
3-13 Benchmark Percentages	• When I read or hear that something is 10%, 25%, 50%, or 75% of an amount, I know what fraction of that amount they are referring to.
3-14 Solving Percentage Problems	• I can choose and create diagrams to help me solve problems about percentages.
3-15 Finding This Percent of That	• I can solve different problems like "What is 40% of 60?" by dividing and multiplying.

Lesson	Learning Target(s)
3-16 Finding the Percentage	• I can solve different problems like "60 is what percentage of 40?" by dividing and multiplying.
3-17 Painting a Room	• I can apply what I have learned about unit rates and percentages to predict how long it will take and how much it will cost to paint all the walls in a room.

(continued on the next page)

Notes:

Unit 4

Dividing Fractions

Yana Gayvoronskaya/123RF

How do fractions and division relate to jars of jam? In this unit, you'll use digital tools, diagrams, and equations to answer this question.

Topics

- Making Sense of Division
- Meanings of Fraction Division
- Algorithm for Fraction Division
- Fractions in Lengths, Areas, and Volumes
- Let's Put It to Work

Unit 4

Dividing Fractions

Lesson 4-1

Size of Divisor and Size of Quotient

NAME ______________________ DATE ____________ PERIOD ________

Learning Goal Let's explore quotients of different sizes.

Warm Up

1.1 Number Talk: Size of Dividend and Divisor

Find the value of each expression mentally.

1. 5,000 ÷ 5 **2.** 5,000 ÷ 2,500 **3.** 5,000 ÷ 10,000 **4.** 5,000 ÷ 500,000

Activity

1.2 All Stacked Up

1. Here are several types of objects. For each type of object, estimate how many are in a stack that is 5 feet high. Be prepared to explain your reasoning.

Cardboard Boxes

Bricks

Notebooks

Coins

2. A stack of books is 72 inches tall. Each book is 2 inches thick. Which expression tells us how many books are in the stack? Be prepared to explain your reasoning.

$72 \cdot 2$ $\qquad$ $72 - 2$ $\qquad$ $2 \div 72$ $\qquad$ $72 \div 2$

3. Another stack of books is 43 inches tall. Each book is $\frac{1}{2}$-inch thick. Write an expression that represents the number of books in the stack.

Activity

1.3 All in Order

Your teacher will give you two sets of papers with division expressions.

1. Without computing, estimate the quotients in each set and order them from greatest to least. Be prepared to explain your reasoning.

Pause here for a class discussion.

Record the expressions in each set in order from the greatest value to the least.

a. Set 1

b. Set 2

NAME ____________________ DATE __________ PERIOD __________

2. Without computing, estimate the quotients and sort them into the following three groups. Be prepared to explain your reasoning.

$30 \div \frac{1}{2}$	$9 \div 10$	$18 \div 19$	$15{,}000 \div 1{,}500{,}000$
$30 \div 0.45$	$9 \div 10{,}000$	$18 \div 0.18$	$15{,}000 \div 14{,}500$

Close to 0 **Close to 1** **Much larger than 1**

Are you ready for more?

Write 10 expressions of the form $12 \div ?$ in a list ordered from least to greatest. Can you list expressions that have value near 1 without equaling 1? How close can you get to the value 1?

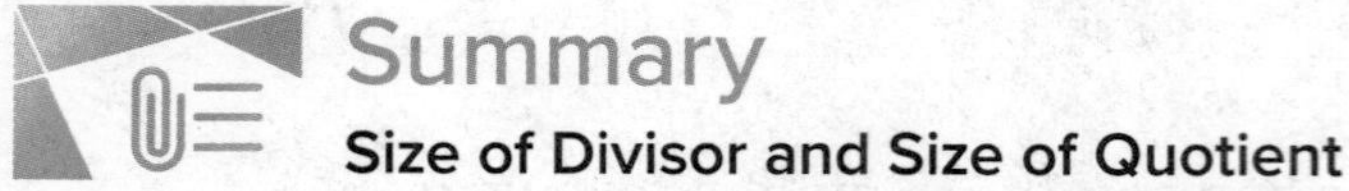

Summary

Size of Divisor and Size of Quotient

Here is a division expression:

$60 \div 4.$

In this division, we call 60 the *dividend* and 4 the *divisor.* The result of the division is the quotient.

In this example, the quotient is 15, because $60 \div 4 = 15$.

We don't always have to make calculations to have a sense of what a quotient will be. We can reason about it by looking at the size of the dividend and the divisor.

Let's look at some examples.

- In $100 \div 11$ and in $18 \div 2.9$ the dividend is larger than the divisor.
  $100 \div 11$ is very close to $99 \div 11$, which is 9.
  The quotient $18 \div 2.9$ is close to $18 \div 3$ or 6.

  In general, when a larger number is divided by a smaller number, the quotient is greater than 1.

- In $99 \div 101$ and in $7.5 \div 7.4$ the dividend and divisor are very close to each other. $99 \div 101$ is very close to $99 \div 100$, which is $\frac{99}{100}$ or 0.99.
  The quotient $7.5 \div 7.4$ is close to $7.5 \div 7.5$, which is 1.

  In general, when we divide two numbers that are nearly equal to each other, the quotient is close to 1.

- In $10 \div 101$ and in $50 \div 198$ the dividend is smaller than the divisor.
  $10 \div 101$ is very close to $10 \div 100$, which is $\frac{10}{100}$ or 0.1.
  The division $50 \div 198$ is close to $50 \div 200$, which is $\frac{1}{4}$ or 0.25.

  In general, when a smaller number is divided by a larger number, the quotient is less than 1.

NAME ______________________ DATE ____________ PERIOD ____________

Practice

Size of Divisor and Size of Quotient

1. Order from smallest to largest:

- Number of pennies in a stack that is 1 ft high
- Number of books in a stack that is 1 ft high
- Number of dollar bills in a stack that is 1 ft high
- Number of slices of bread in a stack that is 1 ft high

2. Use each of the numbers 4, 40, and 4,000 once to complete the sentences.

a. The value of __________ ÷ 40.01 is close to 1.

b. The value of __________ ÷ 40.01 is much less than 1.

c. The value of __________ ÷ 40.01 is much greater than 1.

3. Without computing, decide whether the value of each expression is much smaller than 1, close to 1, or much greater than 1.

a. $100 \div \frac{1}{1{,}000}$

b. $50\frac{1}{3} \div 50\frac{1}{4}$

c. $4.7 \div 5.2$

d. $2 \div 7{,}335$

e. $2{,}000{,}001 \div 9$

f. $0.002 \div 2{,}000$

4. A rocking horse has a weight limit of 60 pounds. **(Lesson 3-16)**

a. What percentage of the weight limit is 33 pounds?

b. What percentage of the weight limit is 114 pounds?

c. What weight is 95% of the limit?

5. Compare using >, =, or <. **(Lesson 3-15)**

a. 0.7 ______ 0.70

b. $0.03 + \frac{6}{10}$ ______ $0.30 + \frac{6}{100}$

c. 0.9 ______ 0.12

6. Diego has 90 songs on his playlist. How many songs are there for each genre? **(Lesson 3-14)**

a. 40% rock

b. 10% country

c. 30% hip-hop

d. The rest is electronica.

7. A garden hose emits 9 quarts of water in 6 seconds. At this rate: **(Lesson 3-8)**

a. How long will it take the hose to emit 12 quarts?

b. How much water does the hose emit in 10 seconds?

Lesson 4-2

Meanings of Division

NAME ______________________ DATE ______________ PERIOD ______________

Learning Goal Let's explore ways to think about division.

Warm Up

2.1 A Division Expression

Here is an expression: $20 \div 4$.
What are some ways to think about this expression?
Describe at least two meanings you think it could have.

Activity

2.2 Bags of Almonds

A baker has 12 pounds of almonds. She puts them in bags, so that each bag has the same weight.

Clare and Tyler drew diagrams and wrote equations to show how they were thinking about $12 \div 6$.

	12	
6		6

$\_\_\_ \cdot 6 = 12$

Clare's diagram and equation

12					
2	2	2	2	2	2

$6 \cdot \_\_\_ = 12$

Tyler's diagram and equation

1. How do you think Clare and Tyler thought about $12 \div 6$? Explain what each diagram and the parts of each equation could mean about the situation with the bags of almonds. Make sure to include the meaning of the missing number.

   Pause here for a class discussion.

2. Explain what each division expression could mean about the situation with the bags of almonds. Then draw a diagram and write a multiplication equation to show how you are thinking about the expression.

a. $12 \div 4$

b. $12 \div 2$

c. $12 \div \frac{1}{2}$

Are you ready for more?

A loaf of bread is cut into slices.

1. If each slice is $\frac{1}{2}$ of a loaf, how many slices are there?

2. If each slice is $\frac{1}{5}$ of a loaf, how many slices are there?

3. What happens to the number of slices as each slice gets smaller?

4. What would dividing by 0 mean in this situation about slicing bread?

NAME ______________________ DATE ______________ PERIOD ______________

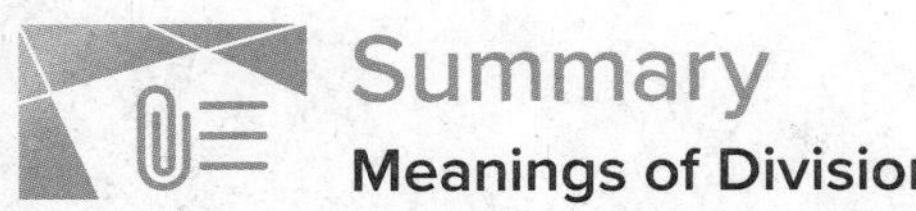

Summary

Meanings of Division

Suppose 24 bagels are being distributed into boxes.
The expression $24 \div 3$ could be understood in two ways:

- 24 bagels are distributed equally into 3 boxes, as represented by this diagram.

24		
8	8	8

- 24 bagels are distributed into boxes, 3 bagels in each box, as represented by this diagram.

24							
3	3	3	3	3	3	3	3

In both interpretations, the quotient is the same ($24 \div 3 = 8$), but it has different meanings in each case.

- In the first case, the 8 represents the number of bagels in each of the 3 boxes.
- In the second, it represents the number of boxes that were formed with 3 bagels in each box.

These two ways of seeing division are related to how 3, 8, and 24 are related in a multiplication. Both $3 \cdot 8$ and $8 \cdot 3$ equal 24.

- $3 \cdot 8 = 24$ can be read as "3 groups of 8 make 24."
- $8 \cdot 3 = 24$ can be read as "8 groups of 3 make 24."

If 3 and 24 are the only numbers given, the multiplication equations would be:

$3 \cdot ? = 24$

$? \cdot 3 = 24$

In both cases, the division $24 \div 3$ can be used to find the value of the "?"
But now we see that it can be interpreted in more than one way, because the "?" can refer to *the size of a group* (as in "3 groups of what number make 24?"), or to *the number of groups* (as in "How many groups of 3 make 24?").

Practice

Meanings of Division

1. Twenty pounds of strawberries are being shared equally by a group of friends. The equation $20 \div 5 = 4$ represents the division of strawberries.
   a. If the 5 represents the number of people, what does the 4 represent?

   b. If the 5 represents the pounds of strawberries per person, what does the 4 represent?

2. A sixth-grade science club needs $180 to pay for the tickets to a science museum. All tickets cost the same amount.

   What could $180 \div 15$ mean in this situation? Describe two different possible meanings of this expression. Then, find the quotient and explain what it means in each case.

3. Write a multiplication equation that corresponds to each division equation.
   a. $10 \div 5 = ?$

   b. $4.5 \div 3 = ?$

   c. $\frac{1}{2} \div 4 = ?$

NAME ______________________ DATE ____________ PERIOD ____________

4. Write a division or multiplication equation that represents each situation. Use a "?" for the unknown quantity.

a. 2.5 gallons of water are poured into 5 equally sized bottles. How much water is in each bottle?

b. A large bucket of 200 golf balls is divided into 4 smaller buckets. How many golf balls are in each small bucket?

c. Sixteen socks are put into pairs. How many pairs are there?

5. Find a value for a that makes each statement true. **(Lesson 4-1)**

a. $a \div 6$ is greater than 1

b. $a \div 6$ is equal to 1

c. $a \div 6$ is less than 1

d. $a \div 6$ is equal to a whole number

6. Complete the table. Write each percentage as a percent of 1. **(Lesson 3-14)**

Fraction	Decimal	Percentage
$\frac{1}{4}$	0.25	25% of 1
	0.1	
		75% of 1
$\frac{1}{5}$		
	1.5	
		140% of 1

7. Jada walks at a speed of 3 miles per hour. Elena walks at a speed of 2.8 miles per hour. If they both begin walking along a walking trail at the same time, how much farther will Jada walk after 3 hours? Explain your reasoning. **(Lesson 3-8)**

Lesson 4-3

Interpreting Division Situations

NAME ______________________ DATE ______________ PERIOD ____________

Learning Goal Let's explore situations that involve division.

Warm Up

3.1 Dot Image: Properties of Multiplication

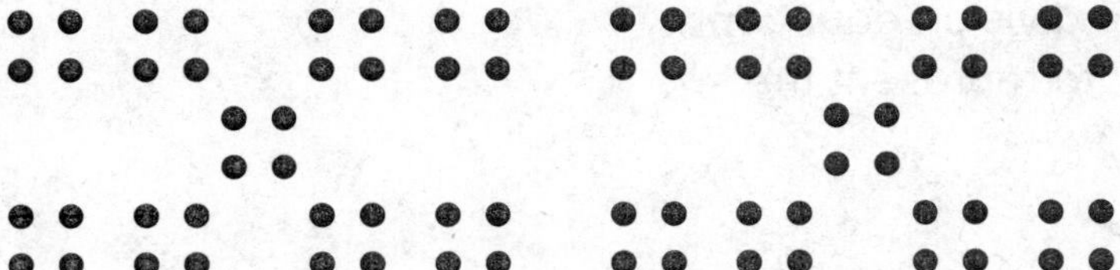

Activity

3.2 Homemade Jams

Draw a diagram, and write a multiplication equation to represent each situation. Then answer the question.

1. Mai had 4 jars. In each jar, she put $2\frac{1}{4}$ cups of homemade blueberry jam. Altogether, how many cups of jam are in the jars?

2. Priya filled 5 jars, using a total of $7\frac{1}{2}$ cups of strawberry jam. How many cups of jam are in each jar?

3. Han had some jars. He put $\frac{3}{4}$ cup of grape jam in each jar, using a total of $6\frac{3}{4}$ cups. How many jars did he fill?

Activity
3.3 Making Granola

1. Consider the problem: To make 1 batch of granola, Kiran needs 26 ounces of oats. The only measuring tool he has is a 4-ounce scoop. How many scoops will it take to measure 26 ounces of oats?

   a. Will the answer be more than 1 or less than 1?

   b. Write a multiplication equation and a division equation that represent this situation. Use "?" to represent the unknown quantity.

   c. Find the unknown quantity. If you get stuck, consider drawing a diagram.

2. The recipe calls for 14 ounces of mixed nuts. To get that amount, Kiran uses 4 bags of mixed nuts.

   a. Write a mathematical question that might be asked about this situation.

   b. What might the equation $14 \div 4 = ?$ represent in Kiran's situation?

   c. Find the quotient. Show your reasoning. If you get stuck, consider drawing a diagram.

NAME ______________________ DATE ______________ PERIOD ____________

Summary

Interpreting Division Situations

If a situation involves equal-sized groups, it is helpful to make sense of it in terms of the number of groups, the size of each group, and the total amount. Here are three examples to help us better understand such situations.

- Suppose we have 3 bottles with $6\frac{1}{2}$ ounces of water in each, and the total amount of water is not given. Here we have 3 groups, $6\frac{1}{2}$ ounces in each group, and an unknown total, as shown in this diagram.

  ?

$6\frac{1}{2}$	$6\frac{1}{2}$	$6\frac{1}{2}$

  We can express this situation as a multiplication problem. The unknown is the product, so we can simply multiply the 2 known numbers to find it.

  $$3 \cdot 6\frac{1}{2} = ?$$

- Next, suppose we have 20 ounces of water to fill 6 equal-sized bottles, and the amount in each bottle is not given.

  Here we have 6 groups, an unknown amount in each, and a total of 20. We can represent it like this:

  20

?	?	?	?	?	?

  This situation can also be expressed using multiplication, but the unknown is a factor, rather than the product:

  $$6 \cdot ? = 20$$

  To find the unknown, we cannot simply multiply, but we can think of it as a division problem.

  $$20 \div 6 = ?$$

- Now, suppose we have 40 ounces of water to pour into bottles, 12 ounces in each bottle, but the number of bottles is not given.

  Here we have an unknown number of groups, 12 in each group, and a total of 40.

  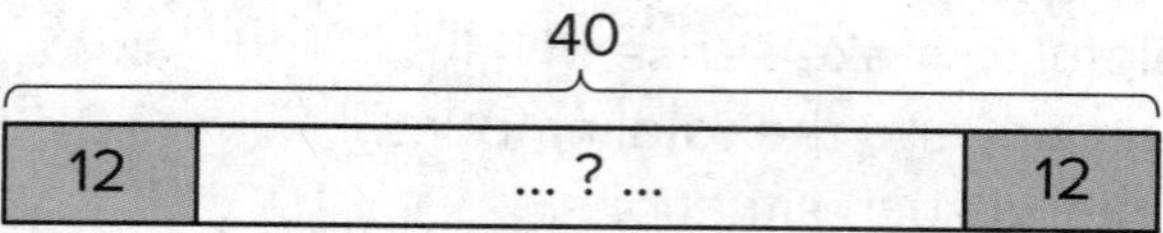

  Again, we can think of this in terms of multiplication, with a different factor being the unknown:

  $$? \cdot 12 = 40$$

  Likewise, we can use division to find the unknown:

  $$40 \div 12 = ?$$

Whenever we have a multiplication situation, one factor tells us *how many groups* there are, and the other factor tells us *how much is in each group*.

Sometimes we want to find...

- the total.
- how many groups there are.
- how much is in each group.

Anytime we want to find out how many groups there are or how much is in each group, we can represent the situation using division.

NAME ______________________ DATE ____________ PERIOD ____________

Practice

Interpreting Division Situations

1. Write a multiplication equation and a division equation that this diagram could represent.

54		
18	18	18

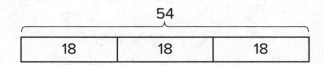

2. Consider the problem: Mai has \$36 to spend on movie tickets. Each movie ticket costs \$4.50. How many tickets can she buy?

   a. Write a multiplication equation and a division equation to represent this situation.

   b. Find the answer. Draw a diagram, if needed.

   c. Use the multiplication equation to check your answer.

3. Kiran said that this diagram can show the solution to $16 \div 8 = ?$ or $16 \div 2 = ?$, depending on how we think about the equations and the "?."

16	
8	8

   Explain or show how Kiran is correct.

4. Write a sentence describing a situation that could be represented by the equation $4 \div 1\frac{1}{3} = ?$. **(Lesson 4-2)**

5. Noah said, "When you divide a number by a second number, the result will always be smaller than the first number."

   Jada said, "I think the result could be larger or smaller, depending on the numbers."

   Do you agree with either of them? Explain or show your reasoning. **(Lesson 4-1)**

6. Mini muffins cost $3.00 per dozen.

   - Andre says, "I have $2.00, so I can afford 8 muffins."
   - Elena says, "I want to get 16 muffins, so I'll need to pay $4.00."

   Do you agree with either of them? Explain your reasoning. **(Lesson 3-7)**

7. A family has a monthly budget of $2,400. How much money is spent on each category? **(Lesson 3-15)**

   a. 44% is spent on housing.

   b. 23% is spent on food.

   c. 6% is spent on clothing.

   d. 17% is spent on transportation.

   e. The rest is put into savings.

Lesson 4-4

How Many Groups? (Part 1)

NAME ______________________ DATE ____________ PERIOD ____________

Learning Goal Let's play with blocks and diagrams to think about division with fractions.

Warm Up

4.1 Equal-sized Groups

Write a multiplication equation and a division equation for each sentence or diagram.

1. Eight $5 bills are worth $40.

2. There are 9 thirds in 3 ones.

3.

$$\overbrace{\begin{array}{|c:c:c:c:c|}\hline \frac{1}{5} & \frac{1}{5} & \frac{1}{5} & \frac{1}{5} & \frac{1}{5} \\ \hline\end{array}}^{1}$$

Activity

4.2 Reasoning with Pattern Blocks

Your teacher will give you pattern blocks as shown here. Use them to answer the questions.

1. If a hexagon represents 1 whole, what fraction does each of the following shapes represent? Be prepared to show or explain your reasoning.

- 1 triangle
- 1 rhombus
- 1 trapezoid
- 4 triangles
- 3 rhombuses
- 2 hexagons
- 1 hexagon and 1 trapezoid

2. Here are Elena's diagrams for $2 \cdot \frac{1}{2} = 1$ and $6 \cdot \frac{1}{3} = 2$. Do you think these diagrams represent the equations? Explain or show your reasoning.

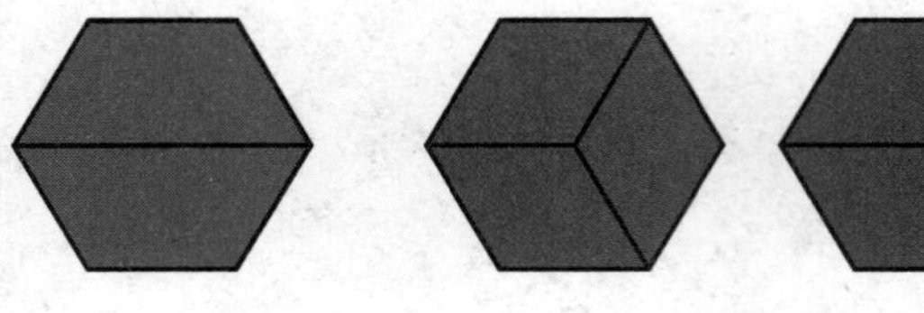

$2 \cdot \frac{1}{2} = 1$ $6 \cdot \frac{1}{3} = 2$

3. Use pattern blocks to represent each multiplication equation. Remember that a hexagon represents 1 whole.

a. $3 \cdot \frac{1}{6} = \frac{1}{2}$

b. $2 \cdot \frac{3}{2} = 3$

4. Answer the questions. If you get stuck, consider using pattern blocks.

a. How many $\frac{1}{2}$s are in 4?

b. How many $\frac{2}{3}$s are in 2?

c. How many $\frac{1}{6}$s are in $1\frac{1}{2}$?

NAME ______________________ DATE ____________ PERIOD ____________

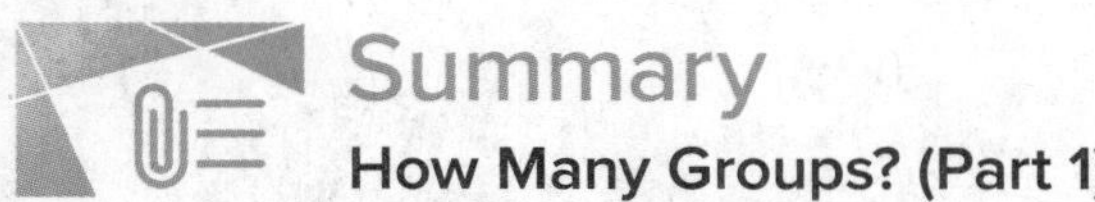

Summary
How Many Groups? (Part 1)

Some problems that involve equal-sized groups also involve fractions.

Here is an example: "How many $\frac{1}{6}$ are in 2?"

We can express this question with multiplication and division equations.

$$? \cdot \frac{1}{6} = 2$$

$$2 \div \frac{1}{6} = ?$$

Pattern-block diagrams can help us make sense of such problems.

Here is a set of pattern blocks.

If the hexagon represents 1 whole, then a triangle must represent $\frac{1}{6}$, because 6 triangles make 1 hexagon. We can use the triangle to represent the $\frac{1}{6}$ in the problem.

Twelve triangles make 2 hexagons, which means there are 12 groups of $\frac{1}{6}$ in 2.

If we write the 12 in the place of the "?" in the original equations, we have:

$$12 \cdot \frac{1}{6} = 2$$

$$2 \div \frac{1}{6} = 12$$

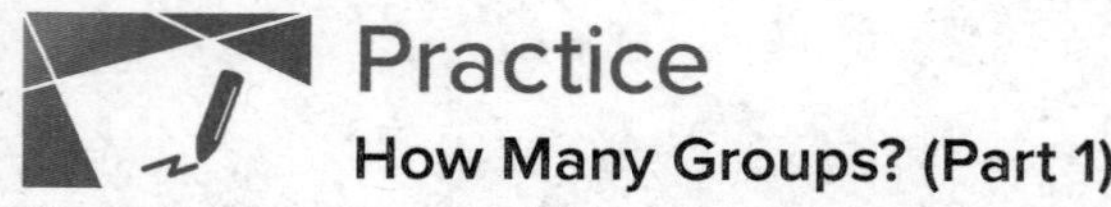

Practice

How Many Groups? (Part 1)

1. Consider the problem: A shopper buys cat food in bags of 3 lbs. Her cat eats $\frac{3}{4}$ lb each week. How many weeks does one bag last?

   a. Draw a diagram to represent the situation and label your diagram so it can be followed by others. Answer the question.

   b. Write a multiplication or division equation to represent the situation.

   c. Multiply your answer in the first question (the number of weeks) by $\frac{3}{4}$. Did you get 3 as a result? If not, revise your previous work.

2. Use the diagram to answer the question: How many $\frac{1}{3}$s are in $1\frac{2}{3}$? The hexagon represents 1 whole. Explain or show your reasoning.

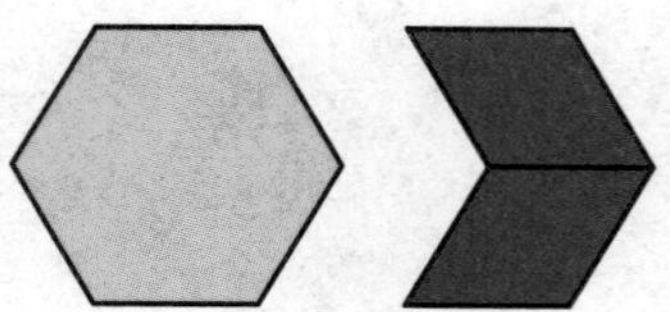

NAME ______________________ DATE ____________ PERIOD ____________

3. Which question can be represented by the equation $? \cdot \frac{1}{8} = 3$?

(A.) How many 3s are in $\frac{1}{8}$?

(B.) What is 3 groups of $\frac{1}{8}$?

(C.) How many $\frac{1}{8}$s are in 3?

(D.) What is $\frac{1}{8}$ of 3?

4. Write two division equations for each multiplication equation.

a. $15 \cdot \frac{2}{5} = 6$

b. $6 \cdot \frac{4}{3} = 8$

c. $16 \cdot \frac{7}{8} = 14$

5. Noah and his friends are going to an amusement park. The total cost of admission for 8 students is $100, and all students share the cost equally. Noah brought $13 for his ticket. Did he bring enough money to get into the park? Explain your reasoning. **(Lesson 4-2)**

6. Write a division expression with a quotient that is: **(Lesson 4-1)**

   a. greater than $8 \div 0.001$

   b. less than $8 \div 0.001$

   c. between $8 \div 0.001$ and $8 \div \frac{1}{10}$

7. Find each unknown number. **(Lesson 3-14)**

   a. 12 is 150% of what number?

   b. 5 is 50% of what number?

   c. 10% of what number is 300?

   d. 5% of what number is 72?

   e. 20 is 80% of what number?

Lesson 4-5

How Many Groups? (Part 2)

NAME ______________________ DATE ____________ PERIOD ____________

Learning Goal Let's use blocks and diagrams to understand more about division with fractions.

Warm Up

5.1 Reasoning with Fraction Strips

Write a fraction or whole number as an answer for each question. If you get stuck, use the fraction strips. Be prepared to share your reasoning.

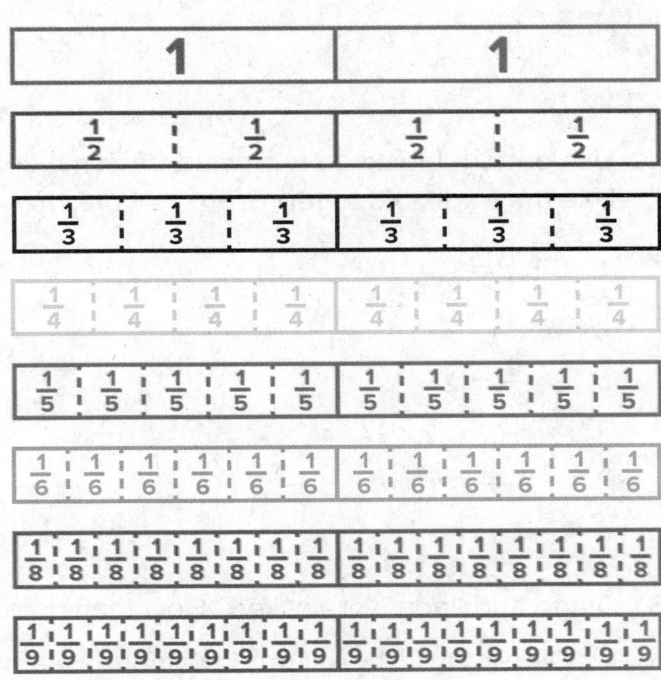

1. How many $\frac{1}{2}$s are in 2?

2. How many $\frac{1}{5}$s are in 3?

3. How many $\frac{1}{8}$s are in $1\frac{1}{4}$?

4. $1 \div \frac{2}{6} = ?$

5. $2 \div \frac{2}{9} = ?$

6. $4 \div \frac{2}{10} = ?$

Activity

5.2 More Reasoning with Pattern Blocks

Your teacher will give you pattern blocks. Use them to answer the questions.

1. If the trapezoid represents 1 whole, what do each of the other shapes represent? Be prepared to show or explain your reasoning.

2. Use pattern blocks to represent each multiplication equation. Use the trapezoid to represent 1 whole.

   a. $3 \cdot \frac{1}{3} = 1$

   b. $3 \cdot \frac{2}{3} = 2$

3. Diego and Jada were asked "How many rhombuses are in a trapezoid?"

   - Diego says, "$1\frac{1}{3}$. If I put 1 rhombus on a trapezoid, the leftover shape is a triangle, which is $\frac{1}{3}$ of the trapezoid."
   - Jada says, "I think it's $1\frac{1}{2}$. Since we want to find out 'how many rhombuses,' we should compare the leftover triangle to a rhombus. A triangle is $\frac{1}{2}$ of a rhombus."

   Do you agree with either of them? Explain or show your reasoning.

NAME ______________________ DATE ____________ PERIOD ____________

4. Select **all** the equations that can be used to answer the question: "How many rhombuses are in a trapezoid?"

A. $\frac{2}{3} \div ? = 1$

B. $? \cdot \frac{2}{3} = 1$

C. $1 \div \frac{2}{3} = ?$

D. $1 \cdot \frac{2}{3} = ?$

E. $? \div \frac{2}{3} = 1$

Activity

5.3 Drawing Diagrams to Show Equal-sized Groups

For each situation, draw a diagram for the relationship of the quantities to help you answer the question. Then write a multiplication equation or a division equation for the relationship. Be prepared to share your reasoning.

1. The distance around a park is $\frac{3}{2}$ miles. Noah rode his bicycle around the park for a total of 3 miles. How many times around the park did he ride?

2. You need $\frac{3}{4}$ yard of ribbon for one gift box. You have 3 yards of ribbon. How many gift boxes do you have ribbon for?

3. The water hose fills a bucket at $\frac{1}{3}$ gallon per minute. How many minutes does it take to fill a 2-gallon bucket?

Are you ready for more?

How many heaping teaspoons are in a heaping tablespoon? How would the answer depend on the shape of the spoons?

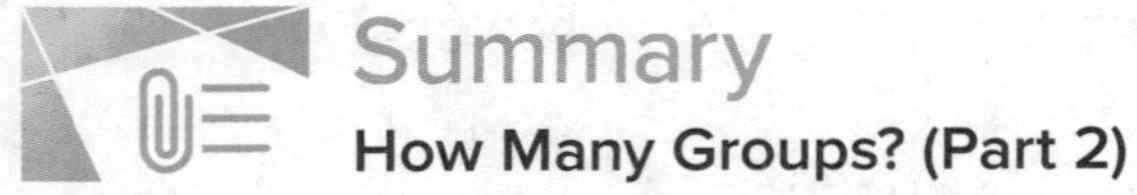

Summary

How Many Groups? (Part 2)

Suppose one batch of cookies requires $\frac{2}{3}$ cup flour. How many batches can be made with 4 cups of flour?

We can think of the question as being: "How many $\frac{2}{3}$s are in 4?" and represent it using multiplication and division equations.

$? \cdot \frac{2}{3} = 4$ $\qquad$ $4 \div \frac{2}{3} = ?$

Let's use pattern blocks to visualize the situation and say that a hexagon is 1 whole.

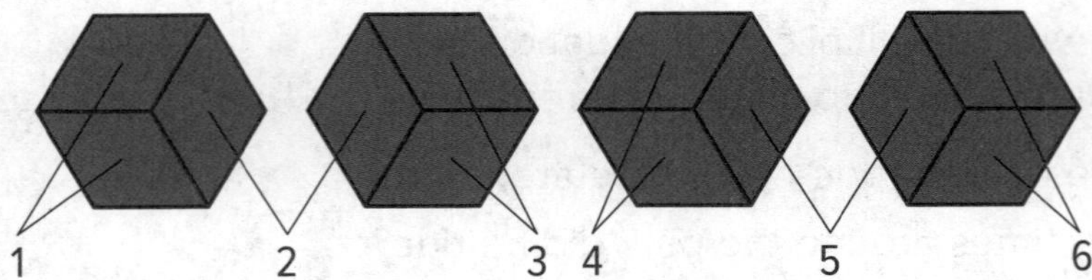

Since 3 rhombuses make a hexagon, 1 rhombus represents $\frac{1}{3}$ and 2 rhombuses represent $\frac{2}{3}$. We can see that 6 pairs of rhombuses make 4 hexagons, so there are 6 groups of $\frac{2}{3}$ in 4.

Other kinds of diagrams can also help us reason about equal-sized groups involving fractions. This example shows how we might reason about the same question from above: "How many $\frac{2}{3}$-cups are in 4 cups?"

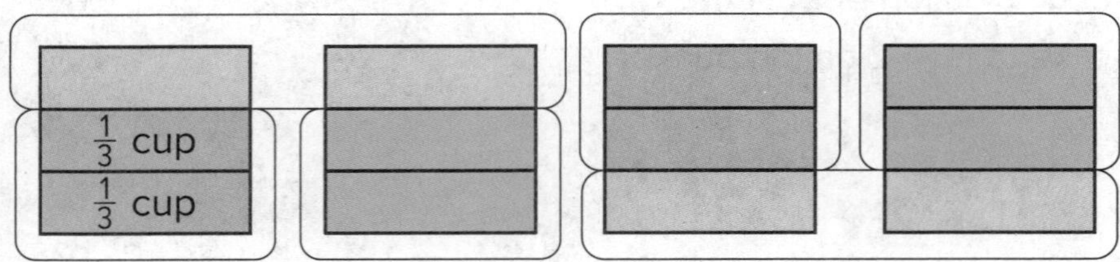

We can see each "cup" partitioned into thirds, and that there are 6 groups of $\frac{2}{3}$-cup in 4 cups. In both diagrams, we see that the unknown value (or the "?" in the equations) is 6. So we can now write:

$6 \cdot \frac{2}{3} = 4$ $\qquad$ $4 \div \frac{2}{3} = 6$

NAME ______________________ DATE ______________ PERIOD ____________

Practice

How Many Groups? (Part 2)

1. Use the tape diagram to find the value of $\frac{1}{2} \div \frac{1}{3}$. Show your reasoning.

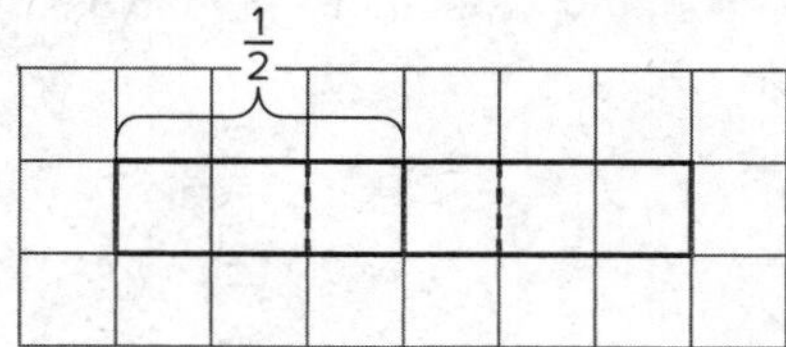

2. What is the value of $\frac{1}{2} \div \frac{1}{3}$? Use pattern blocks to represent and find this value. The yellow hexagon represents 1 whole. Explain or show your reasoning.

3. Use a standard inch ruler to answer each question. Then, write a multiplication equation and a division equation that answer the question.

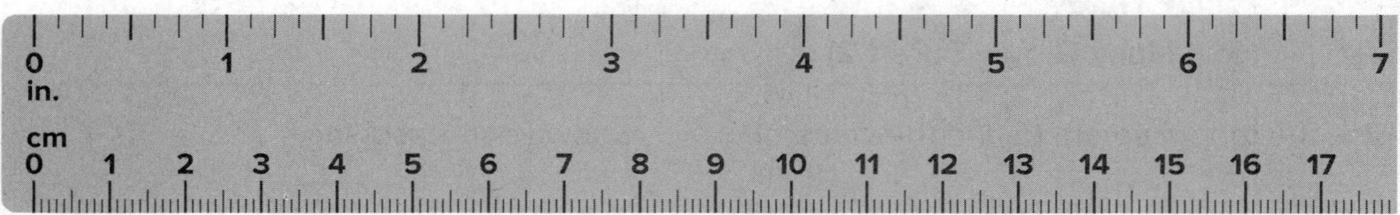

a. How many $\frac{1}{2}$s are in 7?

b. How many $\frac{3}{8}$s are in 6?

c. How many $\frac{5}{16}$s are in $1\frac{7}{8}$?

4. Use the tape diagram to answer the question: How many $\frac{2}{5}$s are in $1\frac{1}{2}$? Show your reasoning.

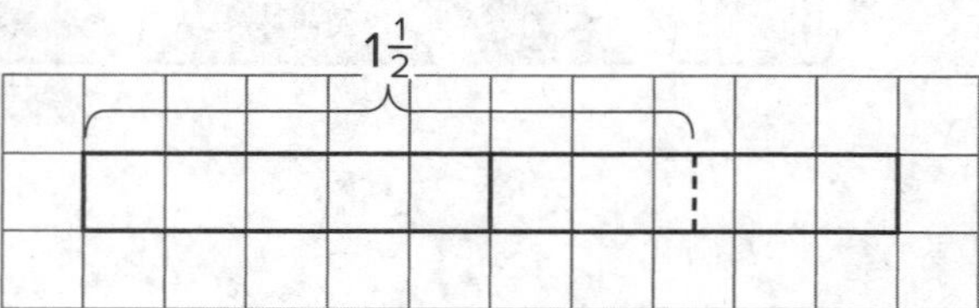

NAME ______________________ DATE ____________ PERIOD ____________

5. Write a multiplication equation and a division equation to represent each sentence or diagram. **(Lesson 4-4)**

a. There are 12 fourths in 3.

b.

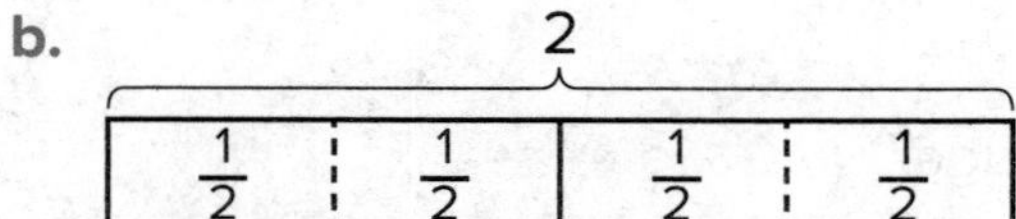

c. How many $\frac{2}{3}$s are in 6?

d.

2

$\frac{2}{5}$	$\frac{2}{5}$	$\frac{2}{5}$	$\frac{2}{5}$	$\frac{2}{5}$

6. At a farmer's market, two vendors sell fresh milk. One vendor sells 2 liters for \$3.80, and another vendor sells 1.5 liters for \$2.70. Which is the better deal? Explain your reasoning. **(Lesson 3-5)**

7. A recipe uses 5 cups of flour for every 2 cups of sugar. **(Lesson 3-6)**
   a. How much sugar is used for 1 cup of flour?

   b. How much flour is used for 1 cup of sugar?

   c. How much flour is used with 7 cups of sugar?

   d. How much sugar is used with 6 cups of flour?

Lesson 4-6

Using Diagrams to Find the Number of Groups

NAME ______________________ DATE ____________ PERIOD ____________

Learning Goal Let's draw tape diagrams to think about division with fractions.

Warm Up

6.1 How Many of These in That?

1. We can think of the division expression $10 \div 2\frac{1}{2}$ as the question: "How many groups of $2\frac{1}{2}$ are in 10?" Complete the tape diagram to represent this question. Then find the answer.

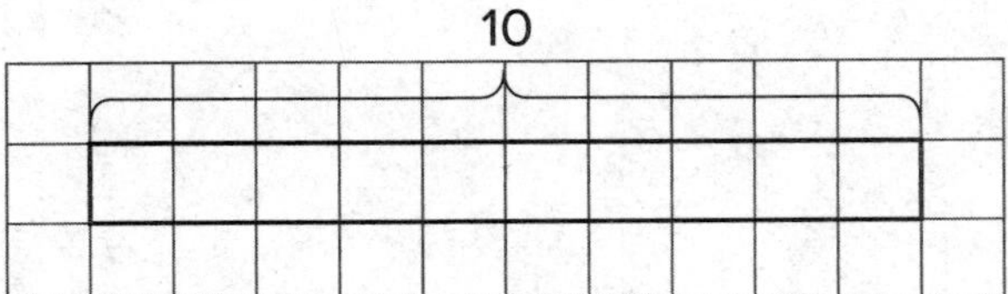

2. Complete the tape diagram to represent the question: "How many groups of 2 are in 7?" Then find the answer.

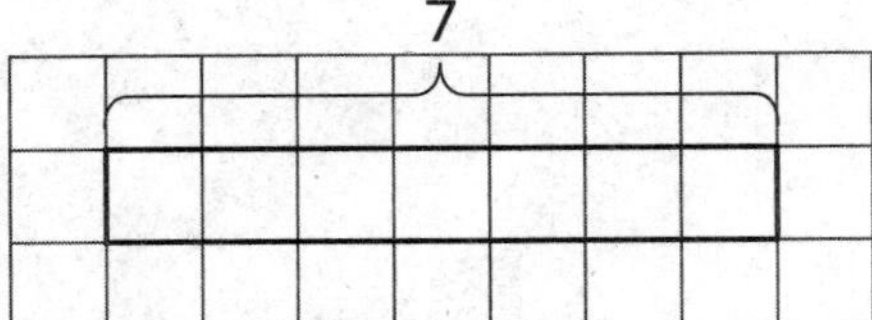

Activity

6.2 Representing Groups of Fractions with Tape Diagrams

To make sense of the question "How many $\frac{2}{3}$s are in 1?," Andre wrote equations and drew a tape diagram.

$$? \cdot \frac{2}{3} = 1$$

$$1 \div \frac{2}{3} = ?$$

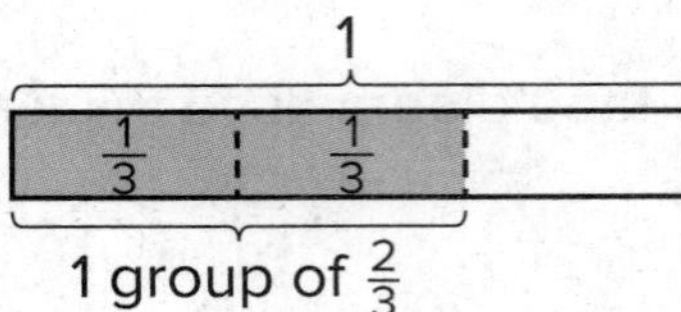

1. In an earlier task, we used pattern blocks to help us solve the equation $1 \div \frac{2}{3} = ?$. Explain how Andre's tape diagram can also help us solve the equation.

2. Write a multiplication equation and a division equation for each question. Then, draw a tape diagram and find the answer.

   a. How many $\frac{3}{4}$s are in 1?

NAME ________________________ DATE ______________ PERIOD ______________

b. How many $\frac{2}{3}$s are in 3?

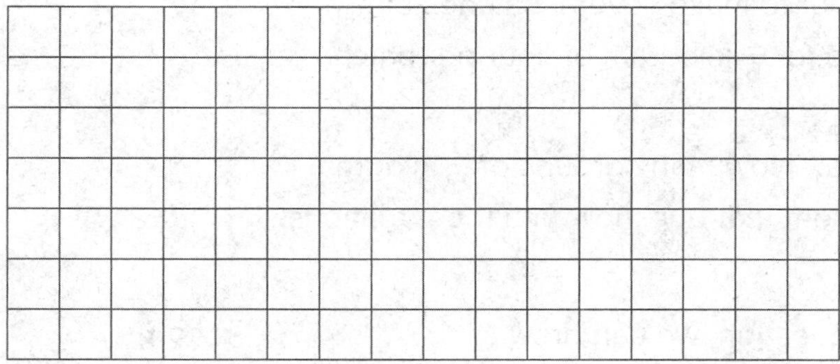

c. How many $\frac{3}{2}$s are in 5?

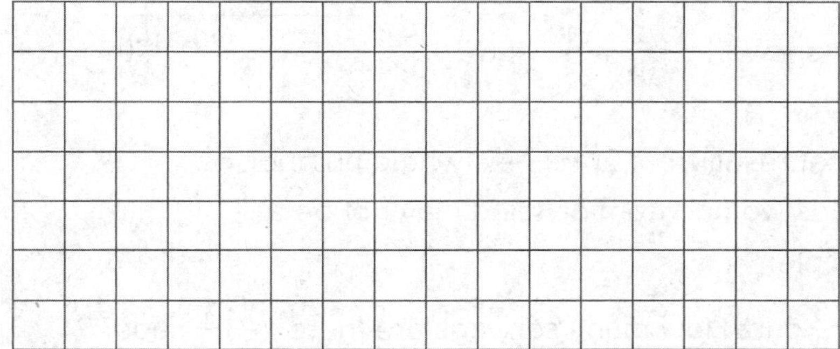

Activity
6.3 Finding Number of Groups

1. Write a multiplication equation or a division equation for each question. Then, find the answer and explain or show your reasoning.
   a. How many $\frac{3}{8}$-inch thick books make a stack that is 6 inches tall?

   b. How many groups of $\frac{1}{2}$ pound are in $2\frac{3}{4}$ pounds?

2. Write a question that can be represented by the division equation $5 \div 1\frac{1}{2} = ?$. Then, find the answer and explain or show your reasoning.

Summary

Using Diagrams to Find the Number of Groups

A baker used 2 kilograms of flour to make several batches of a pastry recipe. The recipe called for $\frac{2}{5}$ kilogram of flour per batch. How many batches did she make?

We can think of the question as: "How many groups of $\frac{2}{5}$ kilogram make 2 kilograms?" and represent that question with the equations:

$$? \cdot \frac{2}{5} = 2 \qquad\qquad 2 \div \frac{2}{5} = ?$$

To help us make sense of the question, we can draw a tape diagram. This diagram shows 2 whole kilograms, with each kilogram partitioned into fifths.

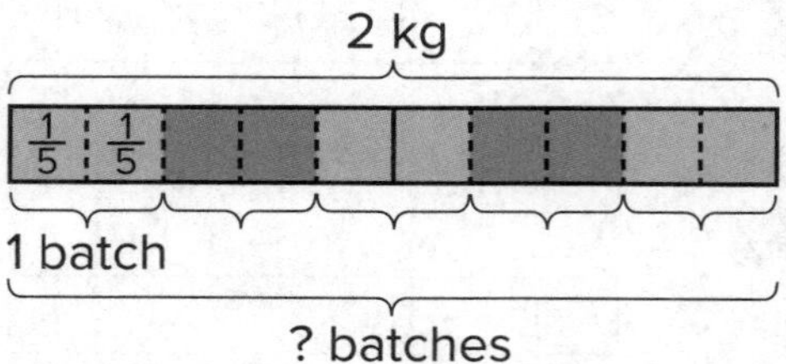

We can see there are 5 groups of $\frac{2}{5}$ in 2. Multiplying 5 and $\frac{2}{5}$ allows us to check this answer: $5 \cdot \frac{2}{5} = \frac{10}{5}$ and $\frac{10}{5} = 2$, so the answer is correct.

Notice the number of groups that result from $2 \div \frac{2}{5}$ is a whole number. Sometimes the number of groups we find from dividing may not be a whole number. Here is an example:

Suppose one serving of rice is $\frac{3}{4}$ cup. How many servings are there in $3\frac{1}{2}$ cups?

$$? \cdot \frac{3}{4} = 3\frac{1}{2} \qquad\qquad 3\frac{1}{2} \div \frac{3}{4} = ?$$

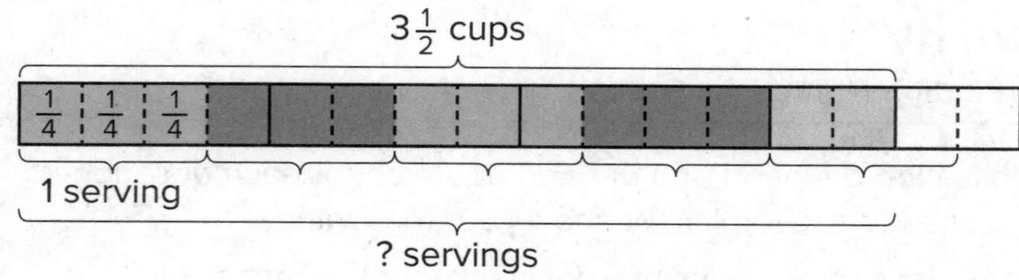

Looking at the diagram, we can see there are 4 full groups of $\frac{3}{4}$, plus 2 fourths. If 3 fourths make a whole group, then 2 fourths make $\frac{2}{3}$ of a group. So, the number of servings (the "?" in each equation) is $4\frac{2}{3}$. We can check this by multiplying $4\frac{2}{3}$ and $\frac{3}{4}$. $4\frac{2}{3} \cdot \frac{3}{4} = \frac{14}{3} \cdot \frac{3}{4}$, and $\frac{14}{3} \cdot \frac{3}{4} = \frac{14}{4}$, which is indeed equivalent to $3\frac{1}{2}$.

NAME ______________________ DATE ______________ PERIOD ______________

Practice

Using Diagrams to Find the Number of Groups

1. We can think of $3 \div \frac{1}{4}$ as the question "How many groups of $\frac{1}{4}$ are in 3?" Draw a tape diagram to represent this question. Then find the answer.

2. Describe how to draw a tape diagram to represent and answer $3 \div \frac{3}{5} = ?$ for a friend who was absent.

3. How many groups of $\frac{1}{2}$ day are in 1 week?

a. Write a multiplication equation or a division equation to represent the question.

b. Draw a tape diagram to show the relationship between the quantities and to answer the question. Use graph paper, if needed.

4. Diego said that the answer to the question "How many groups of $\frac{5}{6}$ are in 1?" is $\frac{6}{5}$ or $1\frac{1}{5}$. Do you agree with him? Explain or show your reasoning.

5. Select **all** the equations that can represent the question: "How many groups of $\frac{4}{5}$ are in 1?" **(Lesson 4-5)**

   A. $? \cdot 1 = \frac{4}{5}$

   B. $1 \cdot \frac{4}{5} = ?$

   C. $\frac{4}{5} \div 1 = ?$

   D. $? \cdot \frac{4}{5} = 1$

   E. $1 \div \frac{4}{5} = ?$

6. Calculate each percentage mentally. **(Lesson 3-14)**

   a. What is 10% of 70?

   b. What is 10% of 110?

   c. What is 25% of 160?

   d. What is 25% of 48?

   e. What is 50% of 90?

   f. What is 50% of 350?

   g. What is 75% of 300?

   h. What is 75% of 48?

Lesson 4-7

What Fraction of a Group?

NAME ______________________ DATE ____________ PERIOD ____________

Learning Goal Let's think about dividing things into groups when we can't even make one whole group.

Warm Up

7.1 Estimating a Fraction of a Number

1. Estimate the quantities.
   a. What is $\frac{1}{3}$ of 7?

   b. What is $\frac{4}{5}$ of $9\frac{2}{3}$?

   c. What is $2\frac{4}{7}$ of $10\frac{1}{9}$?

2. Write a multiplication expression for each of the previous questions.

Activity

7.2 Fractions of Ropes

Here is a diagram that shows four ropes of different lengths.

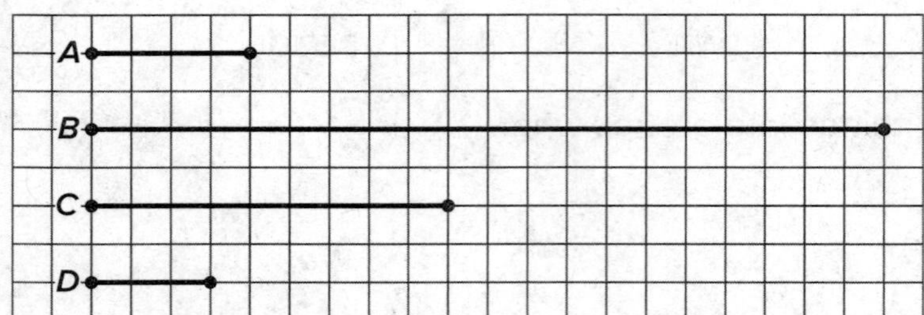

1. Complete each sentence comparing the lengths of the ropes. Then, use the measurements shown on the grid to write a multiplication equation and a division equation for each comparison.

   a. Rope B is ________ times as long as Rope A.

   b. Rope C is ________ times as long as Rope A.

   c. Rope D is ________ times as long as Rope A.

2. Each equation can be used to answer a question about Ropes C and D. What could each question be?

   a. $? \cdot 3 = 9$ and $9 \div 3 = ?$

   b. $? \cdot 9 = 3$ and $3 \div 9 = ?$

NAME ______________________ DATE ______________ PERIOD ____________

Activity
7.3 Fractional Batches of Ice Cream

One batch of an ice cream recipe uses 9 cups of milk. A chef makes different amounts of ice cream on different days. Here are the amounts of milk she used:

- Monday: 12 cups
- Tuesday: $22\frac{1}{2}$ cups
- Thursday: 6 cups
- Friday: $7\frac{1}{2}$ cups

1. How many batches of ice cream did she make on these days? For each day, write a division equation, draw a tape diagram, and find the answer.

a. Monday

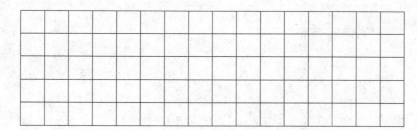

b. Tuesday

2. What fraction of a batch of ice cream did she make on these days? For each day, write a division equation, draw a tape diagram, and find the answer.

a. Thursday

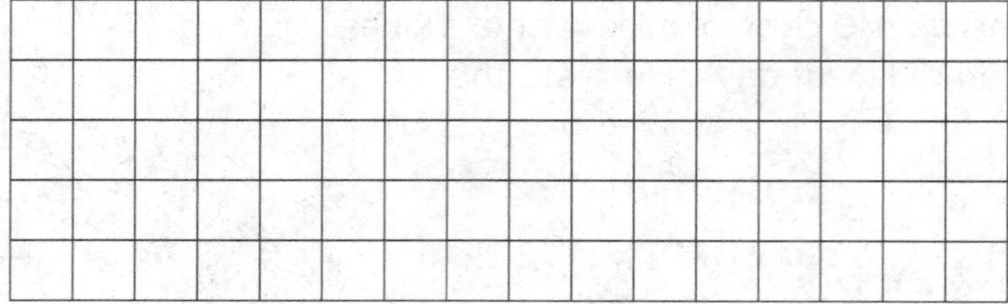

b. Friday

3. For each question, write a division equation, draw a tape diagram, and find the answer.

a. What fraction of 9 is 3?

b. What fraction of 5 is $\frac{1}{2}$?

NAME ______________________ DATE ____________ PERIOD ________

Summary

What Fraction of a Group?

It is natural to think about groups when we have more than one group, but we can also have a *fraction of a group*.

To find the amount in a fraction of a group, we can multiply the fraction by the amount in the whole group. If a bag of rice weighs 5 kg, $\frac{3}{4}$ of a bag would weigh $\left(\frac{3}{4} \cdot 5\right)$ kg.

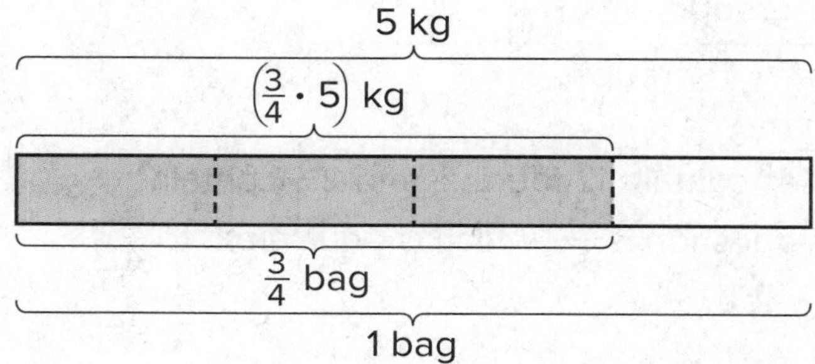

Sometimes we need to find what fraction of a group an amount is.

Suppose a full bag of flour weighs 6 kg. A chef used 3 kg of flour. What fraction of a full bag was used? In other words, what fraction of 6 kg is 3 kg?

This question can be represented by a multiplication equation and a division equation, as well as by a diagram.

Equations

$? \cdot 6 = 3$

$3 \div 6 = ?$

Diagram

6 kg

3 kg

? bag

1 bag

We can see from the diagram that 3 is $\frac{1}{2}$ of 6, and we can check this answer by multiplying: $\frac{1}{2} \cdot 6 = 3$.

In *any* situation where we want to know what fraction one number is of another number, we can write a division equation to help us find the answer.

For example, "What fraction of 3 is $2\frac{1}{4}$?" can be expressed as $? \cdot 3 = 2\frac{1}{4}$, which can also be written as $2\frac{1}{4} \div 3 = ?$

The answer to "What is $2\frac{1}{4} \div 3$?" is also the answer to the original question.

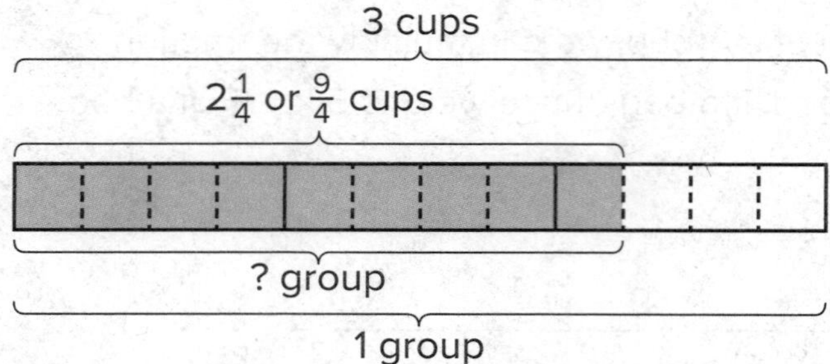

The diagram shows that 3 wholes contain 12 fourths, and $2\frac{1}{4}$ contains 9 fourths, so the answer to this question is $\frac{9}{12}$, which is equivalent to $\frac{3}{4}$.

We can use diagrams to help us solve other division problems that require finding a fraction of a group.

For example, here is a diagram to help us answer the question: "What fraction of $\frac{9}{4}$ is $\frac{3}{2}$?," which can be written as $\frac{3}{2} \div \frac{9}{4} = ?$

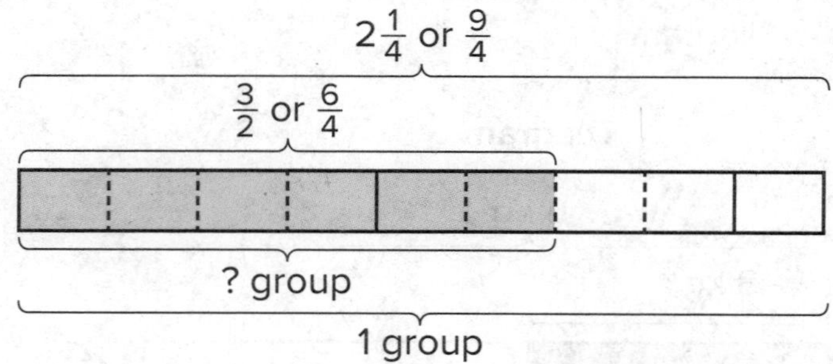

We can see that the quotient is $\frac{6}{9}$, which is equivalent to $\frac{2}{3}$.
To check this, let's multiply.
$\frac{2}{3} \cdot \frac{9}{4} = \frac{18}{12}$, and $\frac{18}{12}$ is, indeed, equal to $\frac{3}{2}$.

NAME ______________________ DATE ______________ PERIOD ______________

Practice

What Fraction of a Group?

1. A recipe calls for $\frac{1}{2}$ lb of flour for 1 batch. How many batches can be made with each of these amounts?

   a. 1 lb b. $\frac{3}{4}$ lb c. $\frac{1}{4}$ lb

2. Whiskers the cat weighs $2\frac{2}{3}$ kg. Piglio weighs 4 kg. For each question, write a multiplication equation and a division equation, decide whether the answer is greater than 1 or less than 1, and then find the answer.

   a. How many times as heavy as Piglio is Whiskers?

   b. How many times as heavy as Whiskers is Piglio?

3. Andre is walking from his home to a festival that is $1\frac{5}{8}$ kilometers away. He walks $\frac{1}{3}$ kilometer and then takes a quick rest. Which question can be represented by the equation $? \cdot 1\frac{5}{8} = \frac{1}{3}$ in this situation?

   (A) What fraction of the trip has Andre completed?

   (B) What fraction of the trip is left?

   (C) How many more kilometers does Andre have to walk to get to the festival?

   (D) How many kilometers is it from home to the festival and back home?

4. Draw a tape diagram to represent the question: What fraction of $2\frac{1}{2}$ is $\frac{4}{5}$? Then find the answer.

5. How many groups of $\frac{3}{4}$ are in each of these quantities? **(Lesson 4-6)**

a. $\frac{11}{4}$

b. $6\frac{1}{2}$

6. Which question can be represented by the equation $4 \div \frac{2}{7} = ?$ **(Lesson 4-4)**

(A.) What is 4 groups of $\frac{2}{7}$?

(B.) How many $\frac{2}{7}$s are in 4?

(C.) What is $\frac{2}{7}$ of 4?

(D.) How many 4s are in $\frac{2}{7}$?

Lesson 4-8

How Much in Each Group? (Part 1)

NAME ______________________ DATE ____________ PERIOD ____________

Learning Goal Let's look at division problems that help us find the size of one group.

Warm Up

8.1 Inventing a Situation

1. Think of a situation with a question that can be represented by the equation $12 \div \frac{2}{3} = ?$. Describe the situation and the question.

2. Trade descriptions with your partner, and answer your partner's question.

Activity

8.2 How Much in One Batch?

To make 5 batches of cookies, 10 cups of flour are required. Consider the question: How many cups of flour does each batch require?

We can write equations and draw a diagram to represent this situation.

Equations

$5 \cdot ? = 10$

$10 \div 5 = ?$

Diagram

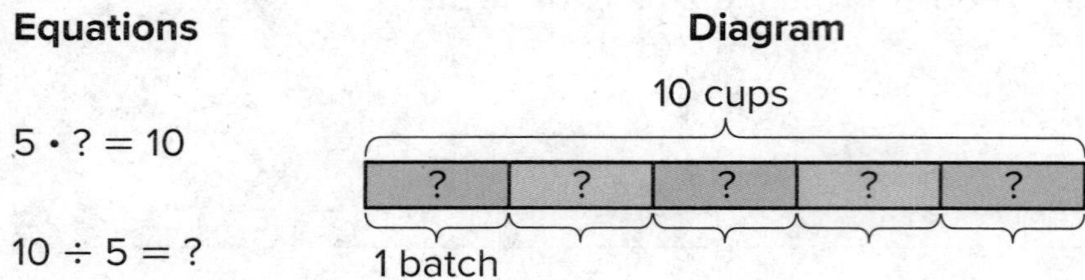

This helps us see that each batch requires 2 cups of flour.

For each question, write a multiplication equation and a division equation, draw a diagram, and find the answer.

1. To make 4 batches of cupcakes, it takes 6 cups of flour. How many cups of flour are needed for 1 batch?

2. To make $\frac{1}{2}$ batch of rolls, it takes $\frac{5}{4}$ cups of flour. How many cups of flour are needed for 1 batch?

3. Two cups of flour make $\frac{2}{3}$ batch of bread. How many cups of flour make 1 batch?

NAME ______________________ DATE ____________ PERIOD ________

Activity

8.3 One Container and One Section of Highway

Here are three tape diagrams that represent situations about filling containers of water.

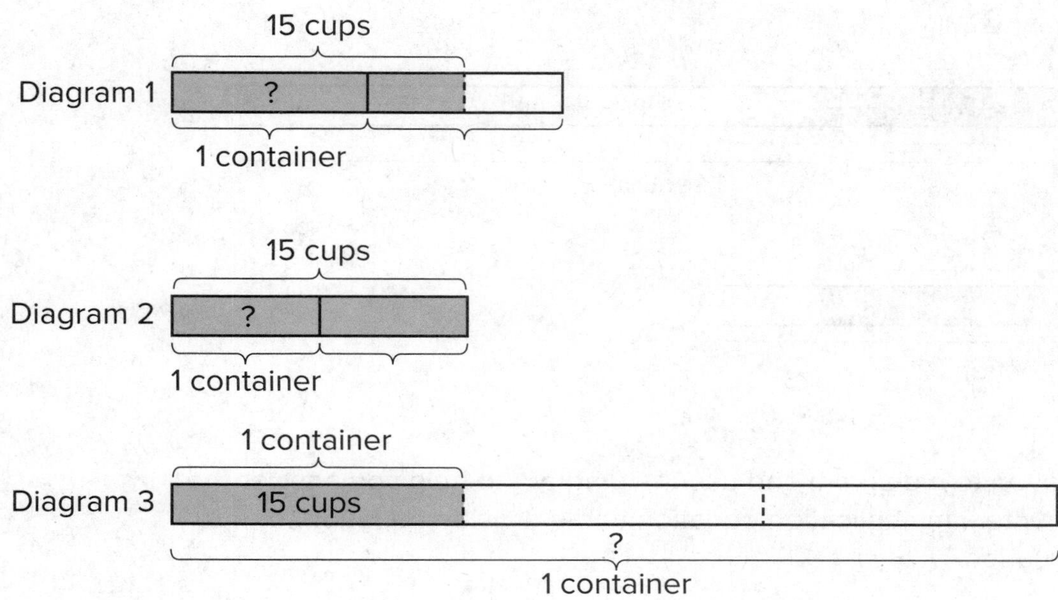

Match each situation to a diagram and use the diagram to help you answer the question. Then, write a multiplication equation and a division equation to represent the situation.

1. Tyler poured a total of 15 cups of water into 2 equal-sized bottles and filled each bottle. How much water was in each bottle?

2. Kiran poured a total of 15 cups of water into equal-sized pitchers and filled $1\frac{1}{2}$ pitchers. How much water was in the full pitcher?

3. It takes 15 cups of water to fill $\frac{1}{3}$ pail. How much water is needed to fill 1 pail?

Here are tape diagrams that represent situations about cleaning sections of highway.

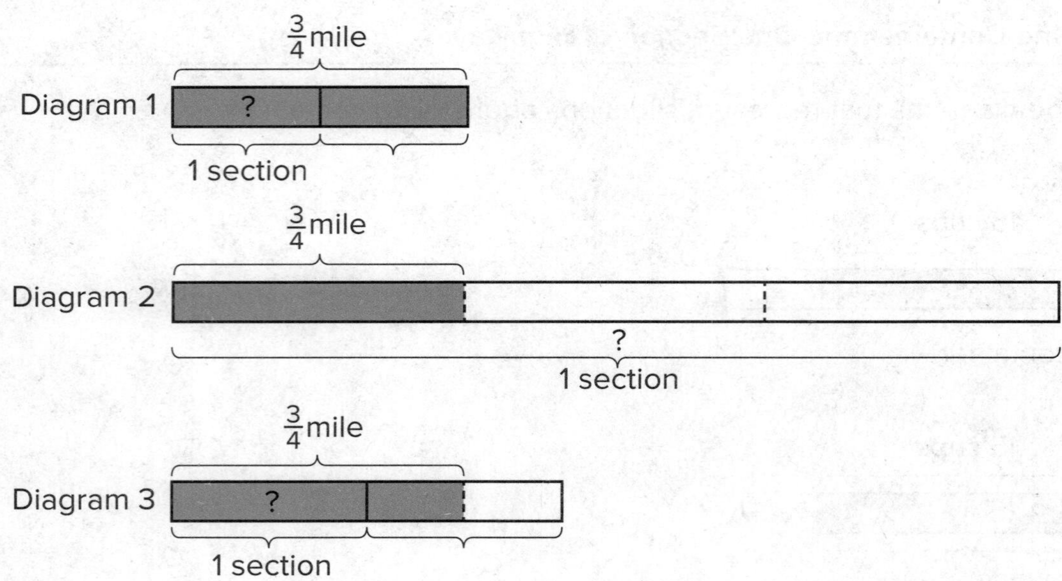

Match each situation to a diagram and use the diagram to help you answer the question. Then, write a multiplication equation and a division equation to represent the situation.

4. Priya's class has adopted two equal sections of a highway to keep clean.

The combined length is $\frac{3}{4}$ of a mile. How long is each section?

5. Lin's class has also adopted some sections of highway to keep clean.

If $1\frac{1}{2}$ sections are $\frac{3}{4}$ mile long, how long is each section?

6. A school has adopted a section of highway to keep clean.

If $\frac{1}{3}$ of the section is $\frac{3}{4}$ mile long, how long is the section?

NAME ______________________ DATE ____________ PERIOD ____________

Are you ready for more?

To make a Cantor ternary set:

- Start with a tape diagram of length 1 unit. This is step 1.
- Color in the middle third of the tape diagram. This is step 2.
- Do the same to each remaining segment that is not colored in. This is step 3.
- Keep repeating this process.

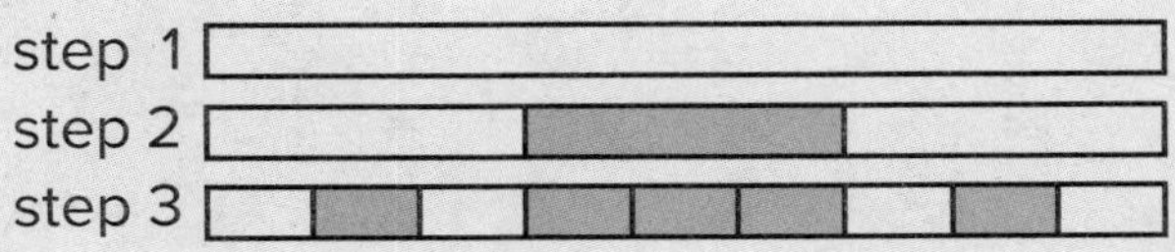

step 1
step 2
step 3

1. How much of the diagram is colored in after Step 2? Step 3? Step 10?

2. If you continue this process, how much of the tape diagram will you color?

3. Can you think of a different process that will give you a similar result?
   For example, color the first fifth instead of the middle third of each strip.

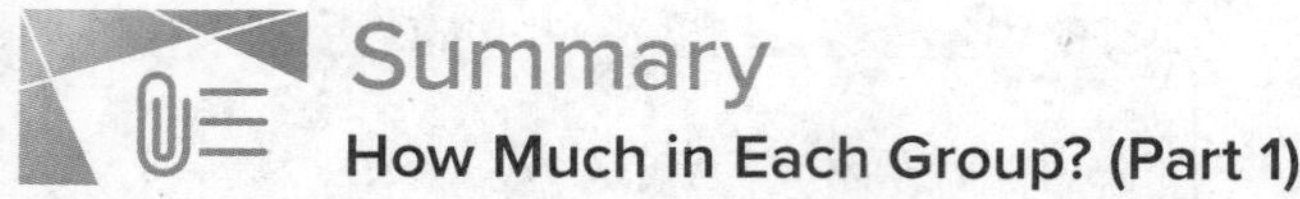

Summary

How Much in Each Group? (Part 1)

Sometimes we know the amount for *multiple* groups, but we don't know how much is in one group. We can use division to find out.

For example, if 5 people share $8\frac{1}{2}$ pounds of cherries equally, how many pounds of cherries does each person get?

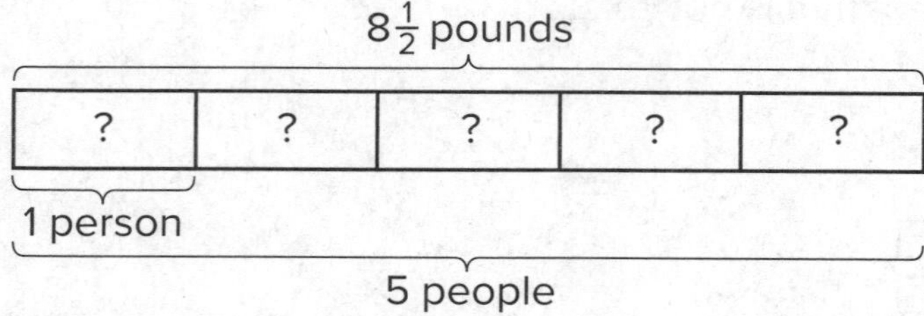

We can represent this situation with a multiplication equation and a division equation.

$$5 \cdot ? = 8\frac{1}{2} \qquad 8\frac{1}{2} \div 5 = ?$$

$8\frac{1}{2} \div 5$ can be written as $\frac{17}{2} \div 5$. Dividing by 5 is equivalent to multiplying by $\frac{1}{5}$, and $\frac{17}{2} \cdot \frac{1}{5} = \frac{17}{10}$. This means each person gets $1\frac{7}{10}$ pounds.

NAME ________________________ DATE ______________ PERIOD ____________

Other times, we know the amount for *a fraction* of a group, but we don't know the size of one whole group. We can also use division to find out.

For example, Jada poured 5 cups of iced tea in a pitcher and filled $\frac{2}{3}$ of the pitcher. How many cups of iced tea fill the entire pitcher?

We can represent this situation with a multiplication equation and a division equation.

$$\frac{2}{3} \cdot ? = 5 \qquad 5 \div \frac{2}{3} = ?$$

The diagram can help us reason about the answer.

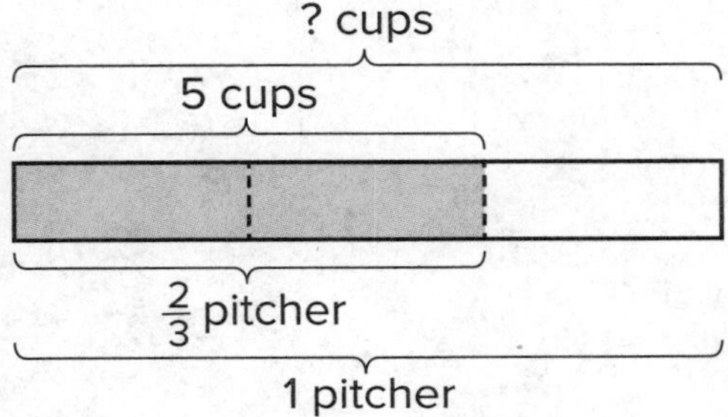

If $\frac{2}{3}$ of a pitcher is 5 cups, then $\frac{1}{3}$ of a pitcher is half of 5, which is $\frac{5}{2}$. Because there are 3 thirds in 1 whole, there would be $\left(3 \cdot \frac{5}{2}\right)$ or $\frac{15}{2}$ cups in one whole pitcher. We can check our answer by multiplying: $\frac{2}{3} \cdot \frac{15}{2} = \frac{30}{6}$, and $\frac{30}{6} = 5$.

Notice that in the first example, the number of groups is greater than 1 (5 people) and in the second, the number of groups is less than 1 $\left(\frac{2}{3}\text{ of a pitcher}\right)$, but the division and multiplication equations for both situations have the same structures.

Practice

How Much in Each Group? (Part 1)

1. For each situation, complete the tape diagram to represent and answer the question.

   a. Mai has picked 1 cup of strawberries for a cake, which is enough for $\frac{3}{4}$ of the cake. How many cups does she need for the whole cake?

   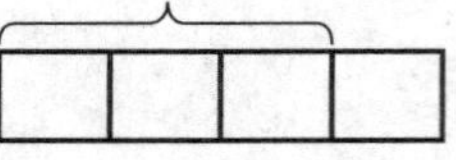

   b. Priya has picked $1\frac{1}{2}$ cups of raspberries, which is enough for $\frac{3}{4}$ of a cake.

   How many cups does she need for the whole cake?

   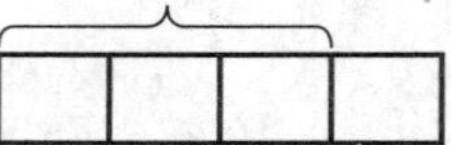

2. Consider the problem: Tyler painted $\frac{9}{2}$ square yards of wall area with 3 gallons of paint. How many gallons of paint does it take to paint each square yard of wall?

   a. Write multiplication and division equations to represent the situation.

   b. Draw a diagram to represent and answer the question.

NAME _______________ DATE _______________ PERIOD _______________

3. Consider the problem: After walking $\frac{1}{4}$ mile from home, Han is $\frac{1}{3}$ of his way to school. What is the distance between his home and school?

a. Write multiplication and division equations to represent this situation.

b. Complete the diagram to represent and answer the question.

4. Here is a division equation: $\frac{4}{5} \div \frac{2}{3} = ?$ **(Lesson 4-7)**

a. Write a multiplication equation that corresponds to the division equation.

b. Draw a diagram to represent and answer the question.

5. Consider the problem.

A set of books that are each 1.5 inches wide are being organized on a bookshelf that is 36 inches wide. How many books can fit on the shelf? **(Lesson 4-3)**

a. Write multiplication and division equations to represent the situation.

b. Find the answer. Draw a diagram, if needed.

c. Use the multiplication equation to check your answer.

6. Respond to each of the following. **(Lesson 4-1)**

a. Without calculating, order the quotients from smallest to largest.

$56 \div 8$
$56 \div 8{,}000{,}000$
$56 \div 0.000008$

b. Explain how you decided the order of the three expressions.

c. Find a number n so that $56 \div n$ is greater than 1 but less than 7.

Lesson 4-9

How Much in Each Group? (Part 2)

NAME ______________________ DATE ____________ PERIOD ____________

Learning Goal Let's practice dividing fractions in different situations.

Warm Up

9.1 Number Talk: Greater Than 1 or Less Than 1?

Decide whether each quotient is greater than 1 or less than 1.

1. $\frac{1}{2} \div \frac{1}{4}$
2. $1 \div \frac{3}{4}$
3. $\frac{2}{3} \div \frac{7}{8}$
4. $2\frac{7}{8} \div 2\frac{3}{5}$

Activity

9.2 Two Water Containers

1. After looking at these pictures, Lin says, "I see the fraction $\frac{2}{5}$." Jada says, "I see the fraction $\frac{3}{4}$." What quantities are Lin and Jada referring to?

2. Consider the problem: How many liters of water fit in the water dispenser?

   a. Write a multiplication equation and a division equation for the question.

   b. Find the answer and explain your reasoning. If you get stuck, consider drawing a diagram.

   c. Check your answer using the multiplication equation.

Activity
9.3 Amount in One Group

Write a multiplication equation and a division equation and draw a diagram to represent each situation. Then, find the answer and explain your reasoning.

1. Jada bought $3\frac{1}{2}$ yards of fabric for \$21. How much did each yard cost?

2. $\frac{4}{9}$ kilogram of baking soda costs \$2. How much does 1 kilogram of baking soda cost?

3. Diego can fill $1\frac{1}{5}$ bottles with 3 liters of water. How many liters of water fill 1 bottle?

4. $\frac{5}{4}$ gallons of water fill $\frac{5}{6}$ of a bucket. How many gallons of water fill the entire bucket?

NAME ______________________ DATE ______________ PERIOD ____________

Are you ready for more?

The largest sandwich ever made weighed 5,440 pounds. If everyone on Earth shares the sandwich equally, how much would you get? What fraction of a regular sandwich does this represent?

Activity

9.4 Inventing Another Situation

1. Think of a situation with a question that can be represented by $\frac{1}{3} \div \frac{1}{4} = ?$ Describe the situation and the question.

2. Trade descriptions with a partner.
   - Review each other's description and discuss whether each question matches the equation.
   - Revise your description based on the feedback from your partner.

3. Find the answer to your question. Explain or show your reasoning. If you get stuck, consider drawing a diagram.

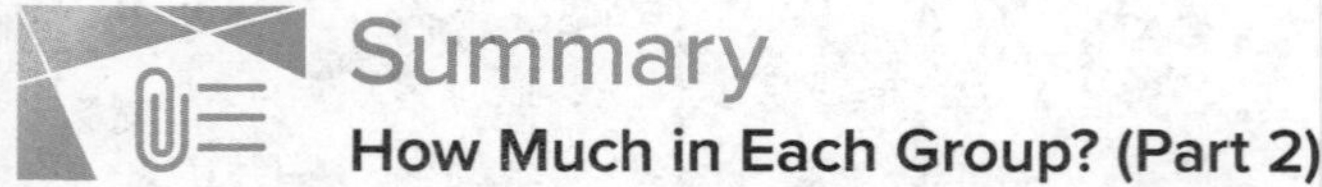

Summary

How Much in Each Group? (Part 2)

Sometimes we have to think carefully about how to solve a problem that involves multiplication and division. Diagrams and equations can help us.

For example, $\frac{3}{4}$ of a pound of rice fills $\frac{2}{5}$ of a container. There are two whole amounts to keep track of here: 1 whole pound and 1 whole container. The equations we write and the diagram we draw depend on what question we are trying to answer.

- How many pounds fill 1 container?

Equations

$\frac{2}{5} \cdot ? = \frac{3}{4}$

$\frac{3}{4} \div \frac{2}{5} = ?$

Diagram

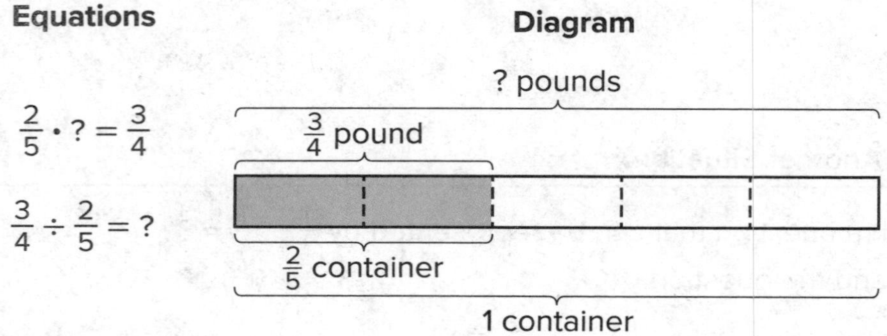

If $\frac{2}{5}$ of a container is filled with $\frac{3}{4}$ pound, then $\frac{1}{5}$ of a container is filled with half of $\frac{3}{4}$, or $\frac{3}{8}$, pound. One whole container then has $5 \cdot \frac{3}{8}$ $\left(\text{or } \frac{15}{8}\right)$ pounds.

- What fraction of a container does 1 pound fill?

Equations

$\frac{3}{4} \cdot ? = \frac{2}{5}$

$\frac{2}{5} \div \frac{3}{4} = ?$

Diagram

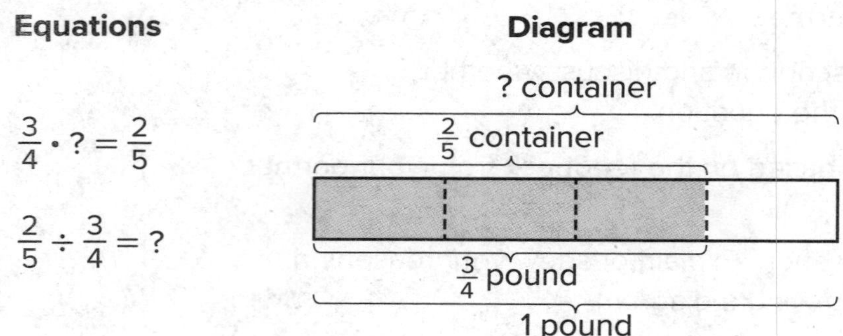

If $\frac{3}{4}$ pound fills $\frac{2}{5}$ of a container, then $\frac{1}{4}$ pound fills a third of $\frac{2}{5}$, or $\frac{2}{15}$, of a container. One whole pound then fills $4 \cdot \frac{2}{15}$ $\left(\text{or } \frac{8}{15}\right)$ of a container.

NAME ______________________ DATE ______________ PERIOD ________

Practice

How Much in Each Group? (Part 2)

1. A group of friends is sharing $2\frac{1}{2}$ pounds of berries.

   a. If each friend received $\frac{5}{4}$ of a pound of berries, how many friends are sharing the berries?

   b. If 5 friends are sharing the berries, how many pounds of berries does each friend receive?

2. $\frac{2}{5}$ kilogram of soil fills $\frac{1}{3}$ of a container. Can 1 kilogram of soil fit in the container? Explain or show your reasoning.

3. After raining for $\frac{3}{4}$ of an hour, a rain gauge is $\frac{2}{5}$ filled. If it continues to rain at that rate for 15 more minutes, what fraction of the rain gauge will be filled?

   a. To help answer this question, Diego wrote the equation $\frac{3}{4} \div \frac{2}{5} = ?$. Explain why this equation does *not* represent the situation.

   b. Write a multiplication equation and a division equation that do represent the situation.

4. 3 tickets to the museum cost $12.75. At this rate, what is the cost of: (Lesson 2-8)

   a. 1 ticket?

   b. 5 tickets?

5. Elena went 60 meters in 15 seconds. Noah went 50 meters in 10 seconds. Elena and Noah both moved at a constant speed. (Lesson 2-9)

   a. How far did Elena go in 1 second?

   b. How far did Noah go in 1 second?

   c. Who went faster? Explain or show your reasoning.

6. The first row in the table shows a recipe for 1 batch of trail mix. Complete the table to show recipes for 2, 3, and 4 batches of the same type of trail mix. (Lesson 2-11)

Number of Batches	Cups of Cereal	Cups of Almonds	Cups of Raisins
1	2	$\frac{1}{3}$	$\frac{1}{4}$
2			
3			
4			

Lesson 4-10

Dividing by Unit and Non-Unit Fractions

NAME ______________________ DATE ____________ PERIOD ____________

Learning Goal Let's look for patterns when we divide by a fraction.

Warm Up

10.1 Dividing by a Whole Number

Work with a partner. One person solves the problems labeled "Partner A" and the other person solves those labeled "Partner B." Write an equation for each question. If you get stuck, consider drawing a diagram.

1. **Partner A:**

   How many 3s are in 12?

   Division equation:

   How many 4s are in 12?

   Division equation:

   How many 6s are in 12?

   Division equation:

Partner B:

What is 12 groups of $\frac{1}{3}$?

Multiplication equation:

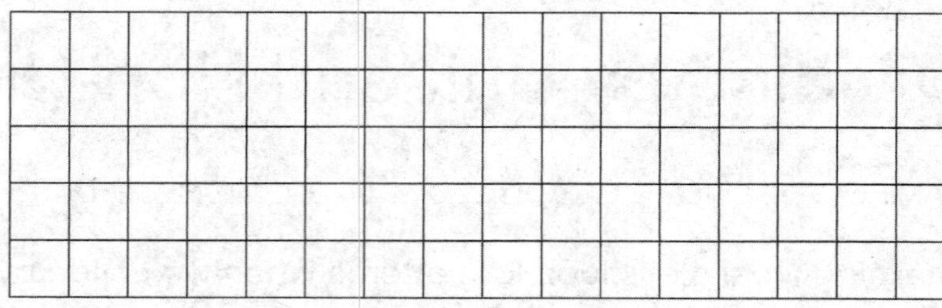

What is 12 groups of $\frac{1}{4}$?

Multiplication equation:

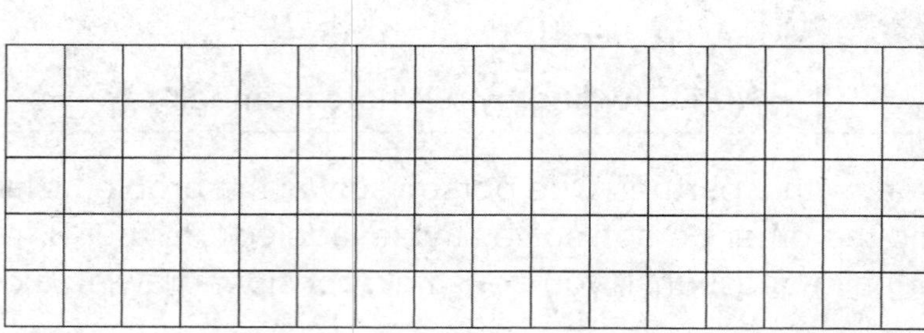

What is 12 groups of $\frac{1}{6}$?

Multiplication equation:

2. What do you notice about the diagrams and equations? Discuss with your partner.

3. Complete this sentence based on what you noticed.

   Dividing by a whole number a produces the same result as multiplying by ________.

NAME ______________________ DATE ____________ PERIOD ________

Activity
10.2 Dividing by Unit Fractions

To find the value of $6 \div \frac{1}{2}$, Elena thought, "How many $\frac{1}{2}$s are in 6?" and then she drew this tape diagram. It shows 6 ones, with each one partitioned into 2 equal pieces.

Expression — **Diagram**

$6 \div \frac{1}{2}$

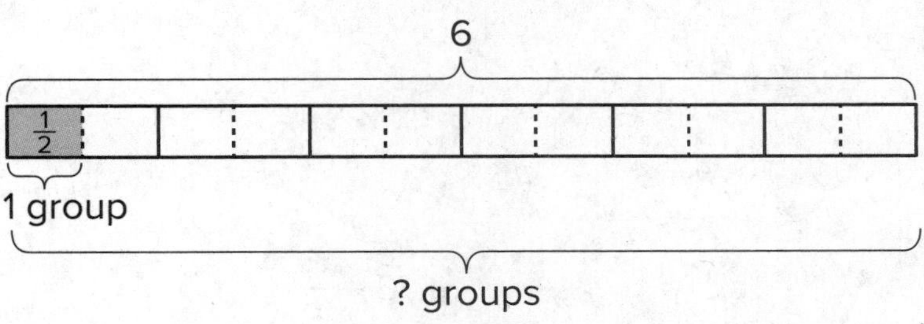

1. For each division expression, complete the diagram using the same method as Elena. Then, find the value of the expression.

   a. $6 \div \frac{1}{3}$

   6

   Value of the expression: ____________

   b. $6 \div \frac{1}{4}$

   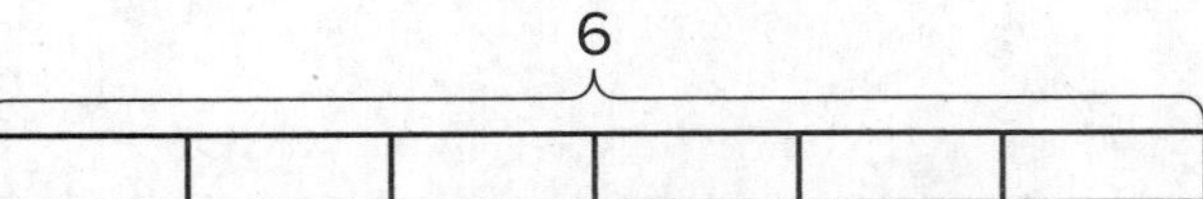

   Value of the expression: ____________

   c. $6 \div \frac{1}{6}$

   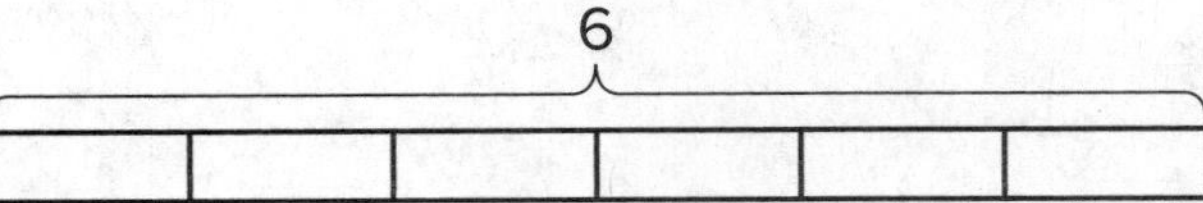

   Value of the expression: ____________

2. Examine the expressions and answers more closely. Look for a pattern. How could you find how many halves, thirds, fourths, or sixths were in 6 without counting all of them? Explain your reasoning.

3. Use the pattern you noticed to find the values of these expressions. If you get stuck, consider drawing a diagram.

a. $6 \div \frac{1}{8}$

b. $6 \div \frac{1}{10}$

c. $6 \div \frac{1}{25}$

d. $6 \div \frac{1}{b}$

4. Find the value of each expression.

a. $8 \div \frac{1}{4}$

b. $12 \div \frac{1}{5}$

c. $a \div \frac{1}{2}$

d. $a \div \frac{1}{b}$

NAME ______________________ DATE ____________ PERIOD ____________

Activity

10.3 Dividing by Non-unit Fractions

1. To find the value of $6 \div \frac{2}{3}$, Elena started by drawing a diagram the same way she did for $6 \div \frac{1}{3}$.

6

$\frac{1}{3}$

a. Complete the diagram to show how many $\frac{2}{3}$s are in 6.

b. Elena says, "To find $6 \div \frac{2}{3}$, I can just take the value of $6 \div \frac{1}{3}$ and then either multiply it by $\frac{1}{2}$ or divide it by 2." Do you agree with her? Explain your reasoning.

2. For each division expression, complete the diagram using the same method as Elena. Then, find the value of the expression. Think about how you could find that value without counting all the pieces in your diagram.

a. $6 \div \frac{3}{4}$

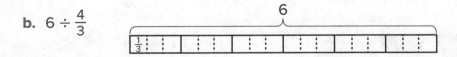

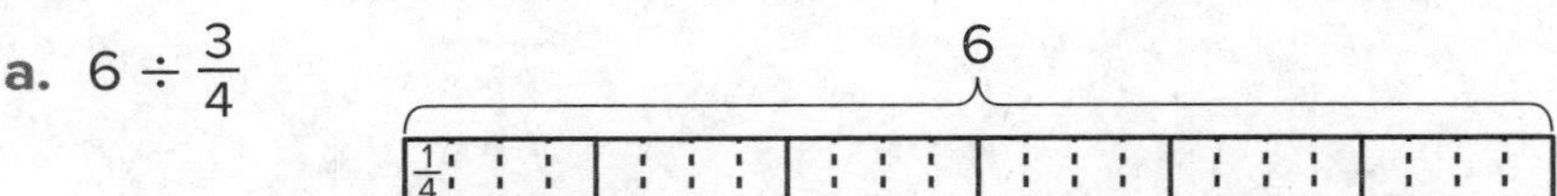

Value of the expression: ____________

b. $6 \div \frac{4}{3}$

6

$\frac{1}{3}$

Value of the expression: ____________

c. $6 \div \frac{4}{6}$

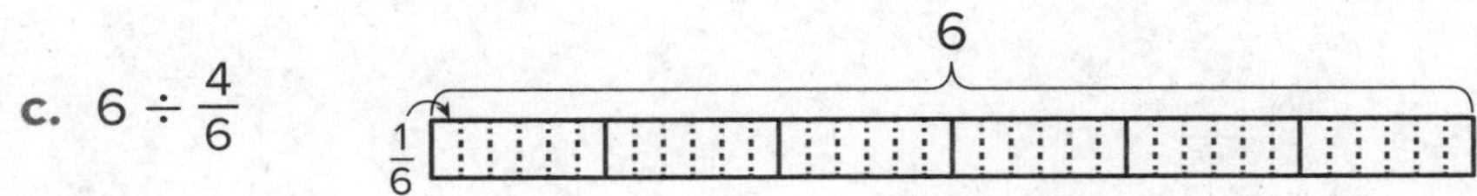

Value of the expression: ____________

3. Elena examined her diagrams and noticed that she always took the same two steps to show division by a fraction on a tape diagram. She said:

"My first step was to divide each 1 whole into as many parts as the number in the denominator. So if the expression is $6 \div \frac{3}{4}$, I would break each 1 whole into 4 parts. Now I have 4 times as many parts.

My second step was to put a certain number of those parts into one group, and that number is the numerator of the divisor. So if the fraction is $\frac{3}{4}$, I would put 3 of the $\frac{1}{4}$s into one group. Then I could tell how many $\frac{3}{4}$s are in 6."

Which expression represents how many $\frac{3}{4}$s Elena would have after these two steps? Be prepared to explain your reasoning.

$6 \div 4 \cdot 3$ $\quad$ $6 \div 4 \div 3$ $\quad$ $6 \cdot 4 \div 3$ $\quad$ $6 \cdot 4 \cdot 3$

4. Use the pattern Elena noticed to find the values of these expressions. If you get stuck, consider drawing a diagram.

   a. $6 \div \frac{2}{7}$

   b. $6 \div \frac{3}{10}$

   c. $6 \div \frac{6}{25}$

Are you ready for more?

Find the missing value.

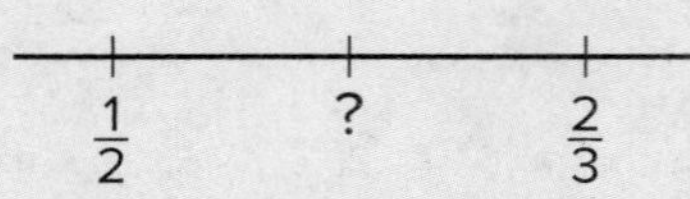

NAME ______________________ DATE ______________ PERIOD ____________

Summary

Dividing by Unit and Non-Unit Fractions

To answer the question "How many $\frac{1}{3}$s are in 4?" or "What is $4 \div \frac{1}{3}$?", we can reason that there are 3 thirds in 1, so there are $(4 \cdot 3)$ thirds in 4.

In other words, dividing 4 by $\frac{1}{3}$ has the same result as multiplying 4 by 3.

$$4 \div \frac{1}{3} = 4 \cdot 3$$

4

$\frac{1}{3}$

1 group

? groups

In general, dividing a number by a unit fraction $\frac{1}{b}$ is the same as multiplying the number by b, which is the **reciprocal** of $\frac{1}{b}$.

How can we reason about $4 \div \frac{2}{3}$?

We already know that there are $(4 \cdot 3)$ or 12 groups of $\frac{1}{3}$s in 4. To find how many $\frac{2}{3}$s are in 4, we need to put together every 2 of the $\frac{1}{3}$s into a group.

Doing this results in half as many groups, which is 6 groups. In other words:

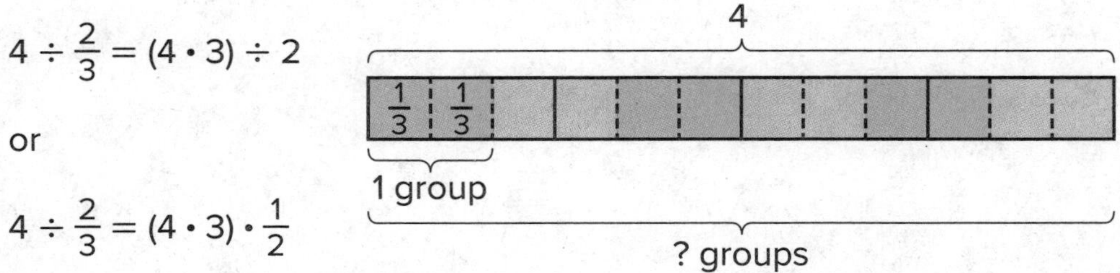

In general, dividing a number by $\frac{a}{b}$, is the same as multiplying the number by b and then dividing by a, or multiplying the number by b and then by $\frac{1}{a}$.

Glossary

reciprocal

Practice

Dividing by Unit and Non-Unit Fractions

1. Priya is sharing 24 apples equally with some friends. She uses division to determine how many people can have a share if each person gets a particular number of apples. For example, $24 \div 4 = 6$ means that if each person gets 4 apples, then 6 people can have apples. Here are some other calculations:

   $24 \div 4 = 6$ $\quad$ $24 \div 2 = 12$ $\quad$ $24 \div 1 = 24$ $\quad$ $24 \div \frac{1}{2} = ?$

   a. Priya thinks the "?" represents a number less than 24. Do you agree? Explain or show your reasoning.

   b. In the case of $24 \div \frac{1}{2} = ?$, how many people can have apples?

2. Here is a centimeter ruler.

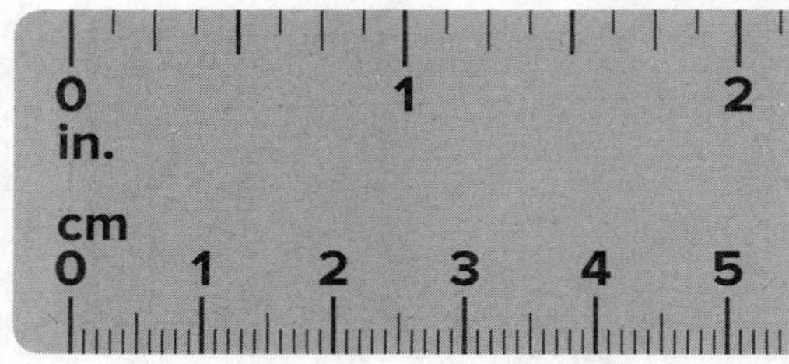

   a. Use the ruler to find $1 \div \frac{1}{10}$ and $4 \div \frac{1}{10}$.

   b. What calculation did you do each time?

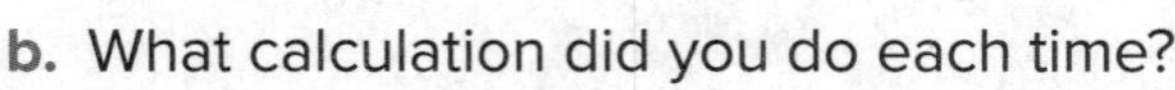

   c. Use this pattern to find $18 \div \frac{1}{10}$.

   d. Explain how you could find $4 \div \frac{2}{10}$ and $4 \div \frac{8}{10}$.

NAME ______________________________ DATE ______________ PERIOD ____________

3. Find each quotient.

a. $5 \div \frac{1}{10}$

b. $5 \div \frac{3}{10}$

c. $5 \div \frac{9}{10}$

4. Use the fact that $2\frac{1}{2} \div \frac{1}{8} = 20$ to find $2\frac{1}{2} \div \frac{5}{8}$. Explain or show your reasoning.

5. Consider the problem: It takes one week for a crew of workers to pave $\frac{3}{5}$ kilometer of a road. At that rate, how long will it take to pave 1 kilometer? **(Lesson 4-9)**

Write a multiplication equation and a division equation to represent the question. Then find the answer and show your reasoning.

6. A box contains $1\frac{3}{4}$ pounds of pancake mix. Jada used $\frac{7}{8}$ pound for a recipe. What fraction of the pancake mix in the box did she use? Explain or show your reasoning. Draw a diagram, if needed. **(Lesson 4-7)**

7. Calculate each percentage mentally. **(Lesson 3-14)**

   a. 25% of 400

   b. 50% of 90

   c. 75% of 200

   d. 10% of 8,000

   e. 5% of 20

Lesson 4-11

Using an Algorithm to Divide Fractions

NAME ______________________ DATE ____________ PERIOD ____________

Learning Goal Let's divide fractions using the rule we learned.

Warm Up

11.1 Multiplying Fractions

Evaluate each expression.

1. $\frac{2}{3} \cdot 27$

2. $\frac{1}{2} \cdot \frac{2}{3}$

3. $\frac{2}{9} \cdot \frac{3}{5}$

4. $\frac{27}{100} \cdot \frac{200}{9}$

5. $\left(1\frac{3}{4}\right) \cdot \frac{5}{7}$

Activity

11.2 Dividing a Fraction by a Fraction

Work with a partner. One person works on the questions labeled "Partner A" and the other person works on those labeled "Partner B."

1. **Partner A:** Find the value of each expression by completing the diagram.

   a. $\frac{3}{4} \div \frac{1}{8}$

   How many $\frac{1}{8}$s in $\frac{3}{4}$?

   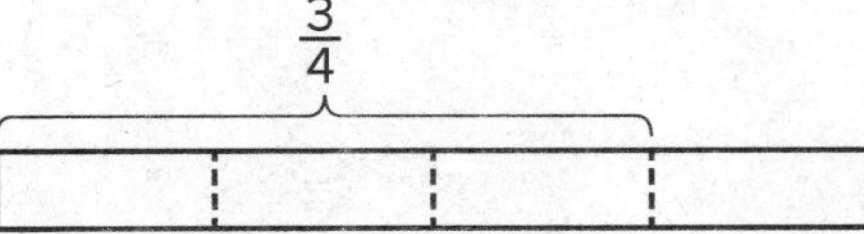

   b. $\frac{9}{10} \div \frac{3}{5}$

   How many $\frac{3}{5}$s in $\frac{9}{10}$?

   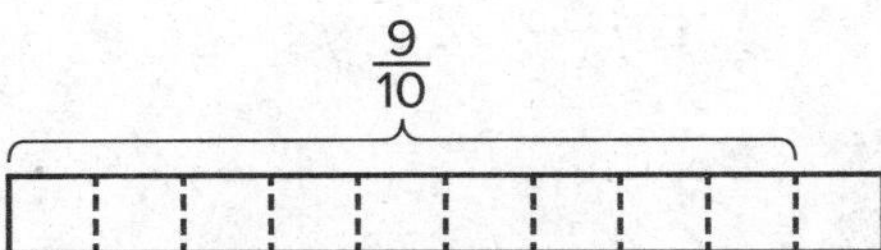

Partner B:

Elena said, "If I want to divide 4 by $\frac{2}{5}$, I can multiply 4 by 5 and then divide it by 2 or multiply it by $\frac{1}{2}$."

Find the value of each expression using the strategy Elena described.

a. $\frac{3}{4} \div \frac{1}{8}$

b. $\frac{9}{10} \div \frac{3}{5}$

2. What do you notice about the diagrams and expressions? Discuss with your partner.

3. Complete this sentence based on what you noticed.

To divide a number n by a fraction $\frac{a}{b}$, we can multiply n by __________ and then divide the product by __________.

4. Select **all** the equations that represent the sentence you completed.

A. $n \div \frac{a}{b} = n \cdot b \div a$

B. $n \div \frac{a}{b} = n \cdot a \div b$

C. $n \div \frac{a}{b} = n \cdot \frac{a}{b}$

D. $n \div \frac{a}{b} = n \cdot \frac{b}{a}$

NAME ______________________ DATE ____________ PERIOD ________

Activity

11.3 Using an Algorithm to Divide Fractions

Calculate each quotient. Show your thinking and be prepared to explain your reasoning.

1. $\frac{8}{9} \div 4$

2. $\frac{3}{4} \div \frac{1}{2}$

3. $3\frac{1}{3} \div \frac{2}{9}$

4. $\frac{9}{2} \div \frac{3}{8}$

5. $6\frac{2}{5} \div 3$

6. After biking $5\frac{1}{2}$ miles, Jada has traveled $\frac{2}{3}$ of the length of her trip. How long (in miles) is the entire length of her trip? Write an equation to represent the situation, and then find the answer.

Are you ready for more?

Suppose you have a pint of grape juice and a pint of milk. Your pour 1 tablespoon of the grape juice into the milk and mix it up. Then you pour 1 tablespoon of this mixture back into the grape juice. Which liquid is more contaminated?

Summary

Using an Algorithm to Divide Fractions

The division $a \div \frac{3}{4} = ?$ is equivalent to $\frac{3}{4} \cdot ? = a$, so we can think of it as meaning "$\frac{3}{4}$ of what number is a?" and represent it with a diagram as shown. The length of the entire diagram represents the unknown number.

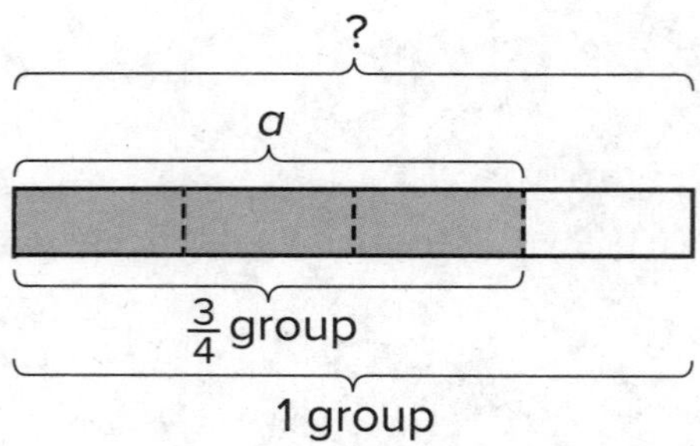

If $\frac{3}{4}$ of a number is a, then to find the number, we can first divide a by 3 to find $\frac{1}{4}$ of the number. Then we multiply the result by 4 to find the number.
The steps above can be written as: $a \div 3 \cdot 4$. Dividing by 3 is the same as multiplying by $\frac{1}{3}$, so we can also write the steps as: $a \cdot \frac{1}{3} \cdot 4$.
In other words: $a \div 3 \cdot 4 = a \cdot \frac{1}{3} \cdot 4$. And $a \cdot \frac{1}{3} \cdot 4 = a \cdot \frac{4}{3}$, so we can say that:

$$a \div \frac{3}{4} = a \cdot \frac{4}{3}$$

In general, dividing a number by a fraction $\frac{c}{d}$ is the same as multiplying the number by $\frac{d}{c}$, which is the reciprocal of the fraction.

NAME ______________________ DATE ____________ PERIOD ________

Practice

Using an Algorithm to Divide Fractions

1. Select **all** the statements that show correct reasoning for finding $\frac{14}{15} \div \frac{7}{5}$.

   (A) Multiplying $\frac{14}{15}$ by 5 and then by $\frac{1}{7}$.

   (B) Dividing $\frac{14}{15}$ by 5, and then multiplying by $\frac{1}{7}$.

   (C) Multiplying $\frac{14}{15}$ by 7, and then multiplying by $\frac{1}{5}$.

   (D) Multiplying $\frac{14}{15}$ by 5 and then dividing by 7.

   (E) Multiplying $\frac{15}{14}$ by 7 and then dividing by 5.

2. Clare said that $\frac{4}{3} \div \frac{5}{2}$ is $\frac{10}{3}$. She reasoned: $\frac{4}{3} \cdot 5 = \frac{20}{3}$ and $\frac{20}{3} \div 2 = \frac{10}{3}$.
   Explain why Clare's answer and reasoning are incorrect.
   Find the correct quotient.

3. Find the value of $\frac{15}{4} \div \frac{5}{8}$. Show your reasoning.

4. Consider the problem: Kiran has $2\frac{3}{4}$ pounds of flour. When he divides the flour into equal-sized bags, he fills $4\frac{1}{8}$ bags. How many pounds fit in each bag? Write a multiplication equation and a division equation to represent the question. Then, find the answer and show your reasoning.

5. Divide $4\frac{1}{2}$ by each of these unit fractions. **(Lesson 4-10)**

   a. $\frac{1}{8}$
   b. $\frac{1}{4}$
   c. $\frac{1}{6}$

6. Consider the problem: After charging for $\frac{1}{3}$ of an hour, a phone is at $\frac{2}{5}$ of its full power. How long will it take the phone to charge completely?

   Decide whether each equation can represent the situation. **(Lesson 4-9)**

   a. $\frac{1}{3} \cdot ? = \frac{2}{5}$
   b. $\frac{1}{3} \div \frac{2}{5} = ?$
   c. $\frac{2}{5} \div \frac{1}{3} = ?$
   d. $\frac{2}{5} \cdot ? = \frac{1}{3}$

7. Elena and Noah are each filling a bucket with water. Noah's bucket is $\frac{2}{5}$ full and the water weighs $2\frac{1}{2}$ pounds. How much does Elena's water weigh if her bucket is full and her bucket is identical to Noah's? **(Lesson 4-8)**

   a. Write multiplication and division equations to represent the question.

   b. Draw a diagram to show the relationship between the quantities and to find the answer.

Lesson 4-12

Fractional Lengths

NAME ______________________ DATE ____________ PERIOD ____________

Learning Goal Let's solve problems about fractional lengths.

Warm Up

12.1 Number Talk: Multiplication Strategies

Find the product mentally.

$19 \cdot 14$

Activity

12.2 Info Gap: How Many Would It Take?

Your teacher will give you either a *problem card* or a *data card*. Do not show or read your card to your partner.

If your teacher gives you the *problem card*:	If your teacher gives you the *data card*:
1. Silently read your card and think about what information you need to be able to answer the question. **2.** Ask your partner for the specific information that you need. **3.** Explain how you are using the information to solve the problem. Continue to ask questions until you have enough information to solve the problem. **4.** Share the *problem card* and solve the problem independently. **5.** Read the *data card* and discuss your reasoning.	**1.** Silently read your card. **2.** Ask your partner *"What specific information do you need?"* and wait for them to *ask* for information. If your partner asks for information that is not on the card, do not do the calculations for them. Tell them you don't have that information. **3.** Before sharing the information, ask "*Why do you need that information?*" Listen to your partner's reasoning and ask clarifying questions. **4.** Read the *problem card* and solve the problem independently. **5.** Share the *data card* and discuss your reasoning.

Are you ready for more?

Lin has a work of art that is 14 inches by 20 inches. She wants to frame it with large paper clips laid end to end.

1. If each paper clip is $1\frac{3}{4}$ inch long, how many paper clips would she need? Show your reasoning and be sure to think about potential gaps and overlaps. Consider making a sketch that shows how the paper clips could be arranged.

2. How many paper clips are needed if the paper clips are spaced $\frac{1}{4}$ inch apart? Describe the arrangement of the paper clips at the corners of the frame.

Activity

12.3 How Many Times as Tall or as Far?

1. A second-grade student is 4 feet tall. Her teacher is $5\frac{2}{3}$ feet tall.
   a. How many times as tall as the student is the teacher?

   b. What fraction of the teacher's height is the student's height?

2. Find each quotient. Show your reasoning and check your answer.
   a. $9 \div \frac{3}{5}$

   b. $1\frac{7}{8} \div \frac{3}{4}$

NAME ______________________ DATE ______________ PERIOD ____________

3. Write a division equation that can help answer each of these questions. Then find the answer. If you get stuck, consider drawing a diagram.

a. A runner ran $1\frac{4}{5}$ miles on Monday and $6\frac{3}{10}$ miles on Tuesday. How many times her Monday distance was her Tuesday distance?

b. A cyclist planned to ride $9\frac{1}{2}$ miles but only managed to travel $3\frac{7}{8}$ miles. What fraction of his planned trip did he travel?

Activity

12.4 Comparing Paper Rolls

The photo shows a situation that involves fractions.

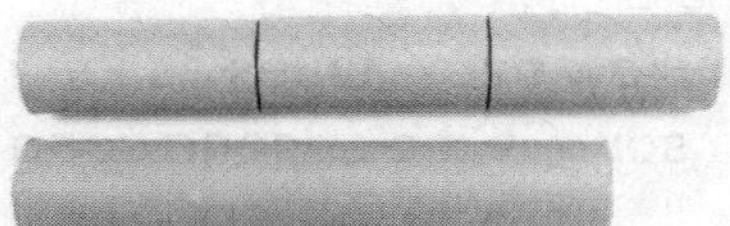

1. Complete the sentences. Be prepared to explain your reasoning.

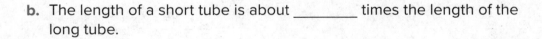

a. The length of the long tube is about ________ times the length of a short tube.

b. The length of a short tube is about ________ times the length of the long tube.

2. If the length of the long paper roll is $11\frac{1}{4}$ inches, what is the length of each short paper roll?

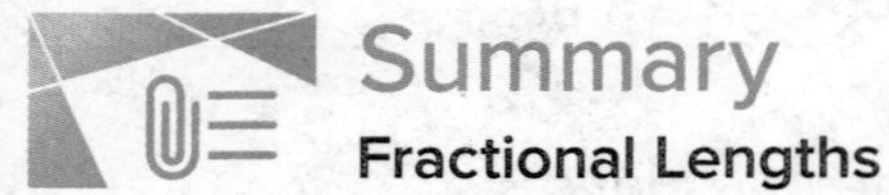

Summary

Fractional Lengths

Division can help us solve comparison problems in which we find out how many times as large or as small one number is compared to another. For example, a student is playing two songs for a music recital. The first song is $1\frac{1}{2}$ minutes long. The second song is $3\frac{3}{4}$ minutes long.

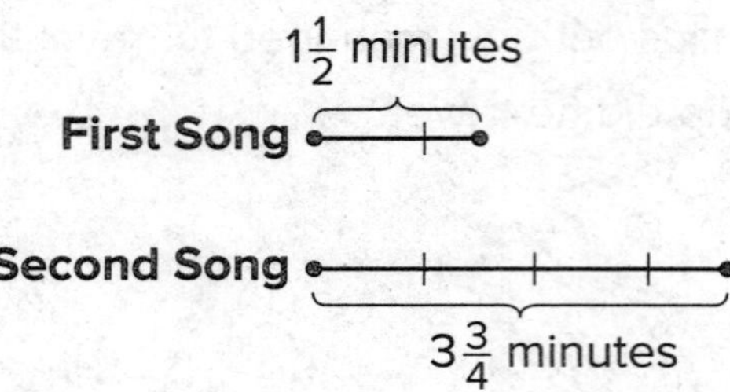

We can ask two different comparison questions and write different multiplication and division equations to represent each question.

1. How many times as long as the first song is the second song?

   $? \cdot 1\frac{1}{2} = 3\frac{3}{4}$ $\quad 3\frac{3}{4} \div 1\frac{1}{2} = ?$

2. What fraction of the second song is the first song?

   $? \cdot 3\frac{3}{4} = 1\frac{1}{2}$ $\quad 1\frac{1}{2} \div 3\frac{3}{4} = ?$

We can use the algorithm we learned to calculate the quotients.

1. $3\frac{3}{4} \div 1\frac{1}{2} = ?$

$$= \frac{15}{4} \div \frac{3}{2}$$

$$= \frac{15}{4} \cdot \frac{2}{3}$$

$$= \frac{30}{12}$$

$$= \frac{5}{2}$$

This means the second song is $2\frac{1}{2}$ times as long as the first song.

2. $1\frac{1}{2} \div 3\frac{3}{4} = ?$

$$= \frac{3}{2} \div \frac{15}{4}$$

$$= \frac{3}{2} \cdot \frac{4}{15}$$

$$= \frac{12}{30}$$

$$= \frac{2}{5}$$

This means the first song is $\frac{2}{5}$ as long as the second song.

NAME ______________________ DATE ______________ PERIOD ____________

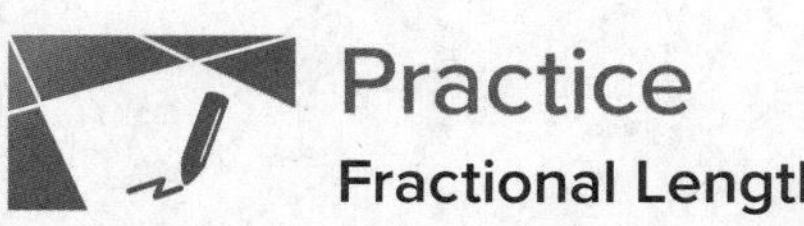

Practice

Fractional Lengths

1. One inch is around $2\frac{11}{20}$ centimeters.

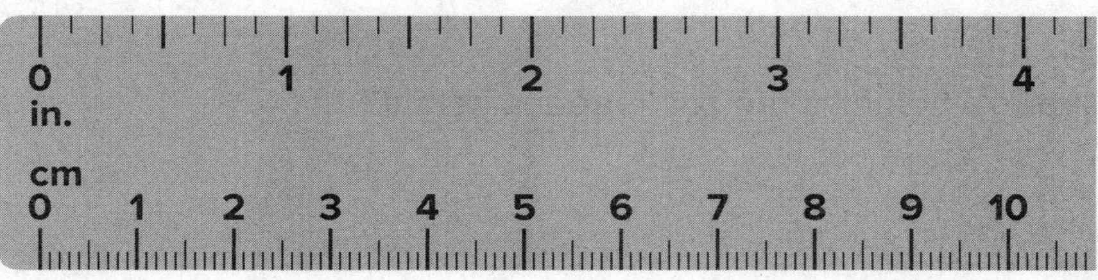

a. How many centimeters long is 3 inches? Show your reasoning.

b. What fraction of an inch is 1 centimeter? Show your reasoning.

c. What question can be answered by finding $10 \div 2\frac{11}{20}$ in this situation?

2. A zookeeper is $6\frac{1}{4}$ feet tall. A young giraffe in his care is $9\frac{3}{8}$ feet tall.

a. How many times as tall as the zookeeper is the giraffe?

b. What fraction of the giraffe's height is the zookeeper's height?

3. A rectangular bathroom floor is covered with square tiles that are $1\frac{1}{2}$ feet by $1\frac{1}{2}$ feet. The length of the bathroom floor is $10\frac{1}{2}$ feet and the width is $6\frac{1}{2}$ feet.

a. How many tiles does it take to cover the length of the floor?

b. How many tiles does it take to cover the width of the floor?

NAME ____________________ DATE ____________ PERIOD ________

4. The Food and Drug Administration (FDA) recommends a certain amount of nutrient intake per day called the "daily value." Food labels usually show percentages of the daily values for several different nutrients—calcium, iron, vitamins, etc.

Consider the problem: In $\frac{3}{4}$ cup of oatmeal, there is $\frac{1}{10}$ of the recommended daily value of iron. What fraction of the daily recommended value of iron is in 1 cup of oatmeal?

Write a multiplication equation and a division equation to represent the question. Then find the answer and show your reasoning. **(Lesson 4-11)**

5. What fraction of $\frac{1}{2}$ is $\frac{1}{3}$? Draw a tape diagram to represent and answer the question. Use graph paper if needed. **(Lesson 4-7)**

6. Noah says, "There are $2\frac{1}{2}$ groups of $\frac{4}{5}$ in 2." Do you agree with him? Draw a tape diagram to show your reasoning. Use graph paper, if needed. **(Lesson 4-6)**

Lesson 4-13

Rectangles with Fractional Side Lengths

NAME ______________________ DATE ______________ PERIOD ______________

Learning Goal Let's explore rectangles that have fractional measurements.

Warm Up

13.1 Areas of Squares

1. What do you notice about the areas of the squares?

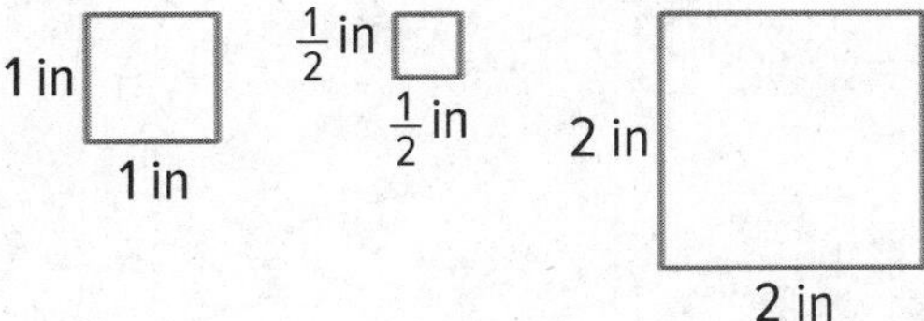

2. Kiran says "A square with side lengths of $\frac{1}{3}$ inch has an area of $\frac{1}{3}$ square inches." Do you agree? Explain or show your reasoning.

Activity

13.2 Areas of Squares and Rectangles

Your teacher will give you graph paper and a ruler.

1. On the graph paper, draw a square with side lengths of 1 inch. Inside this square, draw another square with side lengths of $\frac{1}{4}$ inch.

   Use your drawing to answer the questions.

   a. How many squares with side lengths of $\frac{1}{4}$ inch can fit in a square with side lengths of 1 inch?

   b. What is the area of a square with side lengths of $\frac{1}{4}$ inch? Explain or show your reasoning.

2. On the graph paper, draw a rectangle that is $3\frac{1}{2}$ inches by $2\frac{1}{4}$ inches.

   For each question, write a division expression and then find the answer.

   a. How many $\frac{1}{4}$-inch segments are in a length of $3\frac{1}{2}$ inches?

   b. How many $\frac{1}{4}$-inch segments are in a length of $2\frac{1}{4}$ inches?

3. Use your drawing to show that a rectangle that is $3\frac{1}{2}$ inches by $2\frac{1}{4}$ inches has an area of $7\frac{7}{8}$ square inches.

NAME ________________________ DATE ____________ PERIOD ____________

Activity
13.3 Areas of Rectangles

Each of these multiplication expressions represents the area of a rectangle.

$2 \cdot 4$ $\quad$ $2\frac{1}{2} \cdot 4$ $\quad$ $2 \cdot 4\frac{3}{4}$ $\quad$ $2\frac{1}{2} \cdot 4\frac{3}{4}$

1. All regions shaded in light blue have the same area. Match each diagram to the expression that you think represents its area. Be prepared to explain your reasoning.

Diagram A

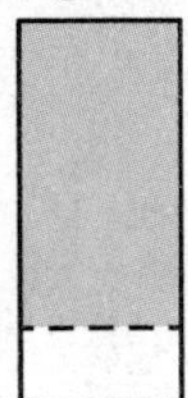

Diagram B

Diagram C

Diagram D

2. Use the diagram that matches $2\frac{1}{2} \cdot 4\frac{3}{4}$ to show that the value of $2\frac{1}{2} \cdot 4\frac{3}{4}$ is $11\frac{7}{8}$.

Are you ready for more?

The following rectangles are composed of squares, and each rectangle is constructed using the previous rectangle. The side length of the first square is 1 unit.

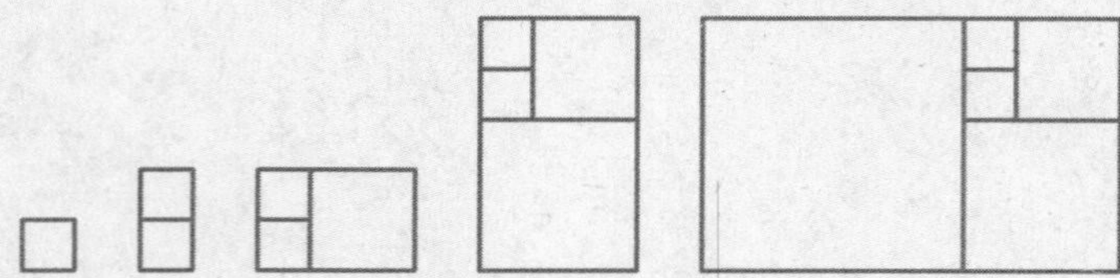

1. Draw the next four rectangles that are constructed in the same way. Then complete the table with the side lengths of the rectangle and the fraction of the longer side over the shorter side.

Short Side	Long Side	$\frac{\text{Long Side}}{\text{Short Side}}$
1		
1		
2		
3		

2. Describe the values of the fraction of the longer side over the shorter side. What happens to the fraction as the pattern continues?

NAME ______________________ DATE ______________ PERIOD ____________

Activity
13.4 How Many Would It Take? (Part 2)

Noah would like to cover a rectangular tray with rectangular tiles. The tray has a width of $11\frac{1}{4}$ inches and an area of $50\frac{5}{8}$ square inches.

1. Find the length of the tray in inches.

2. If the tiles are $\frac{3}{4}$ inch by $\frac{9}{16}$ inch, how many would Noah need to cover the tray completely, without gaps or overlaps? Explain or show your reasoning.

3. Draw a diagram to show how Noah could lay the tiles. Your diagram should show how many tiles would be needed to cover the length and width of the tray, but does not need to show every tile.

Summary

Rectangles with Fractional Side Lengths

If a rectangle has side lengths *a* units and *b* units, the area is $a \cdot b$ square units. For example, if we have a rectangle with $\frac{1}{2}$-inch side lengths, its area is $\frac{1}{2} \cdot \frac{1}{2}$ or $\frac{1}{4}$ square inches.

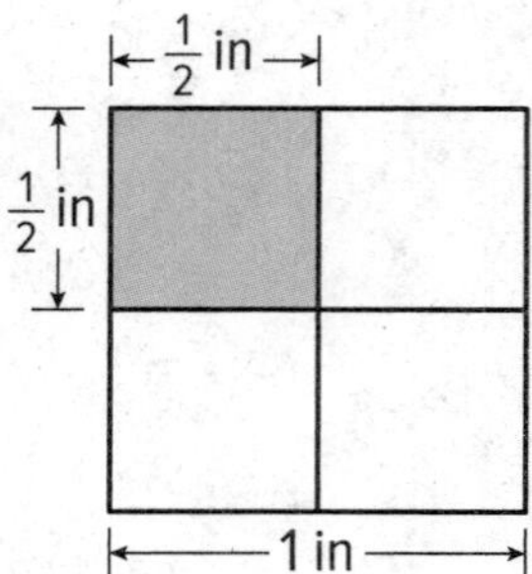

This means that if we know the *area* and *one side length* of a rectangle, we can divide to find the *other* side length.

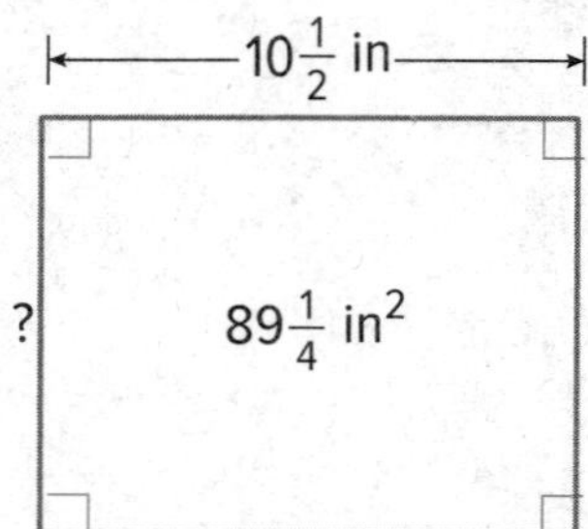

If one side length of a rectangle is $10\frac{1}{2}$ in and its area is $89\frac{1}{4}$ in^2, we can write this equation to show their relationship:

$$\frac{1}{2} \cdot ? = 89\frac{1}{4}$$

Then, we can find the other side length, in inches, using division:

$$89\frac{1}{4} \div 10\frac{1}{2} = ?$$

NAME ______________________ DATE ____________ PERIOD ____________

Practice

Rectangles with Fractional Side Lengths

1. a. Find the unknown side length of the rectangle if its area is 11 m². Show your reasoning.

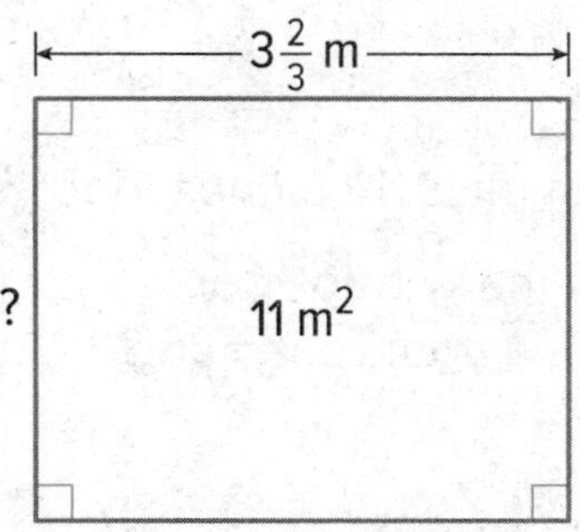

b. Check your answer by multiplying it by the given side length $3\frac{2}{3}$. Is the resulting product 11? If not, revise your previous work.

2. A worker is tiling the floor of a rectangular room that is 12 feet by 15 feet. The tiles are square with side lengths $1\frac{1}{3}$ feet. How many tiles are needed to cover the entire floor? Show your reasoning.

3. A television screen has length $16\frac{1}{2}$ inches, width w inches, and area 462 square inches. Select **all** the equations that represent the relationship of the side lengths and area of the television.

A. $w \cdot 462 = 16\frac{1}{2}$

B. $16\frac{1}{2} \cdot w = 462$

C. $462 \div 16\frac{1}{2} = w$

D. $462 \div w = 16\frac{1}{2}$

E. $16\frac{1}{2} \cdot 462 = w$

4. The area of a rectangle is $17\frac{1}{2}$ in^2 and its shorter side is $3\frac{1}{2}$ in. Draw a diagram that shows this information. What is the length of the longer side?

5. A bookshelf is 42 inches long. **(Lesson 4-12)**

   a. How many books of length $1\frac{1}{2}$ inches will fit on the bookshelf? Explain your reasoning.

   b. A bookcase has 5 of these bookshelves. How many feet of shelf space is there? Explain your reasoning.

6. Find the value of $\frac{5}{32} \div \frac{25}{4}$. Show your reasoning. **(Lesson 4-11)**

7. How many groups of $1\frac{2}{3}$ are in each of these quantities? **(Lesson 4-6)**

   a. $1\frac{5}{6}$

   b. $4\frac{1}{3}$

   c. $\frac{5}{6}$

8. It takes $1\frac{1}{4}$ minutes to fill a 3-gallon bucket of water with a hose. At this rate, how long does it take to fill a 50-gallon tub? If you get stuck, consider using a table. **(Lesson 2-14)**

Lesson 4-14

Fractional Lengths in Triangles and Prisms

NAME ______________________ DATE ____________ PERIOD ____________

Learning Goal Let's explore area and volume when fractions are involved.

Warm Up

14.1 Area of Triangle

Find the area of Triangle A in square centimeters. Show your reasoning.

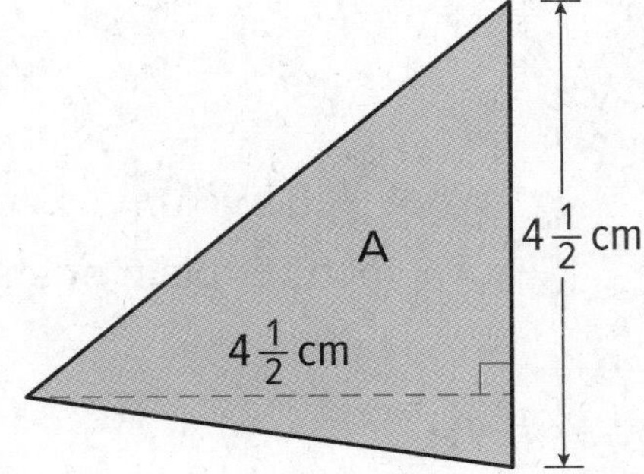

Activity

14.2 Bases and Heights of Triangles

1. The area of Triangle B is 8 square units. Find the length of *b*. Show your reasoning.

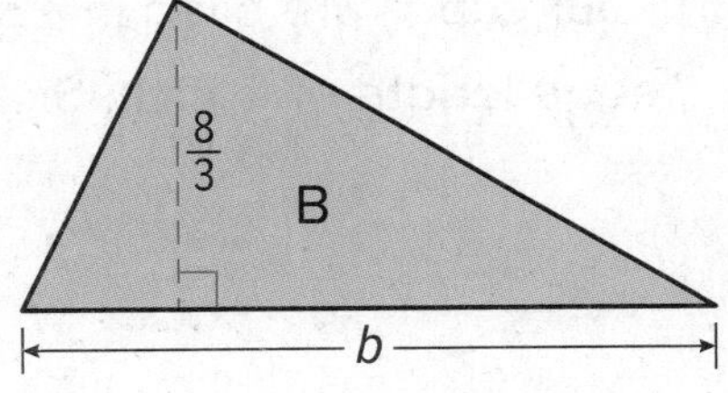

2. The area of Triangle C is $\frac{54}{5}$ square units. What is the length of *h*? Show your reasoning.

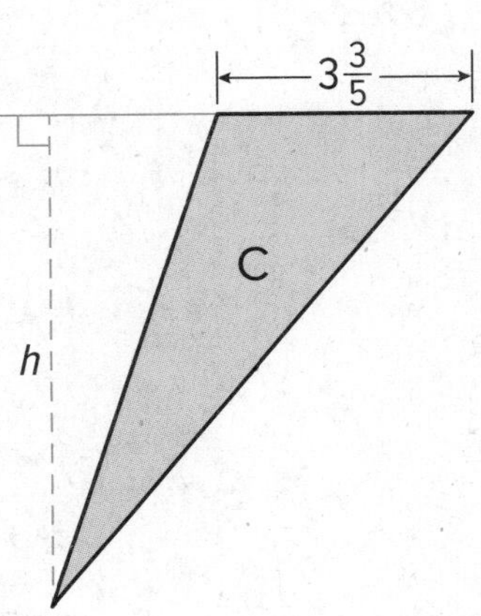

Activity

14.3 Volumes of Cubes and Prisms

Your teacher will give you cubes that have edge lengths of $\frac{1}{2}$ inch.

1. Here is a drawing of a cube with edge lengths of 1 inch.

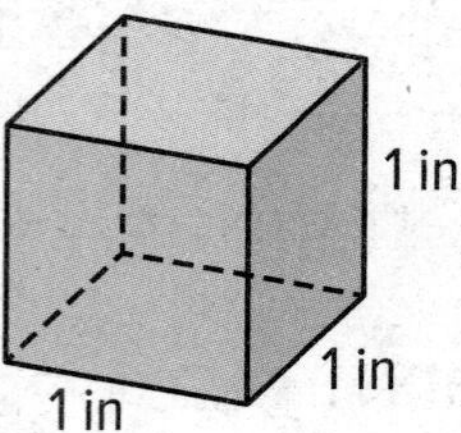

   a. How many cubes with edge lengths of $\frac{1}{2}$ inch are needed to fill this cube?

   b. What is the volume, in cubic inches, of a cube with edge lengths of $\frac{1}{2}$ inch? Explain or show your reasoning.

2. Four cubes are piled in a single stack to make a prism. Each cube has an edge length of $\frac{1}{2}$ inch. Sketch the prism, and find its volume in cubic inches.

NAME ______________________ DATE ______________ PERIOD ______________

3. Use cubes with an edge length of $\frac{1}{2}$ inch to build prisms with the lengths, widths, and heights shown in the table.

a. For each prism, record in the table how many $\frac{1}{2}$-inch cubes can be packed into the prism and the volume of the prism.

Prism Length (in)	Prism Width (in)	Prism Height (in)	Number of $\frac{1}{2}$-inch Cubes in Prism	Volume of Prism (in^3)
$\frac{1}{2}$	$\frac{1}{2}$	$\frac{1}{2}$		
1	1	$\frac{1}{2}$		
2	1	$\frac{1}{2}$		
2	2	1		
4	2	$\frac{3}{2}$		
5	4	2		
5	4	$2\frac{1}{2}$		

b. Examine the values in the table. What do you notice about the relationship between the edge lengths of each prism and its volume?

4. What is the volume of a rectangular prism that is $1\frac{1}{2}$ inches by $2\frac{1}{4}$ inches by 4 inches? Show your reasoning.

Are you ready for more?

A unit fraction has a 1 in the numerator.

These are unit fractions: $\frac{1}{3}, \frac{1}{100}, \frac{1}{1}$.

These are *not* unit fractions: $\frac{2}{9}, \frac{8}{1}, 2\frac{1}{5}$.

1. Find three unit fractions whose sum is $\frac{1}{2}$. An example is: $\frac{1}{8} + \frac{1}{8} + \frac{1}{4} = \frac{1}{2}$ How many examples like this can you find?

2. Find a box whose surface area in square units equals its volume in cubic units. How many like this can you find?

NAME ______________________ DATE ______________ PERIOD ______________

Summary

Fractional Lengths in Triangles and Prisms

If a rectangular prism has edge lengths of 2 units, 3 units, and 5 units, we can think of it as 2 layers of unit cubes, with each layer having $(3 \cdot 5)$ unit cubes in it. So the volume, in cubic units, is:

$$2 \cdot 3 \cdot 5$$

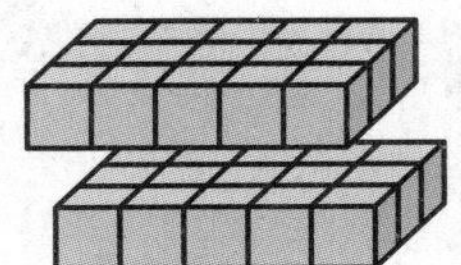

To find the volume of a rectangular prism with fractional edge lengths, we can think of it as being built of cubes that have a unit fraction for their edge length. For instance, if we build a prism that is $\frac{1}{2}$-inch tall, $\frac{3}{2}$-inch wide, and 4 inches long using cubes with a $\frac{1}{2}$-inch edge length, we would have:

- A height of 1 cube, because $1 \cdot \frac{1}{2} = \frac{1}{2}$
- A width of 3 cubes, because $3 \cdot \frac{1}{2} = \frac{3}{2}$
- A length of 8 cubes, because $8 \cdot \frac{1}{2} = 4$

The volume of the prism would be $1 \cdot 3 \cdot 8$, or 24 cubic units. How do we find its volume in cubic inches?

We know that each cube with a $\frac{1}{2}$-inch edge length has a volume of $\frac{1}{8}$ cubic inch, because $\frac{1}{2} \cdot \frac{1}{2} \cdot \frac{1}{2} = \frac{1}{8}$.

Since the prism is built using 24 of these cubes, its volume, in cubic inches, would then be $24 \cdot \frac{1}{8}$, or 3 cubic inches.

The volume of the prism, in cubic inches, can also be found by multiplying the fractional edge lengths in inches:

$$\frac{1}{2} \cdot \frac{3}{2} \cdot 4 = 3$$

Practice

Fractional Lengths in Triangles and Prisms

1. Clare is using little wooden cubes with edge length $\frac{1}{2}$ inch to build a larger cube that has edge length 4 inches. How many little cubes does she need? Explain your reasoning.

2. The triangle has an area of $7\frac{7}{8}$ cm² and a base of $5\frac{1}{4}$ cm. What is the length of *h*? Explain your reasoning.

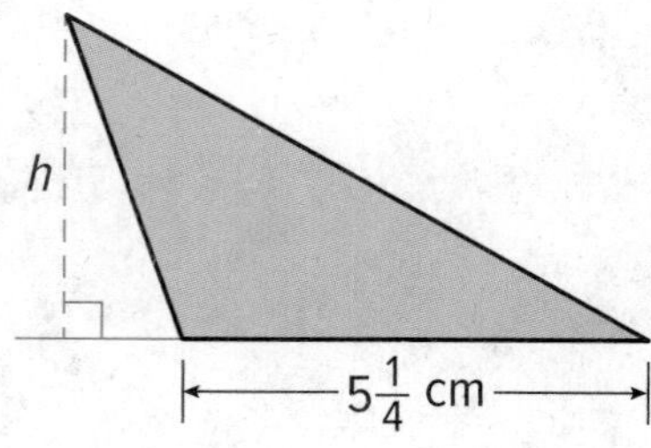

3. a. Which expression can be used to find how many cubes with edge length of $\frac{1}{3}$ unit fit in a prism that is 5 units by 5 units by 8 units? Explain or show your reasoning.
   - $\left(5 \cdot \frac{1}{3}\right) \cdot \left(5 \cdot \frac{1}{3}\right) \cdot \left(8 \cdot \frac{1}{3}\right)$
   - $5 \cdot 5 \cdot 8$
   - $(5 \cdot 3) \cdot (5 \cdot 3) \cdot (8 \cdot 3)$
   - $(5 \cdot 5 \cdot 8) \cdot \left(\frac{1}{3}\right)$

   b. Mai says that we can also find the answer by multiplying the edge lengths of the prism and then multiplying the result by 27. Do you agree with her? Explain your reasoning.

NAME ______________________________ DATE ______________ PERIOD ______________

4. A builder is building a fence with $6\frac{1}{4}$-inch-wide wooden boards, arranged side-by-side with no gaps or overlaps. How many boards are needed to build a fence that is 150 inches long? Show your reasoning. **(Lesson 4-12)**

5. Find the value of each expression. Show your reasoning and check your answer. **(Lesson 14-12)**

a. $2\frac{1}{7} \div \frac{2}{7}$

b. $\frac{17}{20} \div \frac{1}{4}$

6. Consider the problem: A bucket contains $11\frac{2}{3}$ gallons of water and is $\frac{5}{6}$ full. How many gallons of water would be in a full bucket?

Write a multiplication and a division equation to represent the situation. Then, find the answer and show your reasoning. **(Lesson 4-11)**

7. There are 80 kids in a gym. 75% are wearing socks. How many are *not* wearing socks? If you get stuck, consider using a tape diagram. **(Lesson 3-12)**

8. Respond to each of the following questions. **(Lesson 3-11)**

   a. Lin wants to save $75 for a trip to the city. If she has saved $37.50 so far, what percentage of her goal has she saved? What percentage remains?

   b. Noah wants to save $60 so that he can purchase a concert ticket. If he has saved $45 so far, what percentage of his goal has he saved? What percentage remains?

Lesson 4-15

Volume of Prisms

NAME ______________________ DATE ____________ PERIOD ____________

Learning Goal Let's look at the volume of prisms that have fractional measurements.

Warm Up

15.1 A Box of Cubes

1. How many cubes with an edge length of 1 inch fill this box?

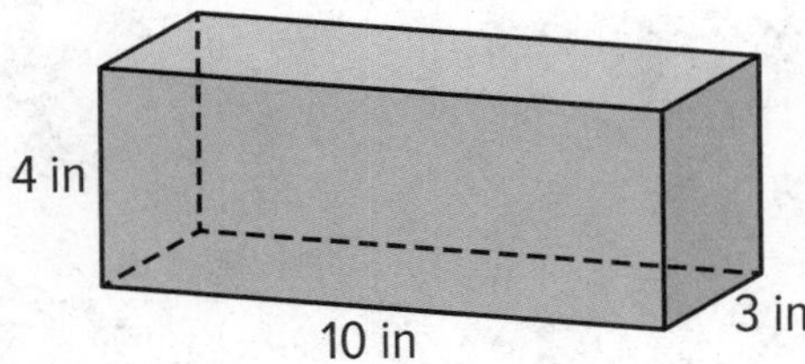

2. If the cubes had an edge length of 2 inches, would you need more or fewer cubes to fill the box? Explain your reasoning.

3. If the cubes had an edge length of $\frac{1}{2}$ inch, would you need more or fewer cubes to fill the box? Explain your reasoning.

Activity

15.2 Cubes with Fractional Edge Lengths

1. Diego says that 108 cubes with an edge length of $\frac{1}{3}$ inch are needed to fill a rectangular prism that is 3 inches by 1 inch by $1\frac{1}{3}$ inch.

   a. Explain or show how this is true. If you get stuck, consider drawing a diagram.

   b. What is the volume, in cubic inches, of the rectangular prism? Explain or show your reasoning.

2. Lin and Noah are packing small cubes into a larger cube with an edge length of $1\frac{1}{2}$ inches. Lin is using cubes with an edge length of $\frac{1}{2}$ inch, and Noah is using cubes with an edge length of $\frac{1}{4}$ inch.

   a. Who would need more cubes to fill the $1\frac{1}{2}$-inch cube? Be prepared to explain your reasoning.

   b. If Lin and Noah each use their small cubes to find the volume of the larger $1\frac{1}{2}$-inch cube, will they get the same answer? Explain or show your reasoning.

NAME ______________________ DATE ______________ PERIOD ______________

Activity

15.3 Fish Tank and Baking Pan

1. A nature center has a fish tank in the shape of a rectangular prism. The tank is 10 feet long, $8\frac{1}{4}$ feet wide, and 6 feet tall.

   a. What is the volume of the tank in cubic feet? Explain or show your reasoning.

   b. The nature center's caretaker filled $\frac{4}{5}$ of the tank with water. What was the volume of the water in the tank, in cubic feet? What was the height of the water in the tank? Explain or show your reasoning.

   c. Another day, the tank was filled with 330 cubic feet of water. The height of the water was what fraction of the height of the tank? Show your reasoning.

vasiliki/iStock/Getty Images

2. Clare's recipe for banana bread won't fit in her favorite pan. The pan is $8\frac{1}{2}$ inches by 11 inches by 2 inches. The batter fills the pan to the very top, and when baking, the batter spills over the sides. To avoid spills, there should be about an inch between the top of the batter and the rim of the pan.

Clare has another pan that is 9 inches by 9 inches by $2\frac{1}{2}$ inches. If she uses this pan, will the batter spill over during baking?

Are you ready for more?

1. Find the area of a rectangle with side lengths $\frac{1}{2}$ and $\frac{2}{3}$.

2. Find the volume of a rectangular prism with side lengths $\frac{1}{2}$, $\frac{2}{3}$, and $\frac{3}{4}$.

3. What do you think happens if we keep multiplying fractions $\frac{1}{2} \cdot \frac{2}{3} \cdot \frac{3}{4} \cdot \frac{4}{5} \cdot \frac{5}{6} \ldots$?

4. Find the area of a rectangle with side lengths $\frac{1}{1}$ and $\frac{2}{1}$.

5. Find the volume of a rectangular prism with side lengths $\frac{1}{1}$, $\frac{2}{1}$, and $\frac{1}{3}$.

6. What do you think happens if we keep multiplying fractions $\frac{1}{1} \cdot \frac{2}{1} \cdot \frac{1}{3} \cdot \frac{4}{1} \cdot \frac{1}{5} \ldots$?

NAME ______________________ DATE ______________ PERIOD ______________

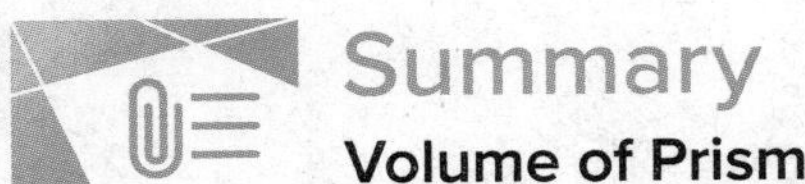

Summary

Volume of Prisms

If a rectangular prism has edge lengths *a* units, *b* units, and *c* units, the volume is the product of *a*, *b*, and *c*.

$$V = a \cdot b \cdot c$$

This means that if we know the *volume* and *two edge lengths*, we can divide to find the *third* edge length.

Suppose the volume of a rectangular prism is $400\frac{1}{2}$ cm^3, one edge length is $\frac{11}{2}$ cm, another is 6 cm, and the third edge length is unknown. We can write a multiplication equation to represent the situation:

$$\frac{11}{2} \cdot 6 \cdot ? = 400\frac{1}{2}$$

We can find the third edge length by dividing:

$$400\frac{1}{2} \div \left(\frac{11}{2} \cdot 6\right) = ?$$

Practice

Volume of Prisms

1. A pool in the shape of a rectangular prism is being filled with water. The length and width of the pool is 24 feet and 15 feet. If the height of the water in the pool is $1\frac{1}{3}$ feet, what is the volume of the water in cubic feet?

2. A rectangular prism measures $2\frac{2}{5}$ inches by $3\frac{1}{5}$ inches by 2 inches.

   a. Priya said, "It takes more cubes with edge length $\frac{2}{5}$ inch than cubes with edge length $\frac{1}{5}$ inch to pack the prism." Do you agree with Priya? Explain or show your reasoning.

   b. How many cubes with edge length $\frac{1}{5}$ inch fit in the prism? Show your reasoning.

   c. Explain how you can use your answer in the previous question to find the volume of the prism in cubic inches.

NAME _______________ DATE _______________ PERIOD _______________

3. Here is a right triangle. **(Lesson 4-14)**

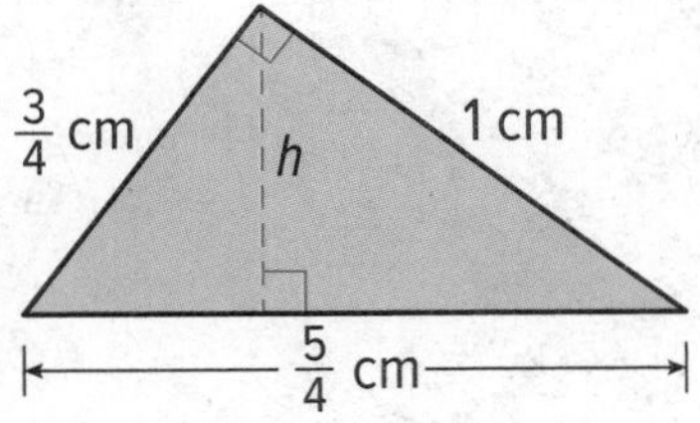

a. What is its area?

b. What is the height *h* for the base that is $\frac{5}{4}$ units long? Show your reasoning.

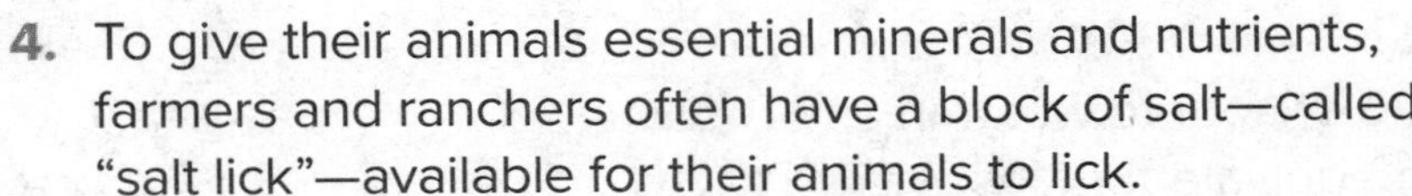

4. To give their animals essential minerals and nutrients, farmers and ranchers often have a block of salt—called "salt lick"—available for their animals to lick.

a. A rancher is ordering a box of cube-shaped salt licks. The edge lengths of each salt lick are $\frac{5}{12}$ foot. Is the volume of one salt lick greater or less than 1 cubic foot? Explain your reasoning.

b. The box that contains the salt lick is $1\frac{1}{4}$ feet by $1\frac{2}{3}$ feet by $\frac{5}{6}$ feet.

How many cubes of salt lick fit in the box? Explain or show your reasoning.

5. a. How many groups of $\frac{1}{3}$ inch are in $\frac{3}{4}$ inch? **(Lesson 4-12)**

b. How many inches are in $1\frac{2}{5}$ groups of $1\frac{2}{3}$ inches?

6. Here is a table that shows the ratio of flour to water in an art paste. Complete the table with values in equivalent ratios. **(Lesson 2-12)**

Cups of Flour	Cups of Water
1	$\frac{1}{2}$
4	
	3
$\frac{1}{2}$	

Lesson 4-16

Solving Problems Involving Fractions

NAME ______________________ DATE ____________ PERIOD ____________

Learning Goal Let's add, subtract, multiply, and divide fractions.

Warm Up

16.1 Operations with Fractions

Without calculating, order the expressions according to their values from least to greatest. Be prepared to explain your reasoning.

$\frac{3}{4}+\frac{2}{3}$ $\qquad$ $\frac{3}{4}-\frac{2}{3}$ $\qquad$ $\frac{3}{4}\cdot\frac{2}{3}$ $\qquad$ $\frac{3}{4}\div\frac{2}{3}$

Activity

16.2 Situations with $\frac{3}{4}$ and $\frac{1}{2}$

Here are four situations that involve $\frac{3}{4}$ and $\frac{1}{2}$.

- Before calculating, decide if each answer is greater than 1 or less than 1.
- Write a multiplication equation or division equation for the situation.
- Answer the question. Show your reasoning. Draw a tape diagram, if needed.

1. There was $\frac{3}{4}$ liter of water in Andre's water bottle. Andre drank $\frac{1}{2}$ of the water. How many liters of water did he drink?

2. The distance from Han's house to his school is $\frac{3}{4}$ kilometers. Han walked $\frac{1}{2}$ kilometers. What fraction of the distance from his house to the school did Han walk?

3. Priya's goal was to collect $\frac{1}{2}$ kilograms of trash. She collected $\frac{3}{4}$ kilograms of trash. How many times her goal was the amount of trash she collected?

4. Mai's class volunteered to clean a park with an area of $\frac{1}{2}$ square mile. Before they took a lunch break, the class had cleaned $\frac{3}{4}$ of the park. How many square miles had they cleaned before lunch?

NAME ______________________ DATE ______________ PERIOD ____________

Activity

16.3 Pairs of Problems

1. Work with a partner to write equations for the following questions. One person works on the questions labeled A1, B1, . . . , E1 and the other person works on those labeled A2, B2, . . . , E2.

A1. Lin's bottle holds $3\frac{1}{4}$ cups of water. She drank 1 cup of water. What fraction of the water in the bottle did she drink?	**A2.** Lin's bottle holds $3\frac{1}{4}$ cups of water. After she drank some, there were $1\frac{1}{2}$ cups of water in the bottle. How many cups did she drink?
B1. Plant A is $\frac{16}{3}$ feet tall. This is $\frac{4}{5}$ as tall as Plant B. How tall is Plant B?	**B2.** Plant A is $\frac{16}{3}$ feet tall. Plant C is $\frac{4}{5}$ as tall as Plant A. How tall is Plant C?
C1. $\frac{8}{9}$ kilogram of berries is put into a container that already has $\frac{7}{3}$ kilogram of berries. How many kilograms are in the container?	**C2.** A container with $\frac{8}{9}$ kilogram of berries is $\frac{2}{3}$ full. How many kilograms can the container hold?
D1. The area of a rectangle is $14\frac{1}{2}$ sq cm and one side is $4\frac{1}{2}$ cm. How long is the other side?	**D2.** The side lengths of a rectangle are $4\frac{1}{2}$ cm and $2\frac{2}{5}$ cm. What is the area of the rectangle?
E1. A stack of magazines is $4\frac{2}{5}$ inches high. The stack needs to fit into a box that is $2\frac{1}{8}$ inches high. How many inches too high is the stack?	**E2.** A stack of magazines is $4\frac{2}{5}$ inches high. Each magazine is $\frac{2}{5}$-inch thick. How many magazines are in the stack?

2. Trade papers with your partner, and check your partner's equations. If you disagree, work to reach an agreement.

3. Your teacher will assign 2 or 3 questions for you to answer. For each question:

   a. Estimate the answer before calculating it.

   b. Find the answer, and show your reasoning.

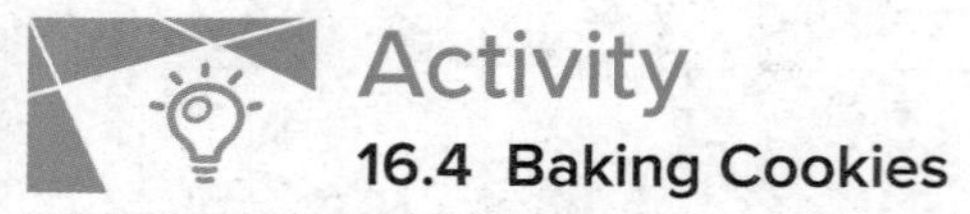

Activity

16.4 Baking Cookies

Mai, Kiran, and Clare are baking cookies together. They need $\frac{3}{4}$ cup of flour and $\frac{1}{3}$ cup of butter to make a batch of cookies. They each brought the ingredients they had at home.

- Mai brought 2 cups of flour and $\frac{1}{4}$ cup of butter.
- Kiran brought 1 cup of flour and $\frac{1}{2}$ cup of butter.
- Clare brought $1\frac{1}{4}$ cups of flour and $\frac{3}{4}$ cup of butter.

If the students have plenty of the other ingredients they need (sugar, salt, baking soda, etc.), how many whole batches of cookies can they make? Explain your reasoning.

NAME ______________________ DATE ____________ PERIOD ____________

Summary

Solving Problems Involving Fractions

We can add, subtract, multiply, and divide both whole numbers and fractions. Here is a summary of how we add, subtract, multiply, and divide fractions.

- To add or subtract fractions, we often look for a common denominator so the pieces involved are the same size. This makes it easy to add or subtract the pieces.

$$\frac{3}{2} - \frac{4}{5} = \frac{15}{10} - \frac{8}{10}$$

- To multiply fractions, we often multiply the numerators and the denominators.

$$\frac{3}{8} \cdot \frac{5}{9} = \frac{3 \cdot 5}{8 \cdot 9}$$

- To divide a number by a fraction $\frac{a}{b}$, we can multiply the number by $\frac{b}{a}$, which is the reciprocal of $\frac{a}{b}$.

$$\frac{4}{7} \div \frac{5}{3} = \frac{4}{7} \cdot \frac{3}{5}$$

Practice

Solving Problems Involving Fractions

1. An orange has about $\frac{1}{4}$ cup of juice. How many oranges are needed to make $2\frac{1}{2}$ cups of juice? Select **all** the equations that represent this question.

   (A.) $? \cdot \frac{1}{4} = 2\frac{1}{2}$

   (B.) $\frac{1}{4} \div 2\frac{1}{2} = ?$

   (C.) $? \cdot 2\frac{1}{2} = \frac{1}{4}$

   (D.) $2\frac{1}{2} \div \frac{1}{4} = ?$

2. Mai, Clare, and Tyler are hiking from a parking lot to the summit of a mountain. They pass a sign that gives distances.

   Parking lot: $\frac{3}{4}$ mile

   Summit: $1\frac{1}{2}$ miles

   - Mai says: "We are one third of the way there."
   - Clare says: "We have to go twice as far as we have already gone."
   - Tyler says: "The total hike is three times as long as what we have already gone."

   Do you agree with any of them? Explain your reasoning.

NAME ______________________ DATE ____________ PERIOD ____________

3. Priya's cat weighs $5\frac{1}{2}$ pounds and her dog weighs $8\frac{1}{4}$ pounds. First, estimate the number that would complete each sentence. Then, calculate the answer. If any of your estimates were not close to the answer, explain why that may be.

a. The cat is __________ as heavy as the dog.

b. Their combined weight is __________ pounds.

c. The dog is __________ pounds heavier than the cat.

4. Before refrigerators existed, some people had blocks of ice delivered to their homes. A delivery wagon had a storage box in the shape of a rectangular prism that was $7\frac{1}{2}$ feet by 6 feet by 6 feet. The cubic ice blocks stored in the box had side lengths $1\frac{1}{2}$ feet. How many ice blocks fit in the storage box? **(Lesson 4-15)**

(A.) 270

(B.) $3\frac{3}{8}$

(C.) 80

(D.) 180

5. Fill in the blanks with 0.001, 0.1, 10, or 1000 so that the value of each quotient is in the correct column.

close to $\frac{1}{100}$	close to 1	greater than 100
a. ____ $\div 9$	b. ____ $\div 0.12$	c. ____ $\div \frac{1}{3}$
d. $12 \div$ ____	e. $\frac{1}{8} \div$ ____	f. $700.7 \div$ ____

6. A school club sold 300 shirts. 31% were sold to fifth graders, 52% were sold to sixth graders, and the rest were sold to teachers. How many shirts were sold to each group—fifth graders, sixth graders, and teachers? Explain or show your reasoning. **(Lesson 3-15)**

7. Jada has some pennies and dimes. The ratio of Jada's pennies to dimes is 2 to 3. **(Lesson 2-15)**

   a. From the information given, can you determine how many coins Jada has?

   b. If Jada has 55 coins, how many of each kind of coin does she have?

   c. How much are her coins worth?

Lesson 4-17

Fitting Boxes into Boxes

NAME ______________________ DATE __________ PERIOD __________

Learning Goal Let's use what we learned about fractions to find shipping costs.

Activity

17.1 Determining Shipping Costs (Part 1)

An artist makes necklaces. She packs each necklace in a small jewelry box that is $1\frac{3}{4}$ inches by $2\frac{1}{4}$ inches by $\frac{3}{4}$ inch.

A department store ordered 270 necklaces. The artist plans to ship the necklaces to the department store using flat-rate shipping boxes from the post office.

1. Consider the problem: Which of the flat-rate boxes should she use to minimize her shipping cost?

What other information would you need to be able to solve the problem?

2. Discuss this information with your group. Make a plan for using this information to find the most inexpensive way to ship the jewelry boxes. Once you have agreed on a plan, write down the main steps.

Activity

17.2 Determining Shipping Costs (Part 2)

Work with your group to find the best plan for shipping the boxes of necklaces. Each member of your group should select a different type of flat-rate shipping box and answer the following questions. Recall that each jewelry box is $1\frac{3}{4}$ inches by $2\frac{1}{4}$ inches by $\frac{3}{4}$ inch, and that there are 270 jewelry boxes to be shipped.

For each type of flat-rate shipping box:

1. Find how many jewelry boxes can fit into the box. Explain or show how the jewelry boxes can be packed in the shipping box. Draw a sketch to show your thinking, if needed.

2. Calculate the total cost of shipping all 270 jewelry boxes in shipping boxes of that type. Show your reasoning and organize your work so it can be followed by others.

Activity

17.3 Determining Shipping Costs (Part 3)

1. Share and discuss your work with the other members of your group. Your teacher will display questions to guide your discussion. Note the feedback from your group so you can use it to revise your work.

2. Using the feedback from your group, revise your work to improve its correctness, clarity, and accuracy. Correct any errors. You may also want to add notes or diagrams, or remove unnecessary information.

3. Which shipping boxes should the artist use? As a group, decide which boxes you recommend for shipping 270 jewelry boxes.
Be prepared to share your reasoning.

Learning Targets

Lesson	Learning Target(s)
4-1 Size of Divisor and Size of Quotient	• When dividing, I know how the size of a divisor affects the quotient.
4-2 Meanings of Division	• I can explain how multiplication and division are related. • I can explain two ways of interpreting a division expression such as 27 ÷ 3. • When given a division equation, I can write a multiplication equation that represents the same situation.
4-3 Interpreting Division Situations	• I can create a diagram or write an equation that represents division and multiplication questions. • I can decide whether a division question is asking “how many groups?” or “how many in each group?”

(continued on the next page)

(continued from the previous page)

Lesson	Learning Target(s)
4-4 How Many Groups? (Part 1)	• I can find how many groups there are when the amount in each group is not a whole number. • I can use diagrams and multiplication and division equations to represent "how many groups?" questions.
4-5 How Many Groups? (Part 2)	• I can find how many groups there are when the number of groups and the amount in each group are not whole numbers.
4-6 Using Diagrams to Find the Number of Groups	• I can use a tape diagram to represent equal-sized groups and find the number of groups.
4-7 What Fraction of a Group?	• I can tell when a question is asking for the number of groups and that number is less than 1. • I can use diagrams and multiplication and division equations to represent and answer "what fraction of a group?" questions.

Lesson	Learning Target(s)
4-8 How Much in Each Group? (Part 1)	• I can tell when a question is asking for the amount in one group. • I can use diagrams and multiplication and division equations to represent and answer "how much in each group?" questions.
4-9 How Much in Each Group? (Part 2)	• I can find the amount in one group in different real-world situations.
4-10 Dividing by Unit and Non-Unit Fractions	• I can divide a number by a non-unit fraction $\frac{a}{b}$ by reasoning with the numerator and denominator, which are whole numbers. • I can divide a number by a unit fraction $\frac{1}{b}$ by reasoning with the denominator, which is a whole number.
4-11 Using an Algorithm to Divide Fractions	• I can describe and apply a rule to divide numbers by any fraction.

(continued on the next page)

(continued from the previous page)

Lesson	Learning Target(s)
4-12 Fractional Lengths	• I can use division and multiplication to solve problems involving fractional lengths.
4-13 Rectangles with Fractional Side Lengths	• I can use division and multiplication to solve problems involving areas of rectangles with fractional side lengths.
4-14 Fractional Lengths in Triangles and Prisms	• I can explain how to find the volume of a rectangular prism using cubes that have a unit fraction as their edge length. • I can use division and multiplication to solve problems involving areas of triangles with fractional bases and heights. • I know how to find the volume of a rectangular prism even when the edge lengths are not whole numbers.
4-15 Volume of Prisms	• I can solve volume problems that involve fractions.

Lesson	Learning Target(s)
4-16 Solving Problems Involving Fractions	• I can use mathematical expressions to represent and solve word problems that involve fractions.
4-17 Fitting Boxes into Boxes	• I can use multiplication and division of fractions to reason about real-world volume problems.

(continued on the next page)

Notes:

Glossary

A

absolute value The absolute value of a number is its distance from 0 on the number line.

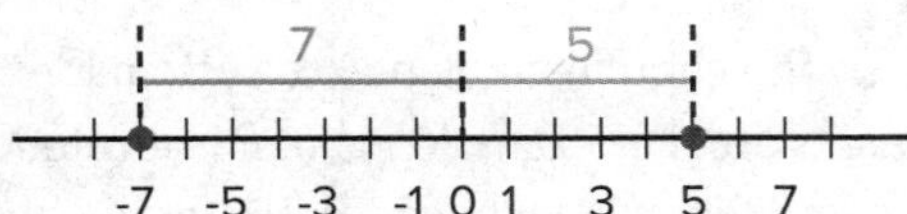

The absolute value of -7 is 7, because it is 7 units away from 0. The absolute value of 5 is 5, because it is 5 units away from 0.

area Area is the number of square units that cover a two-dimensional region, without any gaps or overlaps.

For example, the area of region A is 8 square units. The area of the shaded region of B is $\frac{1}{2}$ square unit.

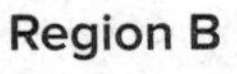

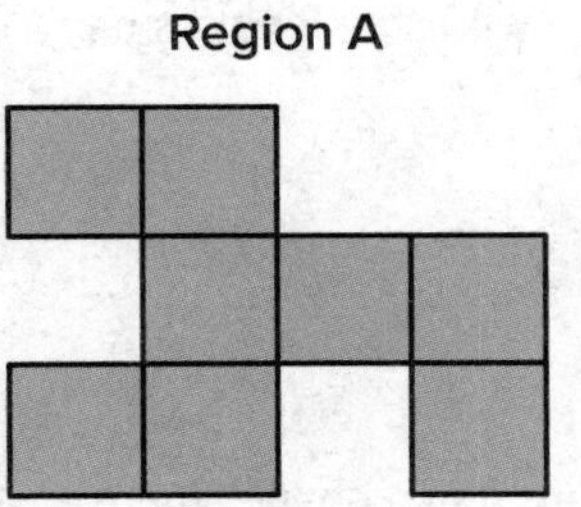

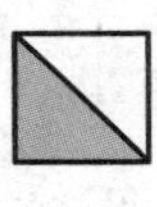

average The average is another name for the mean of a data set.

For the data set 3, 5, 6, 8, 11, 12, the average is 7.5.

$3 + 5 + 6 + 8 + 11 + 12 = 45$

$45 \div 6 = 7.5$

B

base (of a parallelogram or triangle) We can choose any side of a parallelogram or triangle to be the shape's base. Sometimes we use the word *base* to refer to the length of this side.

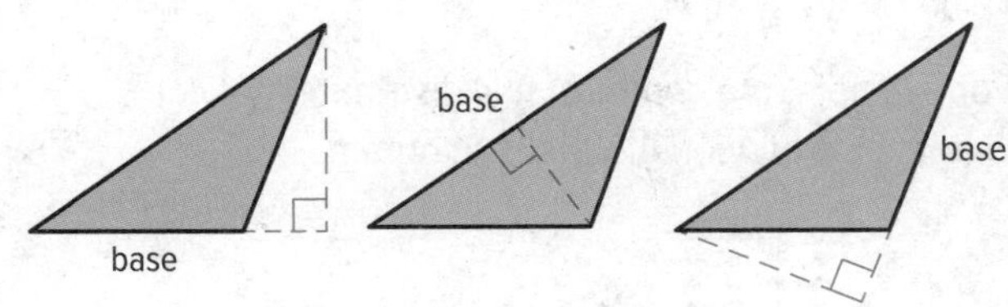

base (of a prism or pyramid) The word *base* can also refer to a face of a polyhedron.

A prism has two identical bases that are parallel. A pyramid has one base.

A prism or pyramid is named for the shape of its base.

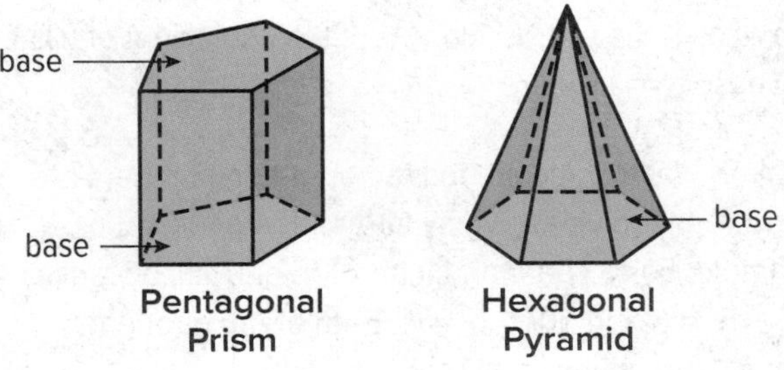

box plot A box plot is a way to represent data on a number line. The data is divided into four sections. The sides of the box represent the first and third quartiles. A line inside the box represents the median. Lines outside the box connect to the minimum and maximum values.

For example, this box plot shows a data set with a minimum of 2 and a maximum of 15. The median is 6, the first quartile is 5, and the third quartile is 10.

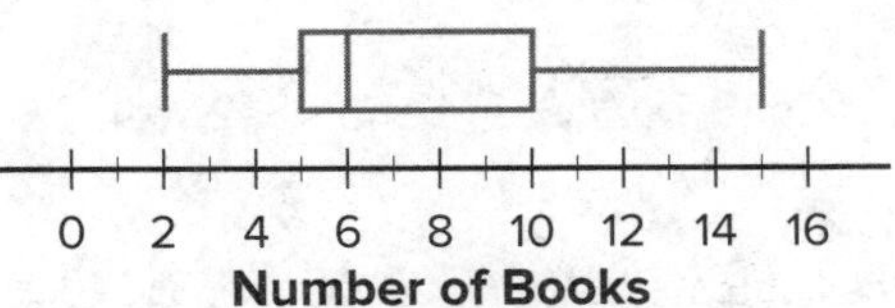

C

categorical data A set of categorical data has values that are words instead of numbers.

For example, Han asks 5 friends to name their favorite color. Their answers are: blue, blue, green, blue, orange.

center The center of a set of numerical data is a value in the middle of the distribution. It represents a typical value for the data set.

For example, the center of this distribution of cat weights is between 4.5 and 5 kilograms.

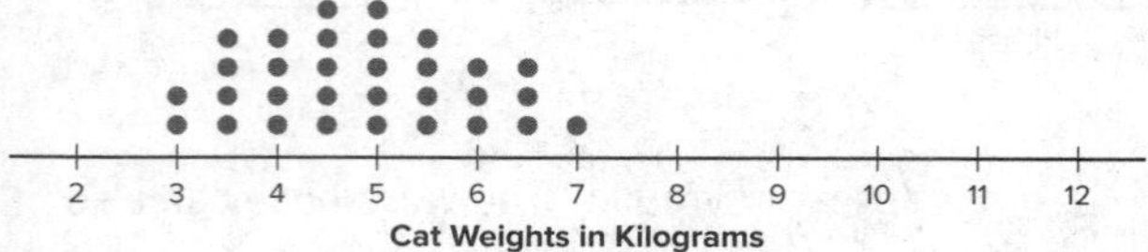

coefficient A coefficient is a number that is multiplied by a variable.

For example, in the expression $3x + 5$, the coefficient of x is 3. In the expression $y + 5$, the coefficient of y is 1, because $y = 1 \cdot y$.

common factor A common factor of two numbers is a number that divides evenly into both numbers. For example, 5 is a common factor of 15 and 20, because $15 \div 5 = 3$ and $20 \div 5 = 4$. Both of the quotients, 3 and 4, are whole numbers.

- The factors of 15 are 1, 3, *5*, and 15.
- The factors of 20 are 1, 2, 4, *5*, 10, and 20.

common multiple A common multiple of two numbers is a product you can get by multiplying each of the two numbers by some whole number. For example, 30 is a common multiple of 3 and 5, because $3 \cdot 10 = 30$ and $5 \cdot 6 = 30$. Both of the factors, 10 and 6, are whole numbers.

- The multiples of 3 are 3, 6, 9, 12, *15*, 18, 21, 24, 27, *30*, 33 . . .
- The multiples of 5 are 5, 10, *15*, 20, 25, *30*, 35, 40 . . .

compose Compose means "put together." We use the word *compose* to describe putting more than one figure together to make a new shape.

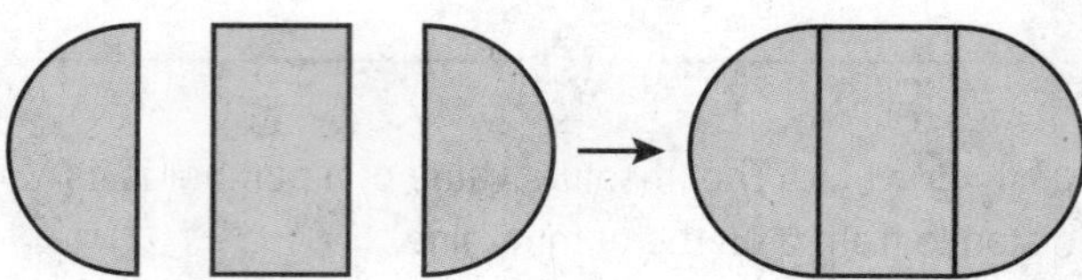

coordinate plane The coordinate plane is a system for telling where points are. For example, point R is located at (3, 2) on the coordinate plane, because it is three units to the right and two units up.

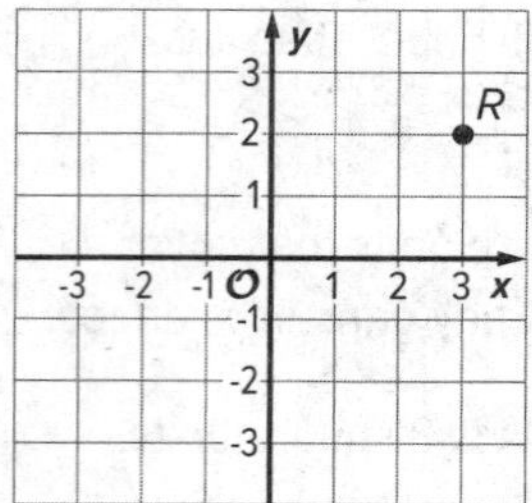

cubed We use the word *cubed* to mean "to the third power." This is because a cube with side length s has a volume of $s \cdot s \cdot s$, or s^3.

D

decompose Decompose means "take apart." We use the word *decompose* to describe taking a figure apart to make more than one new shape.

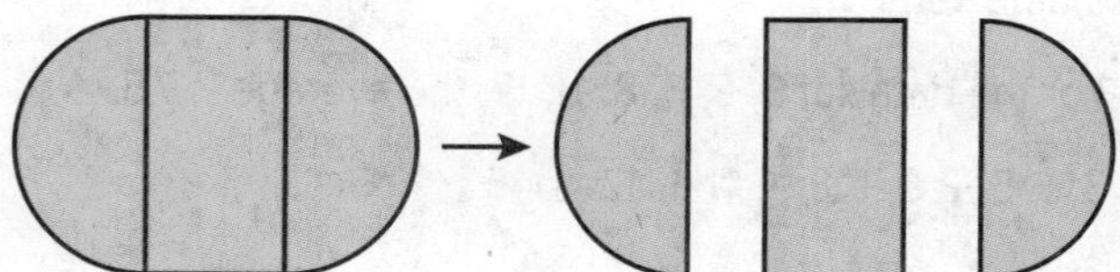

dependent variable The dependent variable is the result of a calculation.

For example, a boat travels at a constant speed of 25 miles per hour. The equation $d = 25t$ describes the relationship between the boat's distance and time. The dependent variable is the distance traveled, because d is the result of multiplying 25 by t.

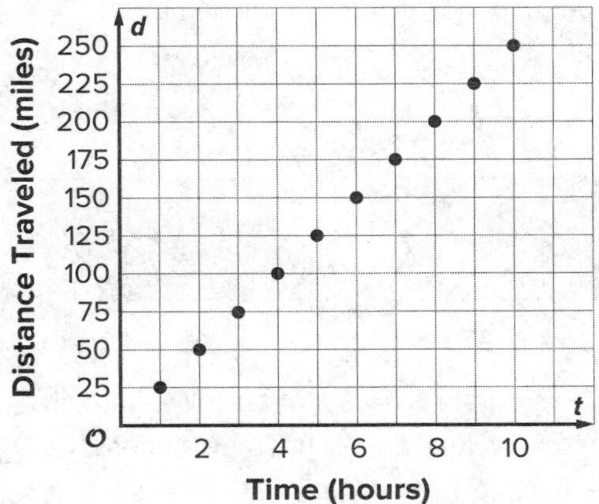

distribution The distribution tells how many times each value occurs in a data set. For example, in the data set blue, blue, green, blue, orange, the distribution is 3 blues, 1 green, and 1 orange.

Here is a dot plot that shows the distribution for the data set 6, 10, 7, 35, 7, 36, 32, 10, 7, 35.

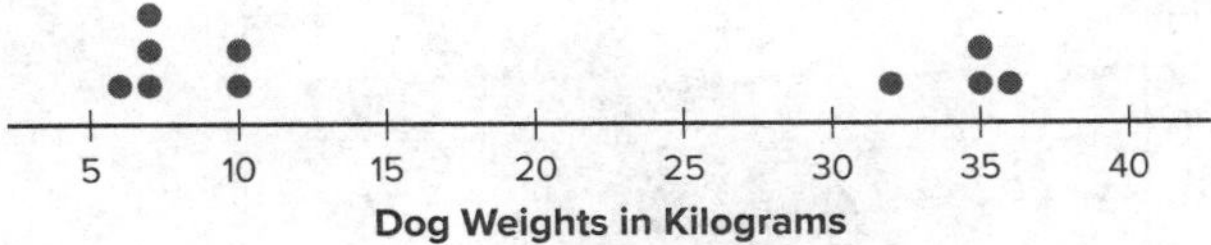

dot plot A dot plot is a way to represent data on a number line. Each time a value appears in the data set, we put another dot above that number on the number line.

For example, in this dot plot there are three dots above the 9. This means that three different plants had a height of 9 cm.

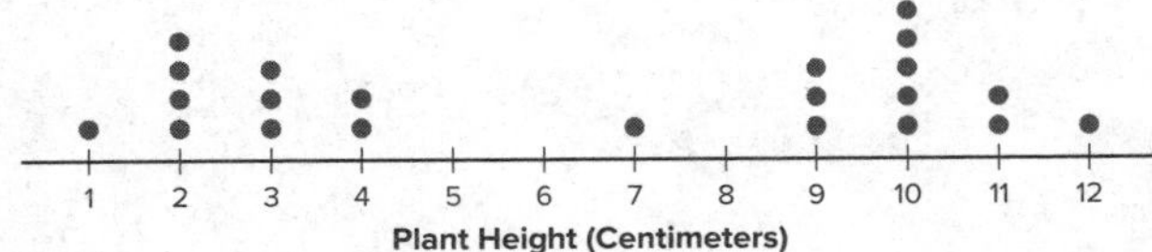

double number line diagram A double number line diagram uses a pair of parallel number lines to represent equivalent ratios. The locations of the tick marks match on both number lines. The tick marks labeled 0 line up, but the other numbers are usually different.

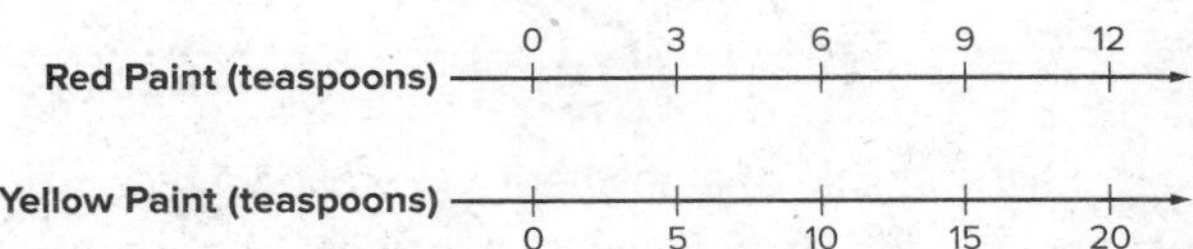

E

edge Each straight side of a polygon is called an edge. For example, the edges of this polygon are segments AB, BC, CD, DE, and EA.

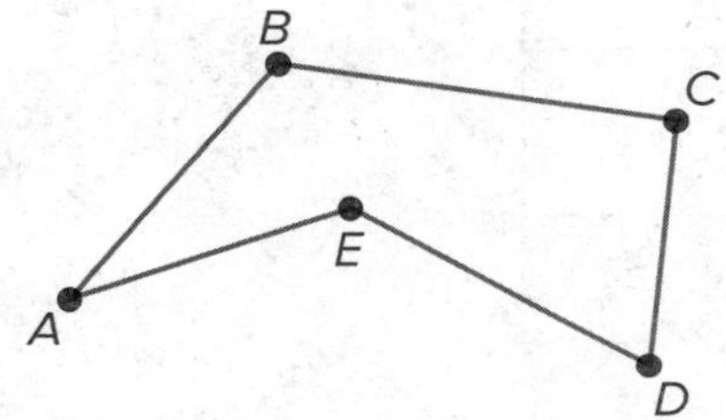

equivalent expressions Equivalent expressions are always equal to each other. If the expressions have variables, they are equal whenever the same value is used for the variable in each expression.

For example, $3x + 4x$ is equivalent to $5x + 2x$. No matter what value we use for x, these expressions are always equal. When x is 3, both expressions equal 21. When x is 10, both expressions equal 70.

equivalent ratios Two ratios are equivalent if you can multiply each of the numbers in the first ratio by the same factor to get the numbers in the second ratio.

For example, 8 : 6 is equivalent to 4 : 3, because $8 \cdot \frac{1}{2} = 4$ and $6 \cdot \frac{1}{2} = 3$.

A recipe for lemonade says to use 8 cups of water and 6 lemons. If we use 4 cups of water and 3 lemons, it will make half as much lemonade. Both recipes taste the same, because 8 : 6 and 4 : 3 are equivalent ratios.

Cups of Water	Number of Lemons
8	6
4	3

exponent In expressions like 5^3 and 8^2, the 3 and the 2 are called exponents. They tell you how many factors to multiply. For example, $5^3 = 5 \cdot 5 \cdot 5$, and $8^2 = 8 \cdot 8$.

F

face Each flat side of a polyhedron is called a face. For example, a cube has 6 faces, and they are all squares.

frequency The frequency of a data value is how many times it occurs in the data set.

For example, there were 20 dogs in a park. The table shows the frequency of each color.

Color	Frequency
White	4
Brown	7
Black	3
Multi-Color	6

G

greatest common factor The greatest common factor of two numbers is the largest number that divides evenly into both numbers. Sometimes we call this the GCF. For example, 15 is the greatest common factor of 45 and 60.

- The factors of 45 are 1, 3, 5, 9, *15*, and 45.
- The factors of 60 are 1, 2, 3, 4, 5, 6, 10, 12, *15*, 20, 30, and 60.

H

height (of a parallelogram or triangle) The height is the shortest distance from the base of the shape to the opposite side (for a parallelogram) or opposite vertex (for a triangle).

We can show the height in more than one place, but it will always be perpendicular to the chosen base.

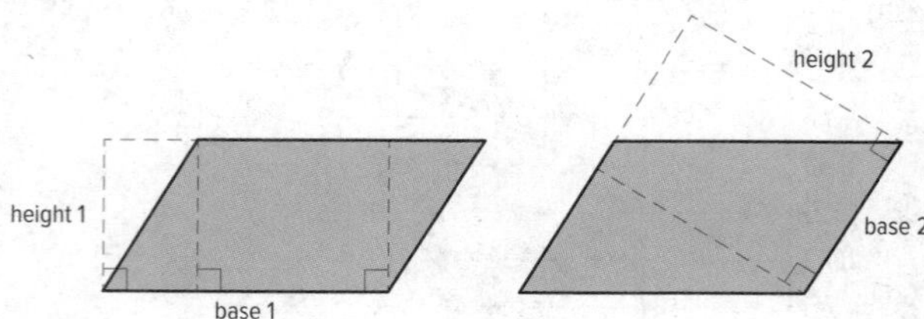

histogram A histogram is a way to represent data on a number line. Data values are grouped by ranges. The height of the bar shows how many data values are in that group.

This histogram shows there were 10 people who earned 2 or 3 tickets. We can't tell how many of them earned 2 tickets or how many earned 3. Each bar includes the left-end value but not the right-end value. (There were 5 people who earned 0 or 1 tickets and 13 people who earned 6 or 7 tickets.)

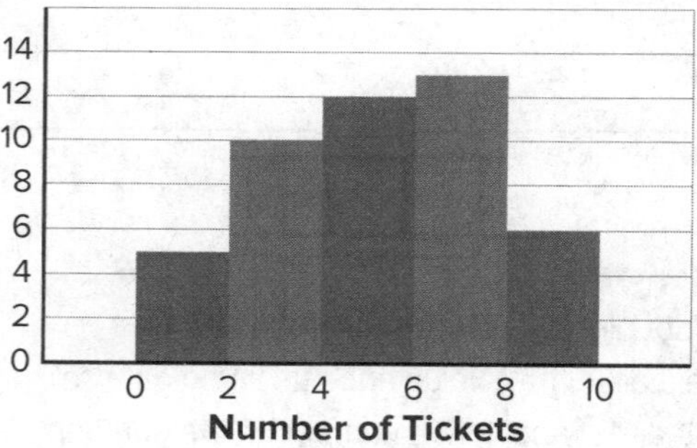

I

independent variable The independent variable is used to calculate the value of another variable.

For example, a boat travels at a constant speed of 25 miles per hour. The equation $d = 25t$ describes the relationship between the boat's distance and time. The independent variable is time, because t is multiplied by 25 to get d.

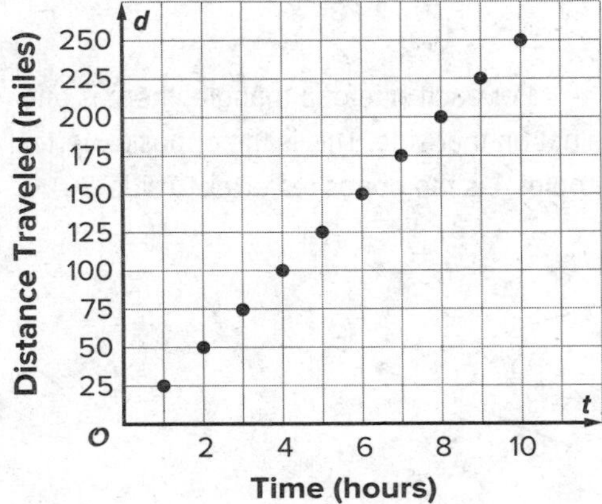

interquartile range (IQR) The interquartile range is one way to measure how spread out a data set is. We sometimes call this the IQR. To find the interquartile range we subtract the first quartile from the third quartile.

For example, the IQR of this data set is 20 because $5 - 30 = 20$.

22	29	30	31	32	43	44	45	50	50	59
		Q1			Q2			Q3		

L

least common multiple The least common multiple of two numbers is the smallest product you can get by multiplying each of the two numbers by some whole number. Sometimes we call this the LCM. For example, 30 is the least common multiple of 6 and 10.

- The multiples of 6 are 6, 12, 18, 24, *30*, 36, 42, 48, 54, 60 . . .
- The multiples of 10 are 10, 20, *30*, 40, 50, 60, 70, 80 . . .

long division Long division is a way to show the steps for dividing numbers in decimal form. It finds the quotient one digit at a time, from left to right. For example, here is the long division for $57 \div 4$.

```
   14.25
 4)57.00
  -4
   17
  -16
    10
    -8
     20
    -20
      0
```

M

mean The mean is one way to measure the center of a data set. We can think of it as a balance point. For example, for the data set 7, 9, 12, 13, 14, the mean is 11.

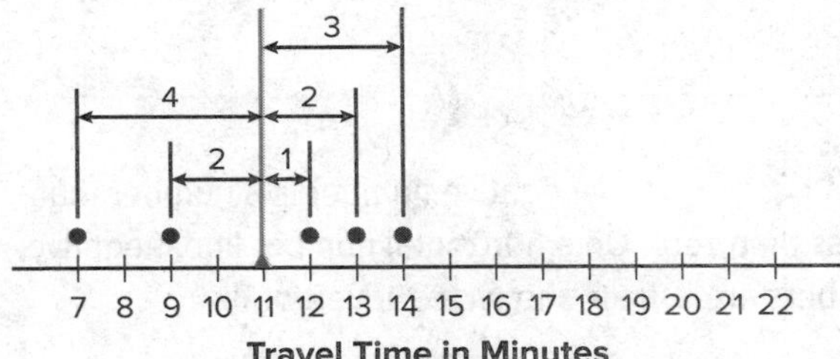

To find the mean, add up all the numbers in the data set. Then, divide by how many numbers there are. $7 + 9 + 12 + 13 + 14 = 55$ and $55 \div 5 = 11$.

mean absolute deviation (MAD) The mean absolute deviation is one way to measure how spread out a data set is. Sometimes we call this the MAD. For example, for the data set 7, 9, 12, 13, 14, the MAD is 2.4. This tells us that these travel times are typically 2.4 minutes away from the mean, which is 11.

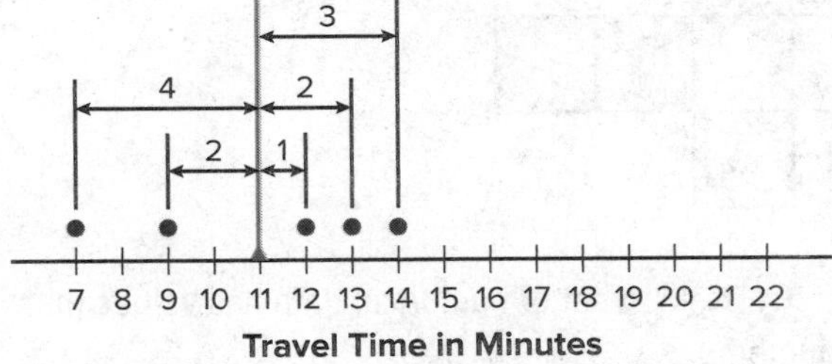

To find the MAD, add up the distance between each data point and the mean. Then, divide by how many numbers there are. $4 + 2 + 1 + 2 + 3 = 12$ and $12 \div 5 = 2.4$.

measure of center A measure of center is a value that seems typical for a data distribution.

Mean and median are both measures of center.

median The median is one way to measure the center of a data set. It is the middle number when the data set is listed in order.

For the data set 7, 9, 12, 13, 14, the median is 12.

For the data set 3, 5, 6, 8, 11, 12, there are two numbers in the middle. The median is the average of these two numbers. $6 + 8 = 14$ and $14 \div 2 = 7$.

meters per second Meters per second is a unit for measuring speed. It tells how many meters an object goes in one second.

For example, a person walking 3 meters per second is going faster than another person walking 2 meters per second.

N

negative number A negative number is a number that is less than zero. On a horizontal number line, negative numbers are usually shown to the left of 0.

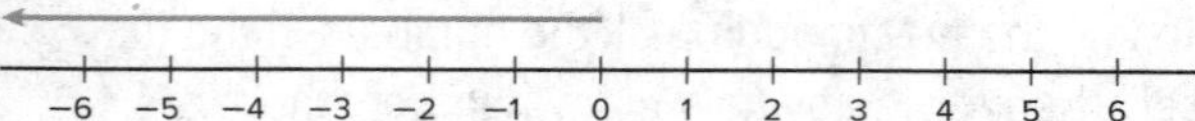

net A net is a two-dimensional figure that can be folded to make a polyhedron.

Here is a net for a cube.

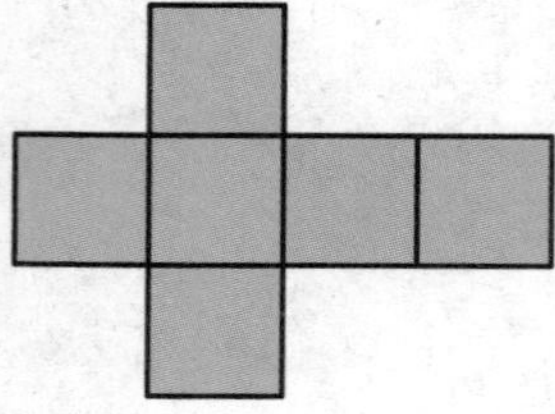

numerical data A set of numerical data has values that are numbers.

For example, Han lists the ages of people in his family: 7, 10, 12, 36, 40, 67.

O

opposite Two numbers are opposites if they are the same distance from 0 and on different sides of the number line.

For example, 4 is the opposite of -4, and -4 is the opposite of 4. They are both the same distance from 0. One is negative, and the other is positive.

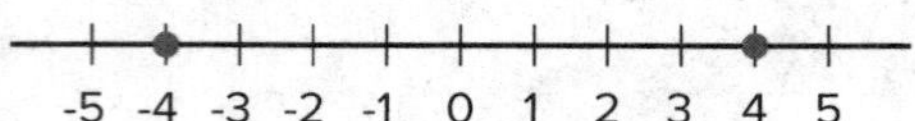

opposite vertex For each side of a triangle, there is one vertex that is not on that side. This is the opposite vertex. For example point A is the opposite vertex to side BC.

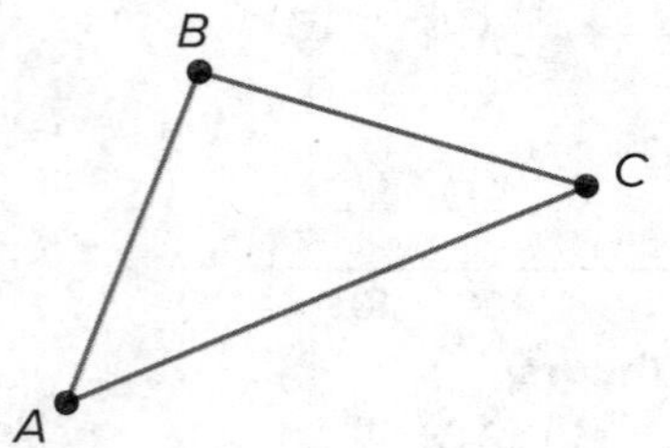

P

pace Pace is one way to describe how fast something is moving. Pace tells how much time it takes the object to travel a certain distance.

For example, Diego walks at a pace of 10 minutes per mile. Elena walks at a pace of 11 minutes per mile. Elena walks slower than Diego, because it takes her more time to travel the same distance.

parallelogram A parallelogram is a type of quadrilateral that has two pairs of parallel sides.

Here are two examples of parallelograms.

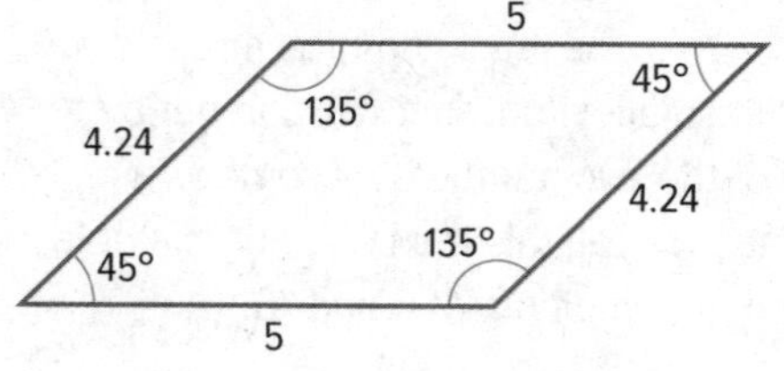

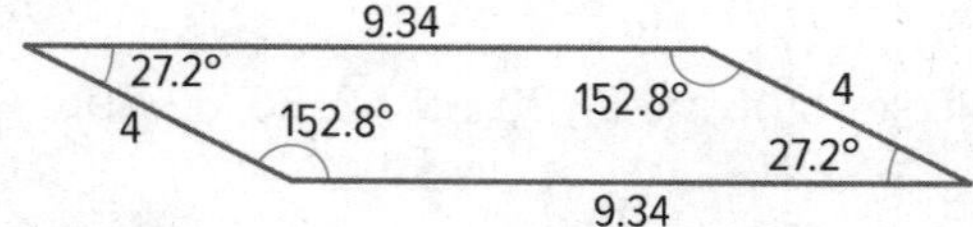

per The word *per* means "for each." For example, if the price is $5 per ticket, that means you will pay $5 *for each* ticket. Buying 4 tickets would cost $20, because $4 \cdot 5 = 20$.

percent The word *percent* means "for each 100." The symbol for percent is %.

For example, a quarter is worth 25 cents, and a dollar is worth 100 cents. We can say that a quarter is worth 25% of a dollar.

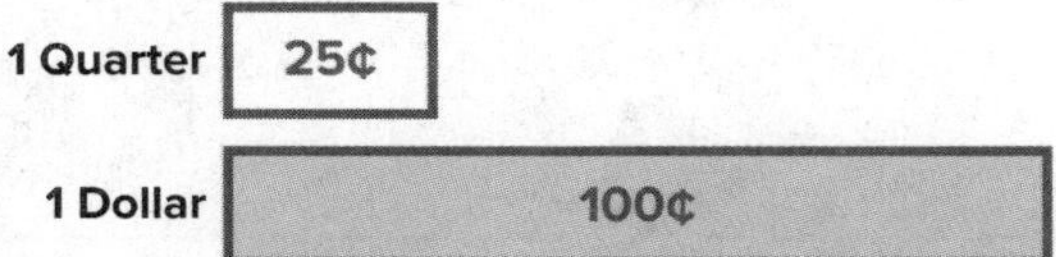

percentage A percentage is a rate per 100.

For example, a fish tank can hold 36 liters. Right now there are 27 liters of water in the tank. The percentage of the tank that is full is 75%.

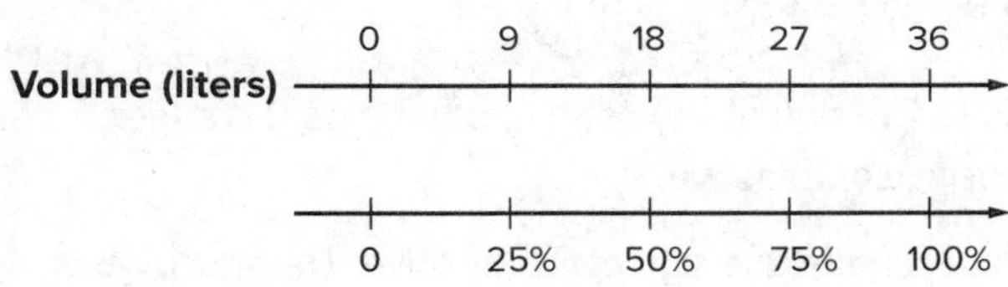

polygon A polygon is a closed, two-dimensional shape with straight sides that do not cross each other. Figure *ABCDE* is an example of a polygon.

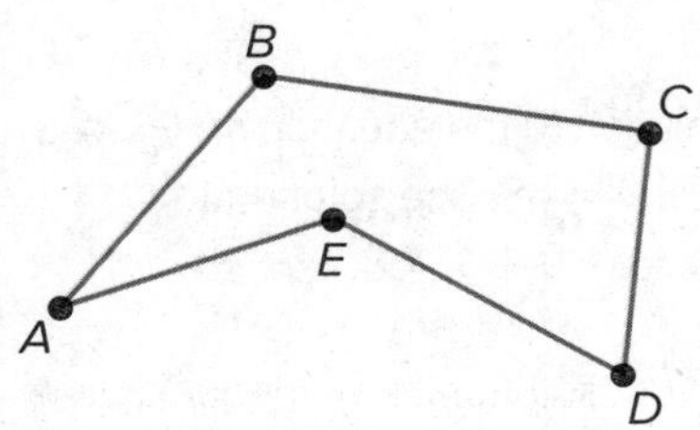

polyhedron (polyhedra) A polyhedron is a closed, three-dimensional shape with flat sides. When we have more than one polyhedron, we call them polyhedra. Here are some drawings of polyhedra.

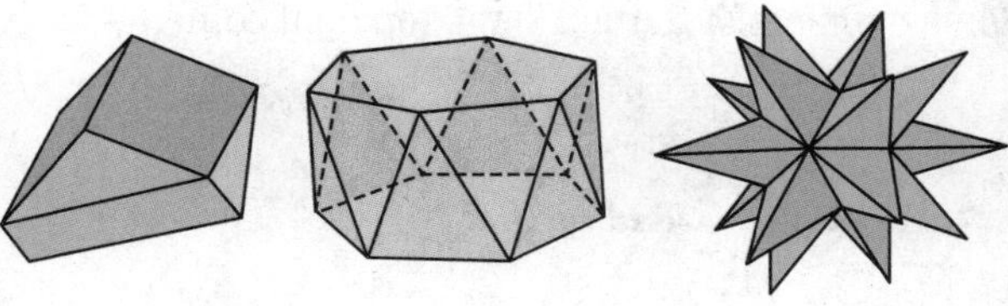

positive number A positive number is a number that is greater than zero. On a horizontal number line, positive numbers are usually shown to the right of 0.

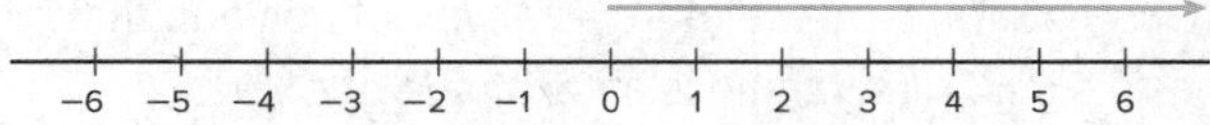

prism A prism is a type of polyhedron that has two bases that are identical copies of each other. The bases are connected by rectangles or parallelograms.

Here are some drawings of some prisms.

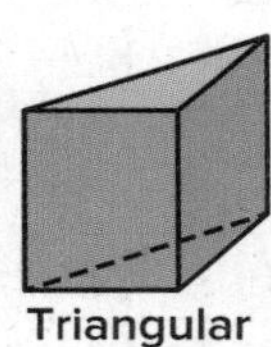
Triangular Prism

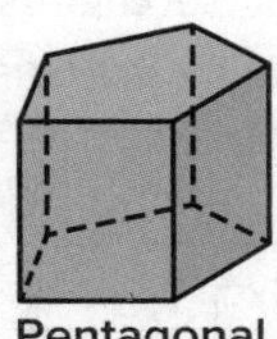
Pentagonal Prism

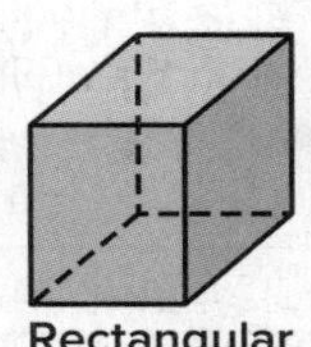
Rectangular Prism

pyramid A pyramid is a type of polyhedron that has one base. All the other faces are triangles, and they all meet at a single vertex

Here are some drawings of pyramids.

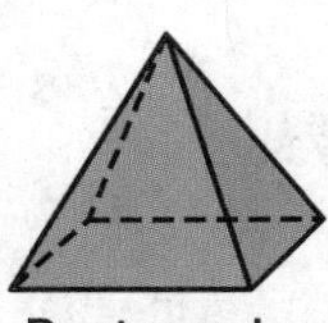
Rectangular Pyramid

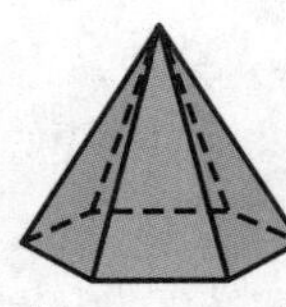
Hexagonal Pyramid

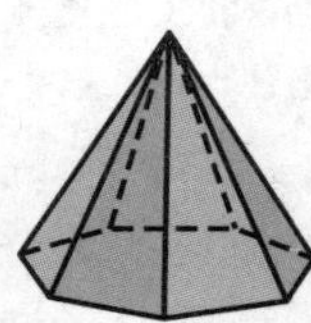
Heptagonal Pyramid

Q

quadrant The coordinate plane is divided into 4 regions called quadrants. The quadrants are numbered using Roman numerals, starting in the top right corner.

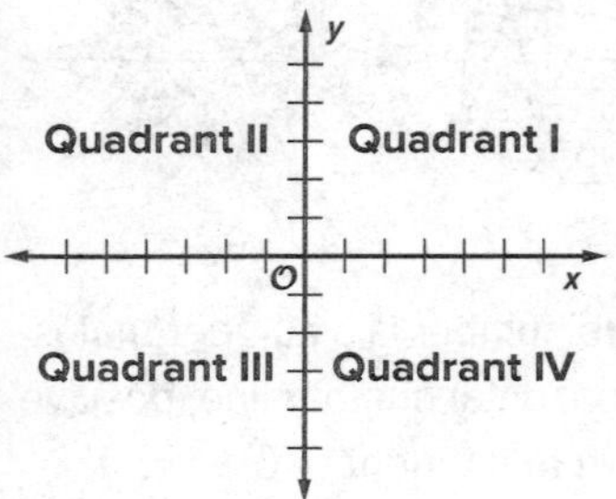

quadrilateral A quadrilateral is a type of polygon that has 4 sides. A rectangle is an example of a quadrilateral. A pentagon is not a quadrilateral, because it has 5 sides.

quartile Quartiles are the numbers that divide a data set into four sections that each have the same number of values.

For example, in this data set the first quartile is 30. The second quartile is the same thing as the median, which is 43. The third quartile is 50.

22	29	30	31	32	43	44	45	50	50	59
		Q1			Q2			Q3		

R

range The range is the distance between the smallest and largest values in a data set. For example, for the data set 3, 5, 6, 8, 11, 12, the range is 9, because $12 - 3 = 9$.

ratio A ratio is an association between two or more quantities.

For example, the ratio 3 : 2 could describe a recipe that uses 3 cups of flour for every 2 eggs, or a boat that moves 3 meters every 2 seconds. One way to represent the ratio 3 : 2 is with a diagram that has 3 blue squares for every 2 green squares.

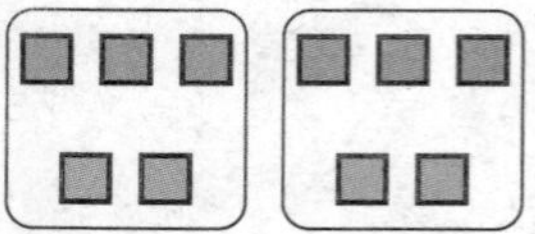

rational number A rational number is a fraction or the opposite of a fraction.

For example, 8 and -8 are rational numbers because they can be written as $\frac{8}{1}$ and, $-\frac{8}{1}$.

Also, 0.75 and -0.75 are rational numbers because they can be written as $\frac{75}{100}$ and, $-\frac{75}{100}$.

reciprocal Dividing 1 by a number gives the reciprocal of that number. For example, the reciprocal of 12 is $\frac{1}{12}$, and the reciprocal of $\frac{2}{5}$ is $\frac{5}{2}$.

region A region is the space inside of a shape. Some examples of two-dimensional regions are inside a circle or inside a polygon. Some examples of three-dimensional regions are the inside of a cube or the inside of a sphere.

S

same rate We use the words *same rate* to describe two situations that have equivalent ratios.

For example, a sink is filling with water at a rate of 2 gallons per minute. If a tub is also filling with water at a rate of 2 gallons per minute, then the sink and the tub are filling at the same rate.

sign The sign of any number other than 0 is either positive or negative.

For example, the sign of 6 is positive. The sign of -6 is negative. Zero does not have a sign, because it is not positive or negative.

solution to an equation A solution to an equation is a number that can be used in place of the variable to make the equation true.

For example, 7 is the solution to the equation $m + 1 = 8$, because it is true that $7 + 1 = 8$. The solution to $m + 1 = 8$ is not 9, because $9 + 1 \neq 8$.

solution to an inequality A solution to an inequality is a number that can be used in place of the variable to make the inequality true.

For example, 5 is a solution to the inequality $c < 10$, because it is true that $5 < 10$. Some other solutions to this inequality are 9.9, 0, and -4.

speed Speed is one way to describe how fast something is moving. Speed tells how much distance the object travels in a certain amount of time.

For example, Tyler walks at a speed of 4 miles per hour. Priya walks at a speed of 5 miles per hour. Priya walks faster than Tyler, because she travels more distance in the same amount of time.

spread The spread of a set of numerical data tells how far apart the values are.

For example, the dot plots show that the travel times for students in South Africa are more spread out than for New Zealand.

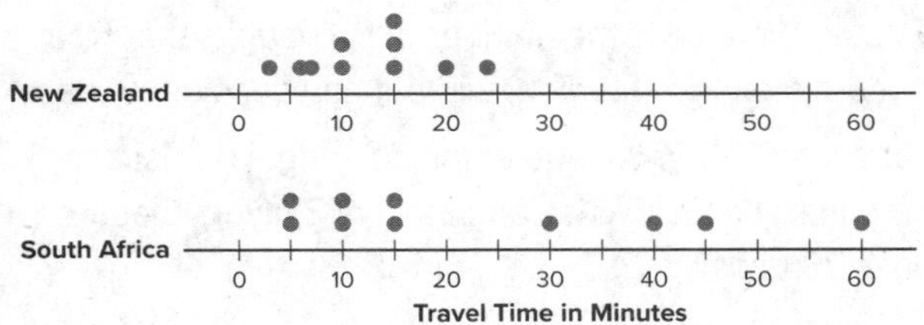

squared We use the word *squared* to mean "to the second power." This is because a square with side length s has an area of $s \cdot s$, or s^2.

statistical question A statistical question can be answered by collecting data that has variability. Here are some examples of statistical questions:

- Who is the most popular musical artist at your school?
- When do students in your class typically eat dinner?
- Which classroom in your school has the most books?

surface area The surface area of a polyhedron is the number of square units that covers all the faces of the polyhedron, without any gaps or overlaps.

For example, if the faces of a cube each have an area of 9 cm^2, then the surface area of the cube is $6 \cdot 9$, or 54 cm^2.

T

table A table organizes information into horizontal rows and vertical columns. The first row or column usually tells what the numbers represent.

For example, here is a table showing the tail lengths of three different pets. This table has four rows and two columns.

Pet	Tail Length (inches)
Dog	22
Cat	12
Mouse	2

tape diagram A tape diagram is a group of rectangles put together to represent a relationship between quantities.

For example, this tape diagram shows a ratio of 30 gallons of yellow paint to 50 gallons of blue paint. If each rectangle were labeled 5, instead of 10, then the same picture could represent the equivalent ratio of 15 gallons of yellow paint to 25 gallons of blue paint.

10	10	10		
10	10	10	10	10

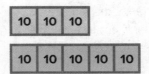

term A term is a part of an expression. It can be a single number, a variable, or a number and a variable that are multiplied together. For example, the expression $5x + 18$ has two terms. The first term is $5x$ and the second term is 18.

U

unit price The unit price is the cost for one item or for one unit of measure. For example, if 10 feet of chain link fencing cost \$150, then the unit price is $150 \div 10$, or \$15 per foot.

unit rate A unit rate is a rate per 1.

For example, 12 people share 2 pies equally. One unit rate is 6 people per pie, because $12 \div 2 = 6$. The other unit rate is $\frac{1}{6}$ of a pie per person, because $2 \div 12 = \frac{1}{6}$.

variability Variability means having different values.

For example, data set B has more variability than data set A. Data set B has many different values, while data set A has more of the same values.

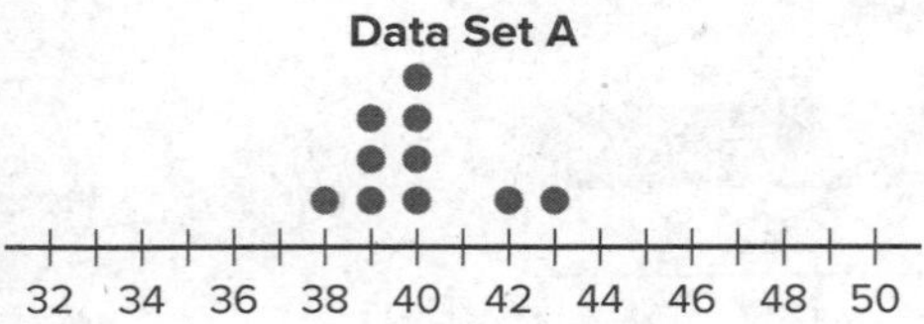

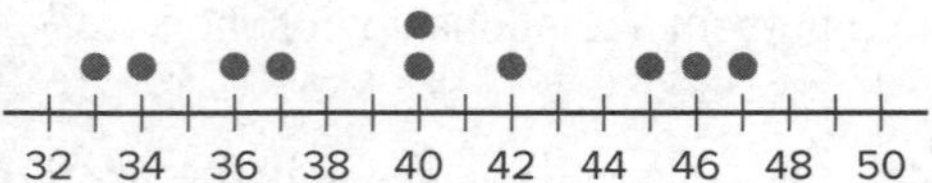

variable A variable is a letter that represents a number. You can choose different numbers for the value of the variable.

For example, in the expression $10 - x$, the variable is x. If the value of x is 3, then $10 - x = 7$, because $10 - 3 = 7$. If the value of x is 6, then $10 - x = 4$, because $10 - 6 = 4$.

vertex (vertices) A vertex is a point where two or more edges meet. When we have more than one vertex, we call them vertices.

The vertices in this polygon are labeled *A*, *B*, *C*, *D*, and *E*.

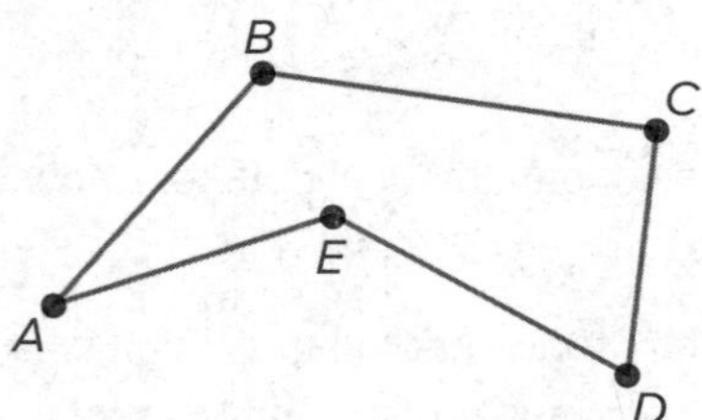

volume Volume is the number of cubic units that fill a three-dimensional region, without any gaps or overlaps.

For example, the volume of this rectangular prism is 60 units^3, because it is composed of 3 layers that are each 20 units^3.

Index

U

V

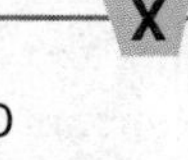
X

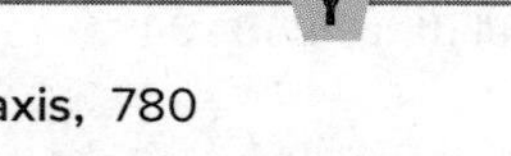
Y

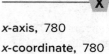